Three-Dimensional Imaging Techniques

Three-Dimensional Imaging Techniques

Takanori Okoshi

Department of Electronic Engineering
University of Tokyo
Tokyo, Japan

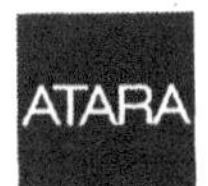

ATARA PRESS *Los Angeles 2011*

To My Late Father

ATARA PRESS, Los Angeles
www.atarapress.com

Second edition: March 2011

Library of Congress Control Number: 2011922903
First published 1976 by Academic Press, New York

ISBN: 978-0-09822251-4-1

Atara Press wishes to express its gratitude to Yasuko Okoshi and Kento Okoshi for their generous assistance in making this edition possible.

Paper sourced from environmentally aware forest product companies.
EDITIONS PRINTED IN THE UNITED KINGDOM AND THE UNITED STATES

Contents

6 Holography

7 Information Recorded in Three-Dimensional Images and Its Reduction

Preface

The reconstruction of spatial images has been a long-held dream. When holography was developed, many people believed that the answer finally had been found. Due to inherent technical limitations, however, holography has not yet become a practical imaging technique. On the other hand, we cannot ignore the technical advantages of older techniques such as lens-sheet pictures, and projection-type displays. Presumably, the most promising direction for future development of three-dimensional imaging techniques may be found in a combination of these various old and new techniques.

The original Japanese edition of this book, entitled "Sanjigen-Gazo Kogaku" (Three-Dimensional Imaging Techniques), was written with the intention of offering a comprehensive treatise on this field, and was published in June 1972 by Sangyo-Tosho, Tokyo. Translation of the Japanese text into English was begun shortly after the original publication, and completed early in 1974. In the course of translation, the author found some sections in the original text unsatisfactory and supplemented these. The length of the English edition, therefore, is probably two or three percent greater than that of the original.

About one-half of the book deals with nonholographic techniques and the other half with holographic techniques. Chapters 4–6 describe in detail the theories and applications of three basic techniques: lens-sheet three-dimensional pictures, projection-type three-dimensional display, and holography, respectively. The preceding three chapters and the following three chapters are devoted to more general discussions. Chapter 1 is an introduction; Chapter 2 a short description of the history of three-dimensional imaging techniques; and Chapter 3 describes the physiological and psychological aspects of three-dimensional sensation (depth perception) in humans. Chapter 7 discusses the quantity and redundancy of information contained in a three-dimensional image and the methods for reducing that redundancy. The discussion in this chapter gives a theoretical basis for three-dimensional image information transmission techniques (three-dimensional television techniques) likely to be developed in the future. In Chapter 8, special three-dimensional imaging techniques, such as those for x-ray images and computer outputs, as well as the technical feasibility of three-dimensional television in the future, are discussed. Concluding remarks are given in Chapter 9.

The author first encountered three-dimensional imaging in 1964. Dr. K.S. Pennington, who did excellent work in the early development of holography, was performing experiments next door to the author's laboratory at the Bell Telephone Laboratories in Murray Hill, New Jersey. Dr. Pennington kindly showed the author some holography experiments, but at that time the author did not recognize the potential of holography.

Later, in 1967, one of the author's friends presented him with a splendid example of a three-dimensional lenticular-sheet picture. (It was a souvenir from the United States, but was later found to have been "printed in Japan.") This piece, together with the memory of Dr. Pennington's experiments, made the author believe that the combination of such conventional techniques as the lenticular sheet with holography might offer new technical possibilities.

In winter 1967, the author investigated for himself the history of various three-dimensional imaging techniques. He also started some research work on three-dimensional imaging; some of this work is described in this book. Meanwhile, he noted that no book had been published covering this field. This fact was the principal motivation for writing the original Japanese edition of this book and then translating it into English.

Acknowledgments

The author is indebted to many people for the work on which this book is based. This work was performed with continuous encouragement from Professors M. Kamiyama, S. Okamura, H. Yanai, and M. Aoki, all of the Department of Electronic Engineering, University of Tokyo. Continuous technical support has been given since the earliest stages of the work by Messrs. H. Suzukawa, J. Mukai, K. Sayanagi, and A. Yano of the Canon Company, and Messrs. R. Yamanashi and S. Hizume of the Toppan Printing Company. There are many others who have helped the author in preparing the photographs and figures used, and in performing the experiments described; their names are mentioned in the text or figure captions, or in the references.

Finally, the author wishes to express his hearty appreciation to his secretary, Miss M. Onozuka, who typed the manuscript and its revisions many times. Without her devoted assistance this translation would never have been completed.

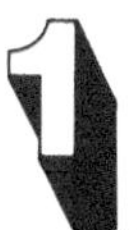

Introduction

The history of information media in human society, the definition and classification of three-dimensional imaging techniques, and the organization of this book are described in this chapter.

1.1 HISTORY OF INFORMATION MEDIA

Information media in human society have progressed steadily over the past centuries. We consider first the history of "direct" information media, those that can be delivered from hand to hand. In this category books and pictures appeared first in the history of man. Afterward, the inventions of printing (1450) and photography (1839) made these media mass-producible and cheap enough for popular use. These two inventions mark the beginnings of important epochs in the history of information. Since then man has become accustomed to expressing his thoughts and sentiments in two-dimensional patterns or images. This is truly a remarkable feature of human culture.

On the other hand, one kind of three-dimensional information medium had been familiar to man since ancient days: sculpture; however, sculpture is often a "thing in itself," and might not deserve to be called an information "medium." Serious efforts by man toward artificial three-dimensional images began in the middle of the nineteenth century, as will be described in the next chapter. Since then this effort has been continued for more than 100 years. In particular, in the past decade new three-dimensional imaging techniques such as holography and lens-sheet pictures have vigorously been developed. Historians of science 100 years from now might characterize our era by the present efforts toward better three-dimensional imaging techniques.

Next we consider the history of telecommunications. In the most primitive stage, letters were forwarded and handed directly to specific individuals. This exchange developed later to the domestic and international mail systems. Late in the eighteenth century, the first high-speed telecommunication system, Chappe's communicator,* appeared. Though it was the most primitive telecommunication system imaginable, it had a great impact on European society. The conquest and rule of European countries by Napoleon is said to be in part due to the use of this communication system.

Telecommunication technology advanced further to bring forth the telegraph (1835), telephone (1876), facsimile (1925), television (1926), and many other versions of these basic techniques. At present, many people believe that the next stage will be three-dimensional television, that is, television transmitting animated, spatial images. It will probably

---

* Towers equipped with pivoted pointing arms for displaying signals were constructed at mile intervals and manned by operators. Each operator moved the pointing arms in the same manner as he observed on the display of the neighboring tower, and thus messages were transmitted.

require an effort of more than 20 years to complete a really practical system of techniques for three-dimensional television. It is, however, one of the last remaining frontiers in the field of telecommunications.

The development of three-dimensional imaging techniques, either still or animated, is an interesting and challenging task; however, it is never an easy one. We might better say that the direction of future development is quite unknown. The purpose of this book is to present the fundamental concepts and available techniques for three-dimensional imaging in as systematic a manner as possible.

1.2 DEFINITION OF THREE-DIMENSIONAL IMAGING

The term *three-dimensional imaging* might require some preliminary comments. In this book, the terms *stereoscopic* and *three-dimensional* are carefully distinguished from one another, except as used in Chapter 3 where physiological and psychological aspects of depth perception are described. A priori there is no essential difference between the meanings of these two terms; however, *three-dimensional* seems to fit better truly three-dimensional, spatial images than does the older *stereoscopic*. When we hear *stereoscopic,* the resulting mental picture is also old and primitive; it is apt to remind us of stereoscopes.

On the other hand, principally discussed in this book are the imaging techniques of truly three-dimensional, spatial images which show gradually their right side when the observer moves rightward, and their left side when he moves leftward. The author believes that what will be promising and interesting in the future are techniques for reproducing such truly spatial images.

Such a truly spatial image is sometimes called an *autostereoscopic* image; however, since this terminology is not universal at present, it will not be used often in this book. Instead, the term *three-dimensional* is used to discuss a spatial image, whereas the term *stereoscopic* is used mainly for binocular images.

This book deals principally with three-dimensional spatial images, but stereoscopic images are also described particularly in two places. One is the historical description of the art in Chapter 2. In the history of three-dimensional imaging, binocular imaging techniques constitute important technical steps. The other exception is in the discussion of binocular television techniques for medical and other special uses. For example, suppose that in a medical school a surgical operation is shown via television to students studying in a classroom adjacent to the operating room. In such cases the final display need not necessarily be truly spatial; binocular imaging with a finer resolution might be superior. In such applications binocular imaging techniques will remain useful.

1.3 CLASSIFICATION OF THREE-DIMENSIONAL IMAGING TECHNIQUES

On the basis of the discussions in the preceding section, we first divide three-dimensional imaging (in the wider sense of the words) into two major groups:

(1) binocular stereoscopic imaging,
(2) three-dimensional spatial imaging (autostereoscopy).

The essential difference between these two kinds of imaging rests upon the quantity of information required to record the images. The quantity of information for a binocular image is only twice that of a planar image, whereas that for a truly spatial image is usually tremendous. This problem will be discussed in detail in Chapter 7.

Each of the above two major groups is further divided into smaller groups according to the principle

TABLE 1.1. Classification of Three-Dimensional Imaging Techniques, I

(A) Binocular stereoscopic imaging	Binocular viewers	A-1
	Parallax stereogram	A-2
	Lenticular-sheet binocular stereoscopic pictures	A-3
	Binocular displays using polaroid glasses	A-4
(B) Three-dimensional spatial imaging (Autostereoscopy)	Those using coherent light: holography	B-1
	Those using incoherent light: parallax panoramagram, lenticular-sheet three-dimensional imaging, integral photography, projection-type three-dimensional displays, etc.	B-2
	Combinations of B-1 and B-2: holocoder-hologram, holographic stereogram, projection-type holography, etc.	B-3

TABLE 1.2. Classification of Three-Dimensional Imaging Techniques, II

	Without transmission	With transmission
Still images	Three-dimensional pictures and three-dimensional projection-type displays (slides)	Three-dimensional facsimile
Animated images	Three-dimensional movies	Three-dimensional television

and techniques used. A coding of such a classification is shown as Table 1.1. This table will not be described further in this chapter, however, because each of the techniques mentioned will be discussed in detail later at least once.

There is another approach to the classification of three-dimensional images: according to their use. An example of such a classification is shown in Table 1.2, which includes some techniques that are not available at present but are expected to be developed in the future.

The descriptions in this book will mainly be restricted to the "still images without transmission" shown in Table 1.2, and some pages will be devoted to the "transmitted animated images," that is, three-dimensional television schemes. Movie techniques can easily be derived from still-picture techniques, and facsimile is nothing but very low-speed television.

1.4 ORGANIZATION OF THIS BOOK

Chapter 2 is a short history of three-dimensional imaging techniques. The history of the art from primitive binocular viewers through the invention of holography will be described. Those techniques that once appeared but later disappeared are described in more detail than those that became practical later; these latter will be described in detail in the following chapters.

Chapter 3 describes the physiological and psychological aspects of depth perception. Chapters 4 to 6 constitute the core of the book; they are devoted to detailed technical discussions of lens-sheet three-dimensional pictures, projection-type three-dimensional displays, and holography, respectively.

Chapter 7 describes the computation of the quantity of information required to record three-dimensional images, and also information-reduction techniques. Such discussions offer a theoretical basis for three-dimensional image transmission techniques to be developed in the future.

Chapter 8 is devoted to the description of various three-dimensional imaging techniques such as x-ray imaging or computer output displays, and also to consideration of future possibilities for three-dimensional television. Concluding remarks are given in Chapter 9.

History of Three-Dimensional Imaging Techniques

A short history of three-dimensional imaging techniques is presented. Starting with Euclid's definition of depth perception, this chapter covers stereoscopes in the nineteenth century, techniques of parallax barriers, integral photography, lenticular-sheet pictures, projection-type displays, and the invention of holography. Emphasis is placed upon those techniques that never became practical because the practical techniques are described in later chapters in more detail.

2.1 PREHISTORY

What is the origin of depth perception? It was very long ago that our ancestors began to wonder about this question. In A.D. 280 Euclid said that depth perception is to receive by means of each eye the simultaneous impression of two dissimilar images of the same object (1).

In modern psychology, depth perception as defined above is called binocular parallax. It is one of the important physiological cues for depth perception. However, in modern psychology, it is accepted that depth perception is based upon as many as ten depth cues. Especially important among these are, in addition to binocular parallax, convergence (the angle made by the viewing axes of a pair of eyes) and accommodation (the adjustment of the focal length of the crystalline lens). Before the invention of holography, however, all of the artificial three-dimensional imaging techniques relied upon only two cues: binocular parallax and convergence.

The first trial of artificial three-dimensional imaging was a stereoscopic drawing technique devised by Giovanni Battista della Porta around the year 1600 (1). It is a technique for drawing two precise pictures of an object observed from two different directions. His drawings do not exist today. Similar stereoscopic pictures, however, were drawn by many people for more than 200 years since della Porta. The invention of photography (early in the nineteenth century) made this sort of technique less significant and prompted the development of the stereoscopes to be described in the next section.

Today stereoscopic drawing is again being investigated in connection with three-dimensional display of computer output. This will be described in Section 8.

2.2 STEREOSCOPES

2.2.1 WHEATSTONE STEREOSCOPE (1)

In 1838, Sir Charles Wheatstone proposed a stereoscopic viewer as shown in Fig.2.1 at a meeting of The Royal Society. This event was 16 years after the first success at photographic imaging by Niepce, and a year before the invention of the daguerrotype by Daguerre. A stereoscopic viewer of this type is now called a Wheatstone stereoscope.

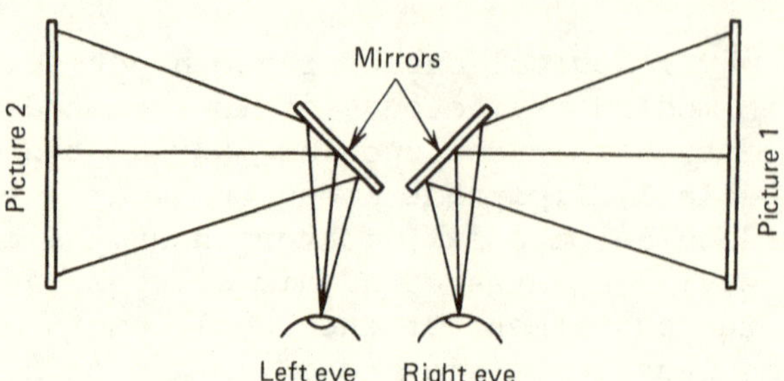

Fig.2.1 Wheatstone stereoscope.

The principle of the Wheatstone stereoscope is probably self-explanatory. Two photographs (or stereoscopic drawings) of an object taken from different directions are placed at the positions of pictures 1 and 2, and observed by a pair of eyes. Such a device was developed for observing stereoscopic pictures larger than the pupil distance (separation of two human eyes), approximately 6.5 cm. When the pictures are small enough, such a device is not needed.

2.2.2 BREWSTER STEREOSCOPE (1)

Following Wheatstone's idea, the first practical stereoscope was constructed by Sir David Brewster of Scotland in 1849. It was a version of the Wheatstone stereoscope in which the mirrors were replaced by prisms as shown in Fig.2.2. A stereoscope of this type is now called a Brewster stereoscope.

2.2.3 HOLMES'S IMPROVEMENT

The Brewster stereoscope was further investigated and improved by Oliver Wendell Holmes, who added convex lenses (eyepieces) to the prisms. At first, his improved stereoscope was used mainly for observing stereoscopic drawings because photography was not yet

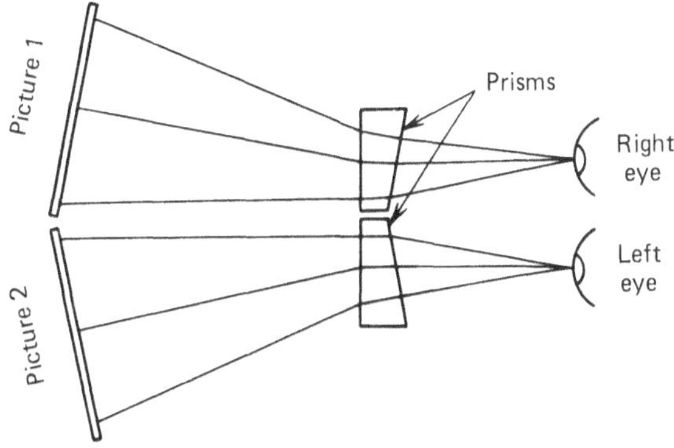

Fig.2.2 Brewster stereoscope.

popular; afterward, however, its use for stereophotographs taken by specially made stereocameras became more and more common, until at last it gained great popularity as the parlor stereoscope, and was also commonly found in amusement parks. Parlor stereoscopes are now seen mostly in museums, but simpler versions can be purchased at toy shops or department stores.

A comment might be needed on the advantage of adding convex lenses. Depth perception is really improved by this addition. The following simplified experiment might be enough to understand the basic principle. We prepare a large convex lens and a portrait, and look at the portrait through the convex lens with one eye, closing the other. When the distances to the lens and picture are appropriate, we shall find excellent depth sensation in that monocular image.

The reason for such a monocular depth sensation may be explained as follows. Let us first assume that we look at the portrait with one eye but without the convex lens. In this case, binocular parallax and convergence are entirely absent; only accommodation is effective. Hence the eye will tell the brain that a planar portrait is being observed. Next, we suppose that a convex lens is inserted so that the virtual

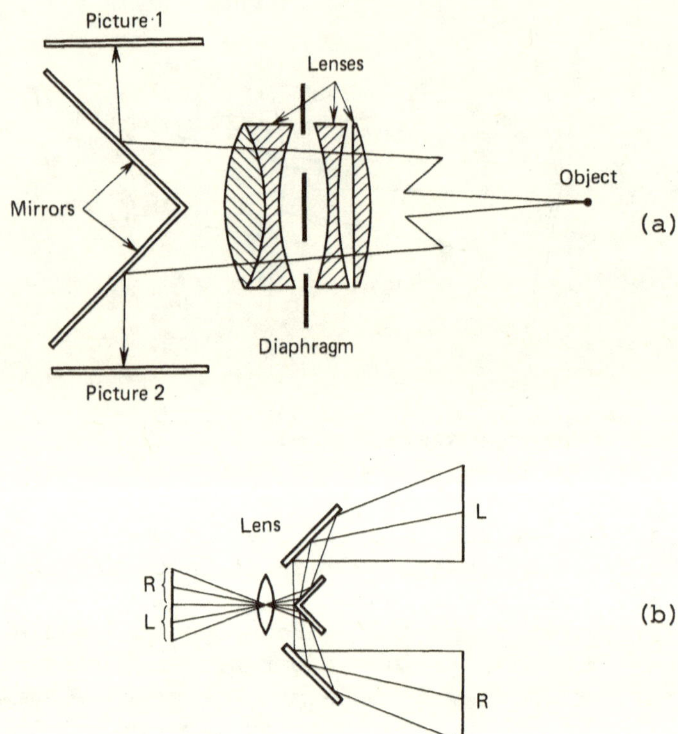

Fig.2.3(a) Kennedy's single-lens stereocamera. (b) Front-lens beam-splitter system.

image of the portrait is now located several meters ahead. Then even accommodation becomes ineffective, and there remains no effective physiological cue for depth perception. Therefore the brain judges, upon empirical grounds, that the person in the portrait must be three-dimensional.

The improvement of depth sensation in stereoscopes is based upon the same principle. In this case, the convex lenses make accommodation less significant, while accommodation is the particular cue telling that what the observer is looking at are just flat pictures. Thus, depth sensation due to binocular parallax and convergence is emphasized.

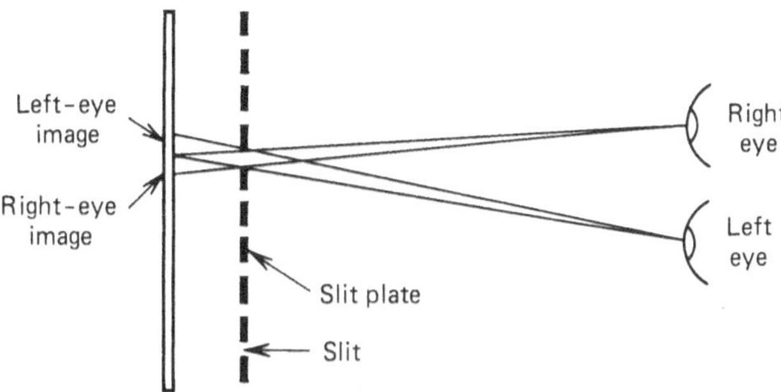

Fig.2.4 Parallax stereogram.

2.2.4 STEREOCAMERAS

The methods of taking binocular stereopictures may be classified into the following three major groups:

(1) Two pictures are taken successively by a single camera.
(2) Two pictures are taken simultaneously by a set of two ganged cameras.
(3) Two pictures are taken simultaneously by a single, special camera.

The special cameras used in methods 2 and 3 are called stereocameras. In method 2, two ordinary cameras may also be used. Method 3 uses more-or-less special cameras. Cameras of this sort can further be classified into two subgroups:

(3a) The single-lens stereocamera as shown in Fig.2.3a, devised by Clarence Kennedy (2). This features a large lens (usually larger than 75 mm) which can collect by itself the necessary binocular image information. A drawback of this method is that the large lens has a long focal length and hence requires a large photographic plate (or film) to assure a reasonable viewing angle.

(3b) The front-lens beam-splitter system as shown in Fig.2.3b collects binocular picture information at

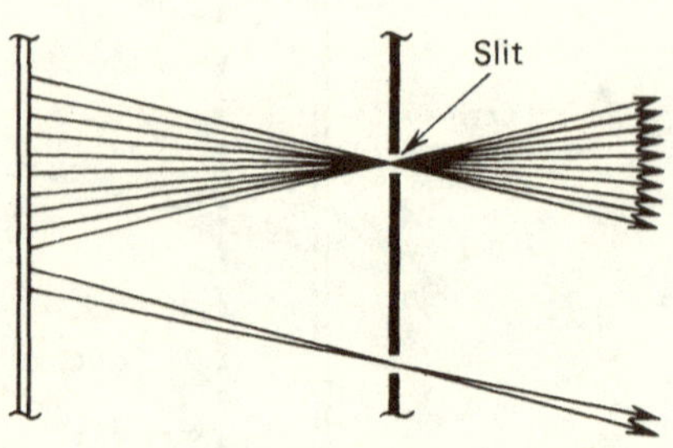

Fig.2.5 Parallax panoramagram.

appropriately separated positions and sends it into an ordinary camera to record the two pictures in half the regular size. At present, such attachments for ordinary cameras are commercially available.

2.3 PARALLAX BARRIERS

It was early in the present century that people began to consider the possibility of three-dimensional images observable from free directions. Two major streams are found in the various proposals made: one consists of those using parallax barriers, and the other consists of integral photography and its variations.

2.3.1 PARALLAX STEREOGRAM

F. E. Ives of the United States, who contributed much to the progress of optics, proposed a method of binocular stereopsis in 1903 and named it the parallax stereogram (3). This was the origin of the first stream mentioned above.

A parallax stereogram consists of a fine, vertical slit plate and a specially prepared picture placed behind the slit plate, as shown in Fig.2.4. The picture consists of the right-eye and left-eye images printed in fine stripes alternately, with approximately the same pitch as the slit plate. The principle of

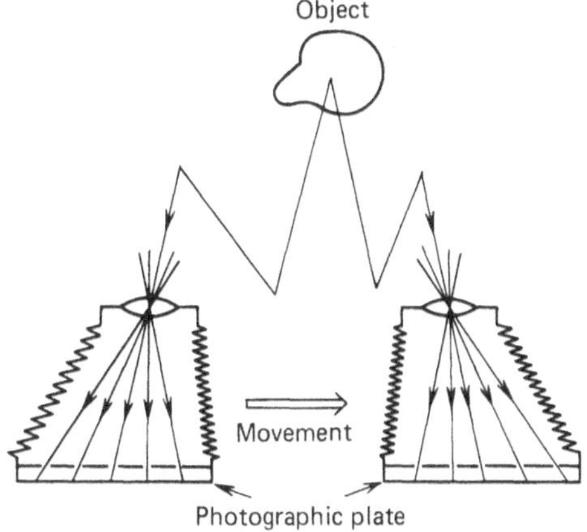

Fig.2.6 Recording of parallax panoramagram, moving camera method.

stereopsis with this device will be self-explanatory.

The idea underlying the use of the slit plate is called the *parallax barrier*. The parallax barrier itself is an older idea; F. E. Ives proposed earlier a device for color photography called a chromolinoscope (4),(5) using a similar aperture plate.

2.3.2 PARALLAX PANORAMAGRAM

The information stored in a parallax stereogram is "binocular," that is, only twice as large as planar, monocular picture information. As a result, there is very little freedom in the viewing position. To overcome this limitation, C. W. Kanolt of the United States proposed a three-dimensional imaging technique provided with a wider viewing angle in 1918 and named it the parallax panoramagram (6).

In a parallax panoramagram, the width-to-pitch ratio of the slit plate is much smaller than in a parallax stereogram, as shown in Fig.2.5; it is typically 1/10 in the panoramagram and 1/2 in the stereogram.

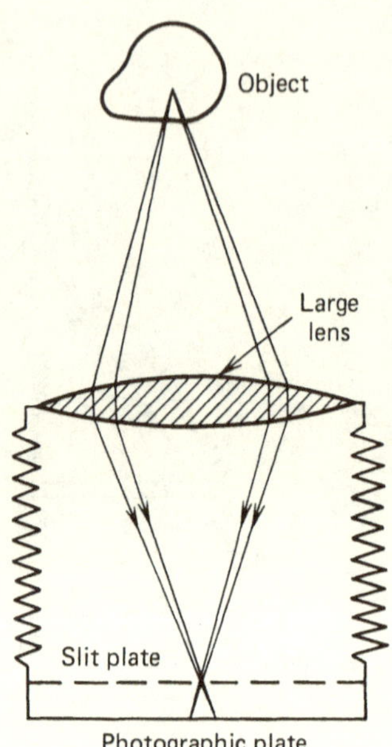

Fig.2.7 Recording of parallax panoramagram, large lens method.

Thus a wider viewing angle and more continuous spatial images can be obtained.

Experiments with the parallax panoramagram were first reported in 1928 by H. E. Ives (7),(8), a son of F. E. Ives who invented the parallax stereogram. The picture sizes were 12.5 × 17.5 cm and the slit pitch was 0.5 mm.

2.3.3 RECORDING OF PARALLAX PANORAMAGRAM--I (SINGLE CAMERA METHODS)

In early experiments with the parallax panoramagram by H. E. Ives, the original picture was prepared using the *moving camera method* shown in Fig.2.6. In this method a camera with a small-diameter lens is used. A photographic plate is placed behind the slit plate in the camera. The camera moves upon a guide rail to collect the picture information of the object viewed from various directions. During the recording, the camera body gradually deforms so that the lens is always positioned upon a line connecting the centers of the object and of the plate.

It was felt that the moving camera method had too many mechanical difficulties. (With present technologies, these difficulties are not serious; similar methods are commonly used in recording lenticular-sheet pictures. See Section 4.2.) To overcome these difficulties, H. E. Ives in 1930 (9) proposed a method using a fixed, large-diameter lens. The optical arrangement of this method is shown in Fig.2.7. If the lens has a diameter larger than the pupil distance, a single lens may collect the necessary picture information of the object viewed from various directions, and this is directly recorded upon the plate behind the slit plate.

The above method seems to work beautifully. In reality, however, the picture thus recorded gives, in the reconstruction process, a pseudoscopic image of the object, that is, an image entirely inverted in depth. This problem will be understood better through the discussion related to integral photography (see the next section).

After the large-diameter lens method, Ives proposed other methods using a fixed, large-diameter concave mirror or a movable, small-diameter concave mirror (10).

2.3.4 RECORDING OF PARALLAX PANORAMAGRAM--II (MULTICAMERA METHODS)

The single camera methods described above have a common disadvantage in that only relatively small objects can be recorded. To facilitate the recording of bigger objects, H. E. Ives proposed a multicamera recording method and also two kinds of projection-type three-dimensional display closely related to multicamera recording (11).

The principle of the multicamera method is shown in Fig.2.8. Photographs of the object are taken from N directions using N cameras (Fig.2.8a). These primary

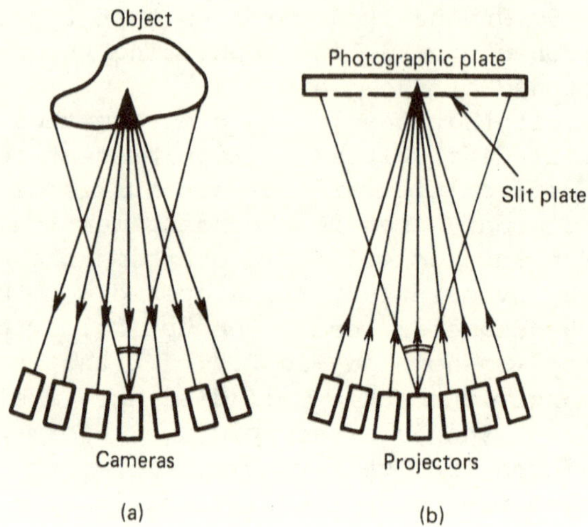

Fig.2.8 Recording of parallax panoramagram, multi-camera method.

photographs may be negative pictures. Then this picture information is transferred to a single photographic plate placed behind a slit plate in the manner shown in Fig.2.8b. The photographic plate is developed, fixed, and again placed behind the slit plate, and a parallax panoramagram is completed.

On the basis of this multicamera recording method, Ives proposed the following projection-type displays. The first one is shown in Fig.2.9 which is apparently identical with Fig.2.8b. In the projection-type display, however, the slit plate is painted dark and the photographic plate is replaced by a white diffusing screen, and the projected image is observed directly. If the two eyes are positioned, for example, as shown in Fig.2.9, the two eyes will observe only pictures projected from projectors 3 and 5, respectively. When the eyes move, the observed pictures will change successively, and thus a spatial image is perceived.

A possible version of the above scheme is a rear-projection-type display using a three-layer composite screen consisting of a slit plate, a translucent

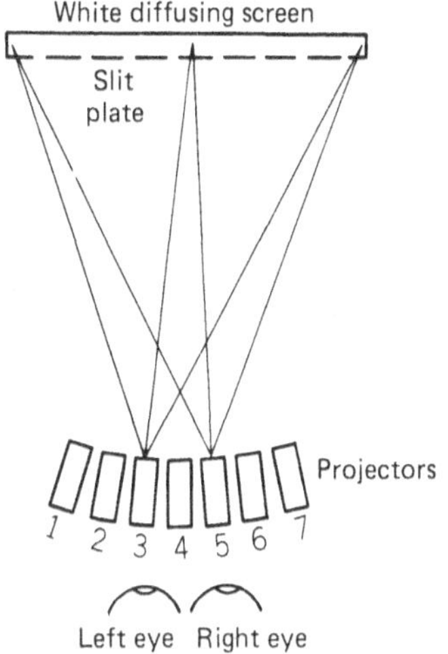

Fig.2.9 Projection-type display developed from the multicamera recording of parallax panoramagram (I).

diffusing plate (like a ground glass plate), and a slit plate (12).

In the second method shown in Fig.2.10, a screen consisting of a slit plate and a translucent diffuser plate is used. The N primary pictures are projected from N directions; they are not observed directly but are again recorded by a camera placed behind the screen (Fig.2.10a). In the final observation, the single picture thus obtained is projected from behind the screen and viewed from the front (Fig.2.10b). In this method, the final picture recorded consists of many fine stripelike images, and registration of the picture is a serious problem.

Projection-type displays will be discussed again in Section 2.7 and Chapter 5.

2.3.5 *DRAWBACKS OF PARALLAX BARRIERS*

The technique of the parallax barrier later showed some progress; multilayer versions and applications to stereoscopic movies were reported by many investigators until late in the 1950s (13). At present, however, the parallax barrier device is just a museum piece

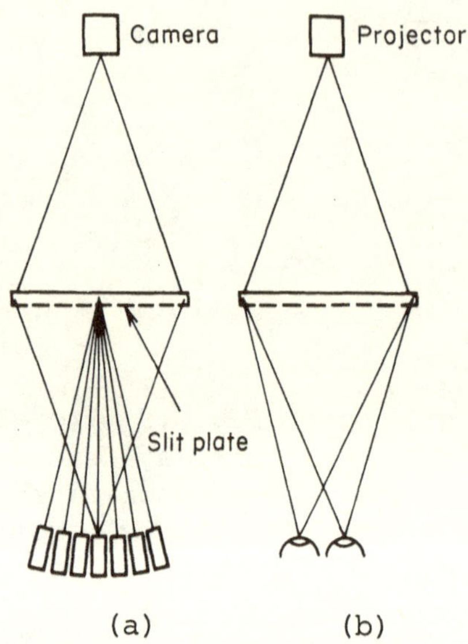

Fig.2.10 Projection-type display developed from the multicamera recording of parallax panoramagram (II).

because it has some essential drawbacks.

The first drawback involves darkening of the images. The second and third stem from the necessity of using very fine slits so that they are not observable. The second is that the very narrow slits bring about a spread of light bundles due to diffraction. The third drawback is a little roundabout; the fine slits must be made on a glass plate, whose surface reflection is often annoying. This reflection is less annoying in the rear-projection system described in the preceding subsection, but in that case the registration of the two slit plates is a difficult problem.

The pitch of the slit p must satisfy, for healthy human eyes,

$$p < (\text{viewing distance})/3500. \tag{2.1}$$

For example, $p < 0.3$ mm for a distance of 1 m and $p < 0.08$ mm for 0.25 m. If the slit width is 1/10 of the pitch, the widths are 30 and 8 μm, respectively. The last value is only about 15 times greater than the wavelength of visible light.

2.4 INTEGRAL PHOTOGRAPHY (IP)

The origin of the second major stream of three-dimensional imaging viewable from free directions is the invention of integral photography by M. G. Lippmann, a famous French optics investigator, in 1908 (14). Integral photography was later simplified to the lenticular-sheet three-dimensional picture; this will be described in the next section.

2.4.1 PRINCIPLE OF INTEGRAL PHOTOGRAPHY

Integral photography is a method of recording a complete spatial image upon a single, flat photographic plate. Since its invention in 1908, it remained almost forgotten for 60 years until a few curious investigators in the mid 1960s noticed the old proposal.

The principle of integral photography is illustrated in Fig.2.11. We prepare a fly's-eye lens sheet, which is a glass or plastic sheet consisting of a tremendous number (for example, 10,000) of small convex lenses. If we place a photographic plate at the focal plane of these lenses and expose it to light from an object, we can record upon the photographic plate a tremendous number of small pictures of the object viewed from various directions.

We develop the photographic plate, convert the negative picture to a positive one if necessary, place the positive picture at exactly the same position as before, and illuminate it from the rear as shown in Fig.2.11b. What will then be observed from the left?

Suppose the object is a person holding a red sign (for example, a red flower) upon his breast. Then 10,000 red spots are recorded on the photographic plate.

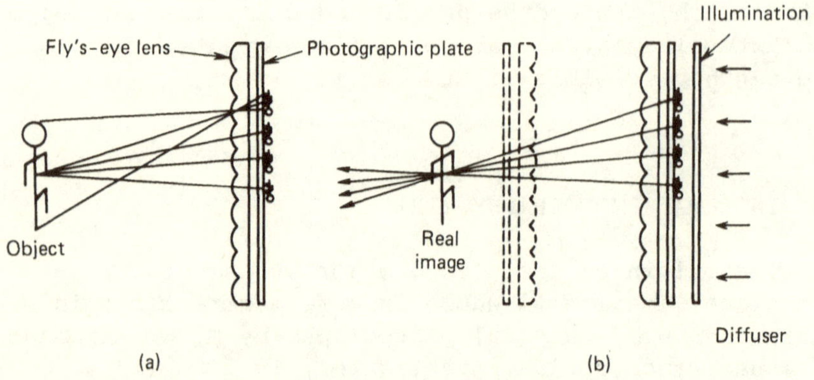

Fig.2.11(a) Recording of an integral photograph. (b) Reconstruction from an integral photograph. The figure in broken lines shows the secondary integral photograph.

In the reconstruction of the image, the rays of the red light scattered upon those red spots on the photographic plate come down along the same paths as in the taking of the picture, but in the opposite directions. Hence, an observer on the left will see something red in the direction toward the original position of the sign. This means that a freely moving observer, or an observer having two eyes, will see something red floating in space where the red sign originally was. The above statement holds not only for the specific point but for all points upon the entire object. Thus a three-dimensional, spatial image is observed.

2.4.2 PROBLEM OF PSEUDOSCOPIC IMAGES

There arises a peculiar difficulty however. The recorded optical information was the shape and color of the object seen from the right. Therefore, in reconstruction, an observer on the left will observe a so-called pseudoscopic image, that is, an image inverted in its depth. For example, what the observer will see as the face of a person is something like the back surface of a death mask painted in the original colors, which is not desirable.

The above difficulty was first pointed out by H. E. Ives in 1931 (15), who also proposed a method for overcoming it. His proposal was to take a second integral photograph of the reconstructed image again from the left, develop, and observe it from the right as shown by broken lines in Fig.2.11b. In such a scheme, although the image is further deteriorated in resolution, it is no longer pseudoscopic but orthoscopic. In this book this scheme of Ives will hereafter be called *two-step integral photography*.

This scheme might look troublesome; however, we should note that it lightens other technical difficulties of integral photography. First, the conversion from a negative to positive picture can be done between the primary and secondary integral photographs; that is, we may simply make the primary picture negative and the secondary positive. Second, as a result of avoiding the independent conversion process as described before, we may provide the photosensitive emulsion layer directly on the rear surface of the lens sheet if desired. When the photographic plate is detached from the lens sheet for development and inversion, as in the description of the principle of integral photography in the preceding subsection, the difficulty of registration arises. In the two-step integral photography scheme, this difficulty can be avoided.

2.4.3 EARLY EXPERIMENTS WITH INTEGRAL PHOTOGRAPHY (PINHOLE INTEGRAL PHOTOGRAPHY)

Experiments with integral photography as described above could not be performed before World War II because good plastic materials were not available. The first experiment with integral photography was performed with a "pinhole sheet" instead of a lens sheet, by A. P. Sokolov of Moscow University in 1911 (16).

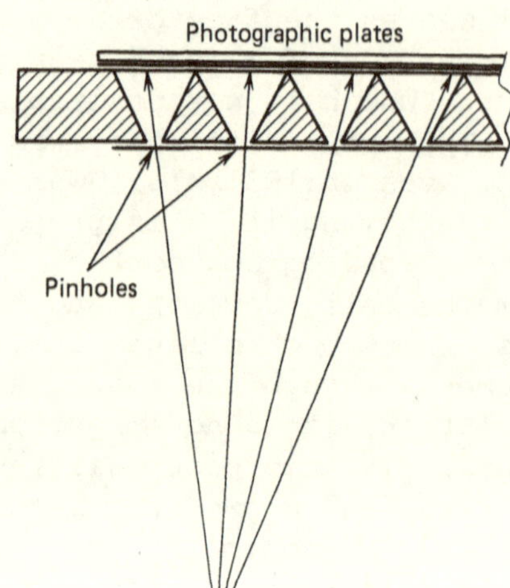

Fig.2.12 The pinhole integral photography by Sokolov.

The original paper by Sokolov (16) is not easily accessible, but his experiment is described by N. A. Valyus (17). Sokolov prepared a "pinhole sheet" provided with 1200 cone-shaped pinholes as shown in Fig. 2.12. The thickness and dimensions of the sheet were 3 mm and 150×200 mm^2, respectively. Such a pinhole sheet is equivalent to a fly's-eye lens sheet except for the brightness of the image obtained. Sokolov reported that he took an integral photograph of the filament of an incandescent lamp and obtained definite depth sensation; however, with only 1200 pinholes, the depth sensation obtained must not have been satisfactory.

2.4.4 RECENT DEVELOPMENTS IN LENS-SHEET INTEGRAL PHOTOGRAPHY

Not many lens-sheet integral photographs have actually been produced. According to Valyus (17), the first experiment with lens-sheet integral photography was performed by S. P. Ivanov and L. V. Akimakina of the Soviet Union in 1948. The lens sheet had 2,000,000 lenslets with diameter and focal length of 0.3 and 0.5 mm, respectively. No quantitative data are available on the picture quality obtained.

On the other hand, in the United States an ardent investigator R. L. de Montebello is said to have made a great contribution to early technological progress in integral photography (18). However, he published no technical papers except for a patent filed in 1966 (19). The first report of the experiment was published by Burckhardt and Doherty (20) of Bell Telephone Laboratories, who used a beaded-plate lens sheet instead of a fly's-eye lens sheet (see Section 4.5).

At the time this book was written, the only manufacturer of integral photography equipment is the MDH Products Corporation of Ann Arbor, Michigan, which is jointly operated by de Montebello and others. They achieved excellent picture quality, with fairly wide viewing angles (almost ±45°) for both horizontal and vertical rotations.

For amateur experiments with multiple photography, which is a simpler version of integral photography, an experimental device for what is called "multiplex recording photography" is said to be commercially available (21). It records 250 small pictures of an object upon a 5 × 5 cm photographic plate.

While integral photography has shown little progress due to technological difficulties, its simplified version, lenticular-sheet three-dimensional picture, has become a more practical imaging technique during the 1960s.

2.5 LENTICULAR-SHEET THREE-DIMENSIONAL PICTURES

About 20 years after the invention of integral photography, many people began to consider simplifications. In an integral photograph, the reconstructed image has so-called vertical parallax information; in other words, the reconstructed image can either be looked at from down or up, from any vertical direction

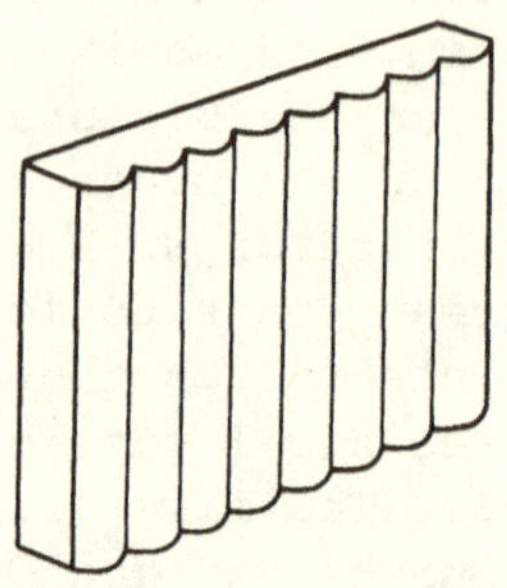

Fig.2.13 A lenticular sheet.

within a limited angle. People reasoned that if they gave up this vertical parallax information, the lens sheet might be simplified from the planar array of convex lenslets to a linear array of cylindrical lenses as shown in Fig.2.13. Such a lens sheet is called a lenticular sheet.*

2.5.1 PRINCIPLE OF LENTICULAR-SHEET THREE-DIMENSIONAL PICTURES

Lenticular-sheet three-dimensional pictures should be called autostereoscopic pictures of the "unidirectional" type, in the sense that only horizontal parallax information is preserved. Their principle is self-explanatory in Fig.2.14. A lenticular sheet is usually made in such a thickness that its rear surface coincides with the focal plane of the lenses. If the images of an object observed from various directions are printed on the rear surface in fine stripes, the right and left eyes will see two different images and depth sensation will be obtained.

---

* The word *lenticular* is originally an adjectival form of *lens*. It is merely a matter of custom that we use this word only for the cylindrical lens sheet, but not for the fly's-eye lens sheet.

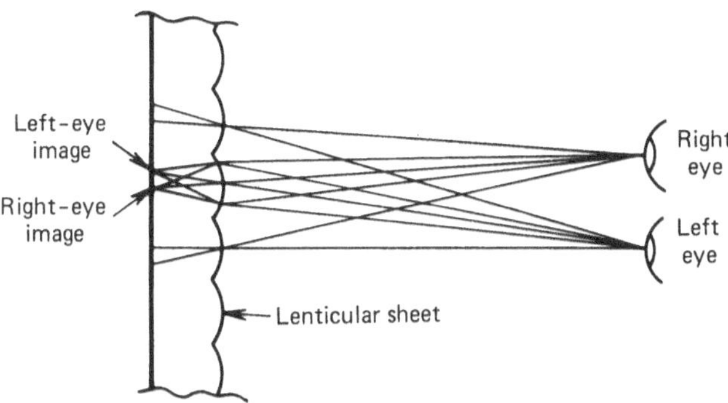

Fig.2.14 A lenticular-sheet three-dimensional picture.

2.5.2 HISTORY OF DEVELOPMENT OF LENTICULAR-SHEET PICTURES

The invention of lenticular-sheet three-dimensional pictures (hereafter sometimes called LS pictures for short) cannot be credited to a single inventor, but to several, including F. E. Ives and his son H. E. Ives (1). Later, further developments were carried out by Vanbenschoeten and Winnek in the United States and Bonnet in France. In the early stage of their development, however, the lack of necessary materials and machining technologies greatly retarded progress.

An epoch was marked when the 1961 Annual Meeting of TAGA (Technical Association of the Graphic Arts) of the United States took up LS picture technology for discussion (22). In addition, progress in the technological backgrounds in the 1950s and 1960s greatly accelerated the developments of LS pictures:

(1) progress in precision machining brought forth lenticular sheets with good optical characteristics;

(2) progress in plastic materials increased freedom of material choice; and

(3) progress in photographic and printing technologies facilitated picture production.

Since a lenticular sheet is much easier to fabricate than a fly's-eye lens sheet, its use showed rapid progress in the 1960s as a forerunner of integral photography. In the United States, the second epoch was marked when LS pictures called Xographs, produced jointly by Eastman Kodak (lens sheet), Eastman Chemicals (photographic material), Harris-Intertype (registration), and Donneley (printing), were published in the April 7, 1964 issue of *Look* magazine, which at that time was printed in millions of copies (23). In Japan, Toppan Printing Company and Dai-Nippon Printing Company both started developing LS pictures around 1960. In France, the studio operated by Bonnet has developed an excellent technique; some samples from Bonnet's studio were presented at the French Pavilion of Expo 1970 held in Osaka, Japan.

2.6 HOLOGRAPHY

In an ordinary photograph, what is recorded at a point upon the photographic plate is the brightness of the image at that point. In an integral photograph, what is recorded at a point is the brightness of the image at that point seen from various directions. Such direction-selective recording of the image is the premise for recording a spatial image upon a flat recording material.

In integral photography, a fine lens sheet is used for direction-selective recording of the image. The holography to be described in this section uses, instead of a lens sheet, a tremendous number of diffraction gratings made upon a flat recording material as interference fringes. In the recording process of holography, a good "coherent" light source is used to make these interference fringes. In the reconstruction process, the diffraction gratings are then illuminated by monochromatic light, and they diffract the illuminating light beam to form a wavefront similar to the originally recorded wavefront from the object.

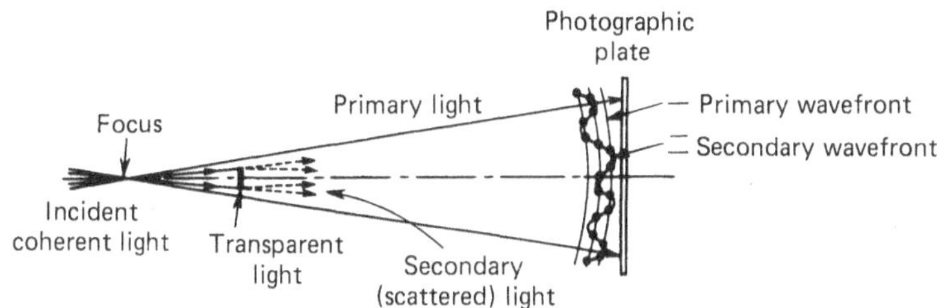

Fig.2.15 An optical model of Gabor's wavefront reconstruction method. On those points (dots) where the primary and secondary wavefronts have the same phase, the light intensity is enhanced.

At present, applications of holography are not limited to spatial imaging. Many people believe that its application to information processing, especially huge-capacity and/or associated-readout digital memory devices might be more promising. However, when D. Gabor of the United Kingdom proposed the original idea of holography in 1948, his aim was neither imaging nor information processing, but the improvement of electron microscopes. The original proposal of Gabor will first be described.

2.6.1 PROPOSAL BY GABOR

An electron microscope uses a very high energy electron beam (hence having a very short wavelength, typically 5×10^{-12} m) instead of light. Because of the very short wavelength of the electron beam, a high resolution is expected. In practice, however, the resolution limit could not be reduced below several angstroms because of spherical aberration of the electron lens.

To overcome this limitation, Gabor devised a new microscopic principle (24),(25) on the basis of Sir Lawrence Bragg's x-ray microscope. As described in the following paragraphs, it is a "lensless" electron

(a)

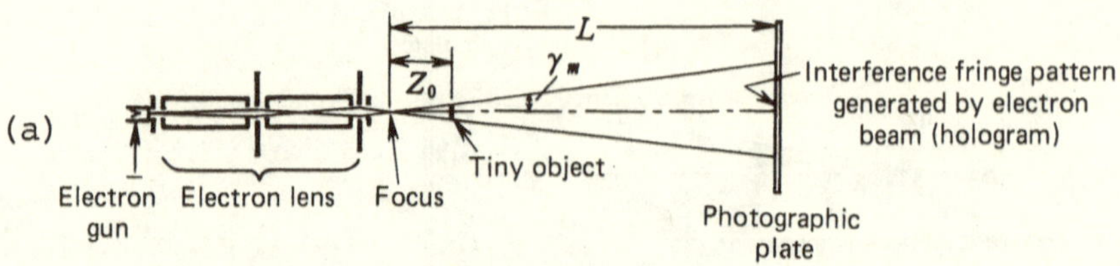

(b)

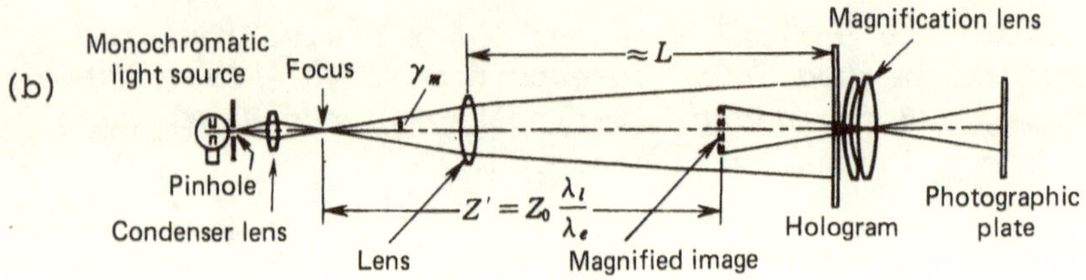

Fig.2.16 Electron microscopy based upon the wavefront reconstruction principle [after Gabor (25)]. (a) Recording of the hologram. (b) Reconstruction of the magnified optical image.

microscopy; hence it is free from the limitation due to lens aberrations.

The principle of Gabor's proposal may better be explained first with respect to an optical system as shown in Fig.2.15. We assume that coherent light is collimated to a pinpoint, while a tiny, transparent object is placed in the diverging light bundle. On the right is a photographic plate. In such an arrangement, the primary, undisturbed light and the secondary light scattered by the object are incident upon the photographic plate. Since the light source is coherent, these primary and secondary light rays interfere with each other, and the resultant interference fringe pattern is recorded upon the photographic plate.

Next we consider an arrangement as shown in Fig. 2.16a, where an electron beam is used instead of light in Fig.2.15. According to wave mechanics, an electron beam may be regarded as a wave, whose wavelength is given as

$$\lambda_e = h/\sqrt{2meV} \quad , \tag{2.2}$$

where h, m, e, and V denote Planck's constant (6.24×10^{-34} J s), the electron mass (9.107×10^{-31} kg), the electron charge (1.602×10^{-19} C), and the accelerating voltage of the electrons. For example, when V = 50 kV, λ_e = 0.005 nm. A photographic plate is usually sensitive to high-energy electron beams. Therefore, in an arrangement like Fig.2.16a, very fine electronic interference fringes are recorded upon the plate; these fringes must record complete image information for the object. Gabor called such a fringe pattern a *hologram,* from Greek words meaning "everything is recorded."

In reconstructing the image from a hologram, an optical arrangement as shown in Fig.2.16b is used. The hologram is illuminated by monochromatic light with wavelength λ_ℓ. It will be shown in Section 6.4.3 that in such an arrangement, a virtual image of the object magnified by a ratio λ_ℓ/λ_e is reconstructed as shown in the figure. We either observe that virtual image directly or take its photograph by using the lens system shown on the right of the figure. For example, when λ_ℓ = 500 nm and λ_e = 0.005 nm, the expected magnification is 100,000.

Gabor called his method of image recording and reconstruction using interference fringe patterns (hologram) "wavefront reconstruction." Later the word *hologram* was combined with *photography* to bring forth the word *holography,* which soon became prevalent. It is not exactly known who used this word for the first time.

2.6.2 EXPERIMENT BY GABOR

Seemingly, at first Gabor was not seriously interested in visible-light holography. He performed his first experiments on wavefront reconstruction using visible light just to "verify its principle," probably because the electron-beam system could not easily be constructed. Since no laser was available at that time, he used in both the recording and reconstruction processes 435.8-nm light from a mercury arc lamp, improving the monochromaticity by letting the light pass through a pinhole having a diameter of 5 μm. In ref. (25), a reconstructed image of a small object (0.65 × 0.5 mm) is shown. The image quality was not excellent because of the poor coherence obtainable with the mercury lamp.

It should be noted that in Ref.(25), Gabor pointed out the possibility of three-dimensional imaging using wavefront reconstruction.

2.6.3 TWO-BEAM METHOD BY LEITH AND UPATNIEKS

The scheme proposed by Gabor had two difficulties. The first is that, in addition to the desired image, another spurious image (the conjugate image) is inevitably seen in the same direction as the desired one. The second difficulty is that the object must be transparent to permit the transmission of the light to be recorded.

In 1962, Leith and Upatnieks of the United States proposed the two-beam method which overcomes both of the above difficulties (26)-(28). They also succeeded in obtaining greatly improved images using the two-beam method and a laser, which had just become available in laboratories, as the light source.

The principle of the two-beam method is shown in Fig.2.17 for the case when a transparent object (a transparency) is recorded. We prepare two light beams from a single coherent light source. One beam is transmitted through the object, partly scattered, and is incident upon the photographic plate. Another beam, after being refracted in a prism, illuminates the photographic plate as a uniform plane wave. The latter beam is called the reference beam or reference wave. The hologram thus recorded will have a pattern consisting of almost parallel fringes produced by the interference between the two beams. The average spacing of the fringes is approximately given as $d = \lambda/\sin\theta$, where

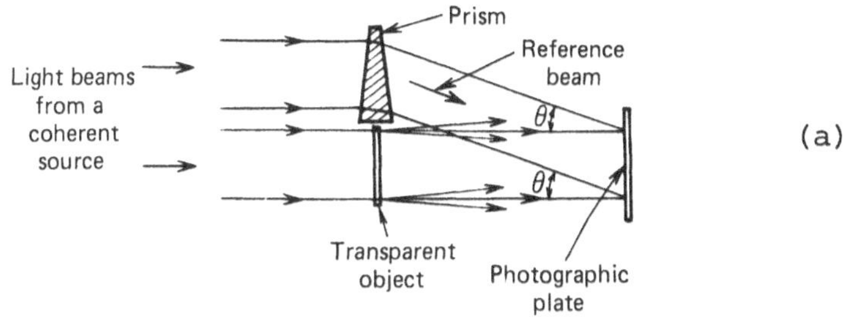

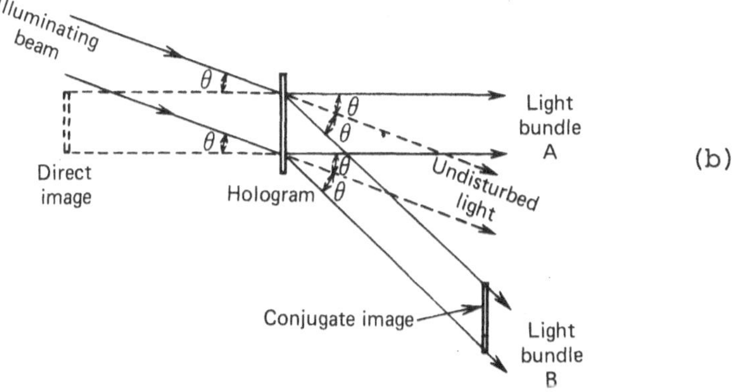

Fig.2.17 Two-beam method of Leith and Upatnieks, I (case of transparent object). (a) Recording of the hologram. (b) Reconstruction [from Leith and Upatnieks (26)].

λ and θ are the wavelength and the incident angle of the reference beam as shown in Fig.2.17, respectively. The spacing and density of the fringes are modulated in accordance with the scattering of the object beam. Since this situation is similar to the modulation of a carrier wave by a signal in electrical communication, the two-beam method is sometimes called the carrier frequency method. (In contrast to that, Gabor's scheme was called the coherent background method, but presently it is more commonly called in-line holography.)

In reconstructing the image, the hologram is illuminated by a beam usually having the same wavefront as the reference beam used in the recording, as shown in

(a)

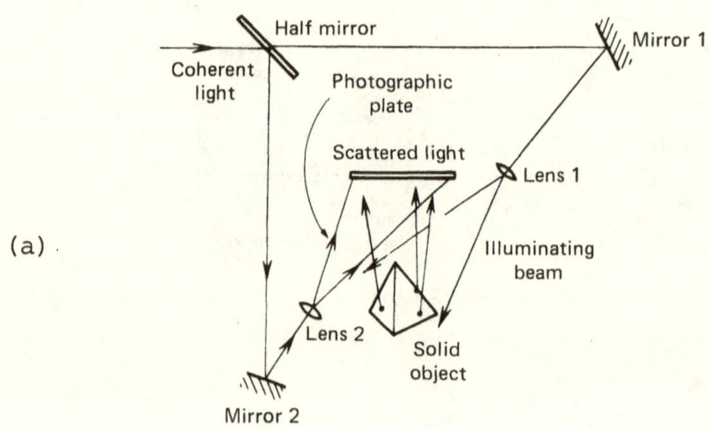

(b)

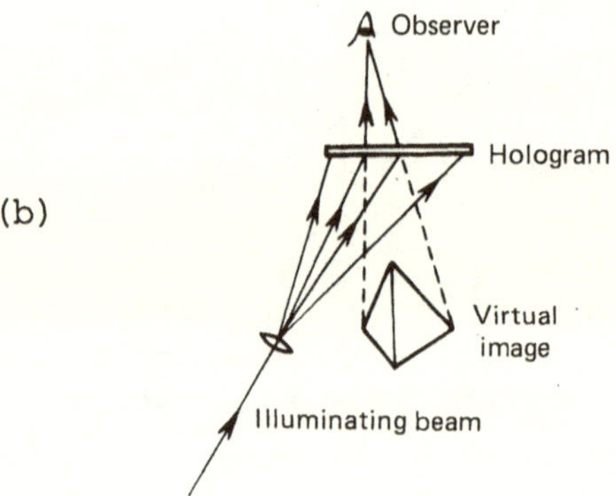

Fig.2.18 Two-beam method of Leith and Upatnieks, II (case of solid object). (a) Recording of the hologram. (b) Reconstruction [from Leith and Upatnieks (27)].

Fig.2.17b. In such a case, the diffraction upon the hologram generates three light components. They are, as shown in Fig.2.17b: (1) light bundle A which has the same wavefront as the original object wave recorded; (2) the undisturbed, directly transmitted light; and (3) light bundle B which produces another image, which is real. The virtual image produced by A is called the direct image, whereas the real image produced by B is called the conjugate image. (Gabor's in-line holography corresponds to the case in which $\theta = 0$ in Fig.2.17. In such a case the direct and conjugate images are not separable.)

The two-beam method can easily be applied to a solid object. The optical arrangements for this case are shown in Fig.2.18. In the recording process illustrated in Fig.2.18a, the light scattered upon the surface of the object (the object wave) interferes with the reference wave everywhere upon the photographic plate, and the resultant fringe pattern is recorded as diffraction gratings upon the hologram. In reconstructing the image, the hologram is illuminated by light having the same wavefront as the reference wave (Fig.2.18b). An observer will see a virtual, spatial image behind the hologram, identical in shape to the object recorded. Thus holography can be used as a three-dimensional imaging technique.

After the proposal of Leith and Upatnieks, holography showed great progress first as an optical imaging technique, contrary to the inventor's early aim. It looked truly promising for three-dimensional imaging, and people even began to talk about the possibility of holographic three-dimensional television in the near future.

2.6.4 HOLOGRAPHY AS A THREE-DIMENSIONAL IMAGING TECHNIQUE

In the history of three-dimensional imaging techniques, the invention of holography was particularly significant because it offered for the first time a method of spatial imaging satisfactory for accommodation. Before holography, all the methods proposed or used relied only upon convergence and parallax.

The holographic process is, however, accompanied by many practical disadvantages, which are mostly attributed to the use of coherent light sources (practically lasers) in the recording and in most of the reconstructing processes. Some of these are:

(1) Recording can be performed only in a darkroom.

(2) The recorded and reconstructed images are monochromatic.

(3) The object must be still in order that a good hologram be produced.

(4) The high resolution plates used in holography are usually very expensive; therefore, holography is not an economical method for large-scale imaging.

The advantages and disadvantages of holography as a three-dimensional imaging technique will be discussed in more detail in Section 6.1.

Vigorous efforts for overcoming the above difficulties have been made. Pole proposed the holocoder-hologram method, and McCrickerd and George proposed the holographic stereogram to overcome difficulties 1 and 3. Color holography techniques were investigated to overcome item 2 mainly by investigators at Bell Laboratories. Item 4 can be avoided by the use of the projection-holography technique. These various proposals and techniques will be described mostly in Chapter 6 and partly in Chapter 7.

2.7 PROJECTION-TYPE THREE-DIMENSIONAL DISPLAYS

Some primitive proposals and experiments of projection-type displays which appeared around 1930 were described in Section 2.3.4. More than 30 years later, in the late 1960s, development of new-type projection screens started again. These recently developed screens, taking advantage of advanced material and machining technologies, show much better characteristics than the older ones; however, they are the main subject of Chapter 5 and are omitted here. In this section, proposals and experiments that appeared in the few years following the invention of the parallax-barrier screen in 1929 (see Section 2.3.4) will be described.

2.7.1 *LENTICULAR-SHEET SCREEN*

As will be discussed in detail in Chapter 5, a lenticular sheet backed by white diffusing paint can be used as a direction-selective screen for three-dimensional imaging. This scheme was first proposed by H. E. Ives in 1930 (29). The first experiment was also performed by H. E. Ives and reported in 1931 (30). He prepared two screens, one with dimensions 12.5 × 17.5 cm^2 and a pitch of 0.5 mm, another with dimensions 60 × 60 cm^2 and a pitch of 3 mm, both painted white on their rear surfaces. Thirty-nine projectors were used in both experiments.

In the same paper, H. E. Ives reported another experiment with the projection system as shown in Fig.2.10, where the parallax-barrier screen is replaced by a lenticular screen (actually consisting of many celluloid rods machined flat on their sides) backed by a translucent diffuser plate (30). Shortly afterward, H. E. Ives also proposed color three-dimensional imaging by combining the above projection system and chromolinoscope (31).

2.7.2 *SPECULAR DIRECTION-SELECTIVE SCREEN*

In another paper, H. E. Ives proposed a specular direction-selective screen consisting of vertically arranged right-angle double mirrors provided with slight, regular unevenness as shown in Fig.2.19 (32). The expected optical properties are similar to those of the diffuser-backed lenticular screen, that is, a direction-selective property in the horizontal direction and a diffusing property in the vertical direction.

2.7.3 *TRANSMISSION-TYPE DIRECTION-SELECTIVE SCREENS*

A transmission-type direction-selective screen consisting of a lenticular sheet, a translucent diffuser

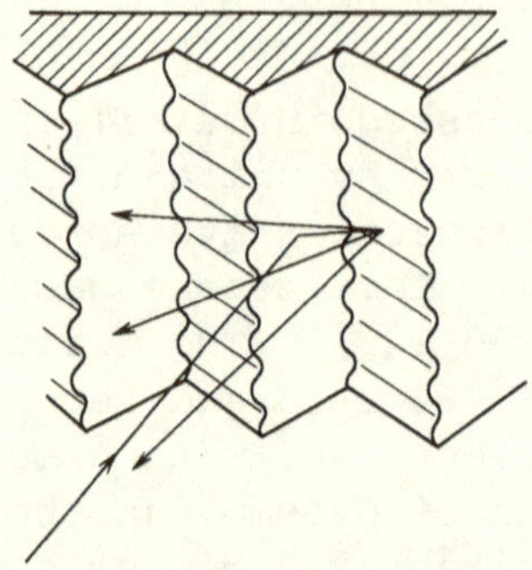

Fig.2.19 Right-angle double-mirror screen with vertically diffusing property [after H. E. Ives (32)].

plate, and a lenticular sheet was invented in 1931 by DeLassus Saint Genies of France (33). (See Section 5.3, where it is referred to as a transmission-type double-lenticular screen.) But, experiments were not performed because of the lack of technological backgrounds.

To summarize, a characteristic feature of the investigations in this period is that the proposals were rarely followed by practical experiments due to lack of materials and machining technologies.

2.8 THREE-DIMENSIONAL MOVIES

Movies will not be dealt with in detail in this book; only a glance at their historical background is given in this section.

The history of three-dimensional movies is as old as the history of ordinary movies. Many proposals appeared and disappeared. The first large-scale stereoscopic movie theater with excellent picture quality appeared in 1939 at the New York World's Fair. At that time a black-and-white film was shown; the projected pictures were polarized in orthogonal directions and the viewers held polaroid glasses. In the following year stereoscopic technicolor films were exhibited in the United States. More than five million people saw those films (1).

In the same period, people in the Soviet Union were also eager to develop stereoscopic movies. Later their efforts brought forth the gigantic movie theater at EXPO 1970 held in Osaka (34). It employed a diffuser-backed radial-lenticular screen, to be described in Section 5.3, and required no glasses.

In the history of three-dimensional (or stereoscopic) movies, the greatest epoch was begun several years after World War II by a new movie technique which itself was not three-dimensional at all. This was "Cinerama" developed by F. Waller in the United States (1).

Cinerama is just an ordinary two-dimensional movie except that the visual field is as wide as 120°. Nevertheless, it succeeded in giving a strong, realistic three-dimensional sensation by surrounding the spectators through a wide angle. Thus, after Cinerama, stereoscopic movie systems relying upon parallax effects became less prevalent; competition among them suddenly disappeared.

The success of Cinerama offers an interesting problem in the psychology of depth perception. In most movie theaters the distance between the spectators and the screen is longer than 10 m. At such a long distance, depth perception due to accommodation is entirely lost. Moreover, depth perception due to binocular parallax and convergence is also much weakened. Therefore, in such big theaters direct physiological depth cues are less important. It is possible to give the spectators a realistic sensation from a superwide screen, taking advantage of their imaginations and of illusion. Such an effect is rarely mentioned in textbooks of psychology, but the presence of the effect is widely accepted by specialists. It might be called the wide-screen effect or the Cinerama effect.

After Cinerama, proposals of stereoscopic or three-dimensional movies are still found in the literature (35)-(37).

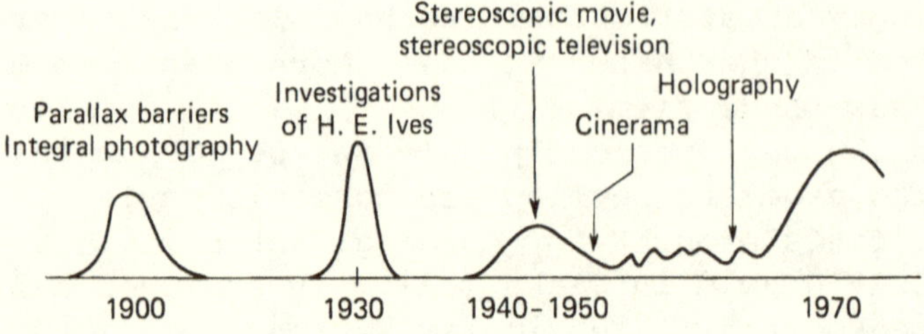

Fig.2.20 History of three-dimensional imaging techniques.

2.9 SUMMARY

The history of three-dimensional imaging techniques described above may be summarized in a graph as shown in Fig.2.20. This figure is featured by several hills and valleys; the present is probably located at a hill. The investigations by H. E. Ives form a narrow peak around 1930. His works were not followed by practical developments because of lack of material and machining technologies.

There are two facts relating to the hill around 1970. First, we have the technique of holography at hand. Second, great progress has been achieved in material and machining technologies in the past 20 years. Future developments will primarily depend upon whether we can take advantage of those merits effectively.

References

(1) Norling, J. A., The Stereoscopic Art--A Reprint, *J. SMPTE 60*, No. 3, 286-308 (March 1953).

(2) Kennedy, C., The Development and Use of Stereo Photography for Educational Purposes, *J. SMPTE 26*, No. 1, 3-17 (Jan. 1936).

(3) Ives, F. E., U.S. Patent, 725,567 (1903).

(4) Ives, F. E., U.S. Patent, 666,424 (1901).

(5) Ives, H. E., The Chromolinoscope Revived, *J. Opt. Soc. Amer. 20,* 343-353 (June 1930).
(6) Kanolt, C. W., U.S. Patent, 1,260,682 (1918).
(7) Ives, H. E., A Camera for Making Parallax Panoramagrams--A Talk Presented at the Michelson Meeting, Sept. 1-3, 1928.
(8) Ives, H. E., A Camera for Making Parallax Panoramagrams, *J. Opt. Soc. Amer. 17,* 435-439 (Dec. 1928).
(9) Ives, H. E., Parallax Panoramagrams Made with a Large Diameter Lens, *J. Opt. Soc. Amer. 20,* 332-342 (June 1930).
(10) Ives, H. E., Parallax Panoramagrams Made with a Large Diameter Concave Mirror, *J. Opt. Soc. Amer. 20,* 597-600 (Nov. 1930).
(11) Ives, H. E., Motion Pictures in Relief, *J. Opt. Soc. Amer. 18,* 118-122 (Feb. 1929).
(12) Okoshi, T., Three-Dimensional Imaging and Television (A Review), *J. IECEJ 51,* No. 10, 1247-1257 (Oct. 1968).
(13) Kaplan, S. H., Theory of Parallax Barriers, *J. SMPTE 59,* No. 7, 11-21 (July 1952).
(14) Lippmann, M. G., Epreuves Reversibles Donnant la Sensation du Relief, *J. Phys. 7,* 4th series, 821-825 (Nov. 1908).
(15) Ives, H. E., Optical Properties of a Lippmann Lenticulated Sheet, *J. Opt. Soc. Amer. 21,* 171-176 (March 1931).
(16) Sokolov, A. P., "Autostereoscopy and Integral Photography by Professor Lippmann's Method." Izd. MGU, Moscow State Univ. Press (1911).
(17) Valyus, N. A., "Stereoscopy." Focal Press, London, 1966.
(18) Burckhardt, C. B., private communication.
(19) DeMontebello, R. L., U.S. Patent, 3,503,315 (1970).
(20) Burckhardt, C. B., and Doherty, E. T., Beaded Plate Recording of Integral Photographs, *Appl. Optics 8,* No. 11, 2329-2331 (Nov. 1969).
(21) Hanson, W., Enter MRP, Photography, 33 (May 1967).
(22) Horiuchi, M., private communication.
(23) Hizume, S., private communication.

(24) Gabor, D., A New Microscopic Principle, *Nature*, No. 161, 777-779 (May 15, 1948).

(25) Gabor, D., Microscopy by Reconstructed Wavefronts, *Proc. Phys. Soc. A194*, 454-487 (1949).

(26) Leith, E. N., and Upatnieks, J., Reconstructed Wavefronts and Communication Theory, *J. Opt. Soc. Amer. 52*, No. 10, 1123-1130 (Oct. 1962).

(27) Leith, E. N., and Upatnieks, J., Wavefront Reconstruction with Continuous-Tone Objects, *J. Opt. Soc. Amer. 53*, No. 12, 1377-1381 (Dec. 1963).

(28) Leith, E. N., and Upatnieks, J., Wavefront Reconstruction with Diffused Illumination and Three-Dimensional Objects, *J. Opt. Soc. Amer. 54*, No. 11, 1295-1301 (Nov. 1964).

(29) Ives, H. E., Parallax Panoramagrams for Viewing by Reflected Light, *J. Opt. Soc. Amer. 20*, 585-592 (Oct. 1930).

(30) Ives, H. E., The Projection of Parallax Panoramagrams, *J. Opt. Soc. Amer. 21*, 397-409 (July 1931).

(31) Ives, H. E., A Method of Projection in Relief and Color, *J. Opt. Soc. Amer. 22*, 191-198 (April 1932).

(32) Ives, H. E., Reflecting Screens for Relief Picture Projection, *J. Opt. Soc. Amer. 21*, 109-118 (Feb. 1931).

(33) Saint Genies, DeLassus, U.S. Patent, 2,139,855 (1934).

(34) Boltiansky, A. G., and Ovsiannikova, N. A., Stereoscopic Cinematography in the Soviet Pavilion at the Expo-70, Cine-Technology and Television (in Russian), 22-27 (Oct. 1970).

(35) Collender, R. B., The Stereoptiplexer, *J. Soc. Photo-opt. Instrum. Eng. 6*, No. 1, 14-18 (Oct./ Nov. 1967).

(36) Collender, R. B., The Stereoscopic Movie System Without Glasses, Information Display, 23-30 (Sept./Oct. 1968).

(37) Collender, R. B., 825 Seat-Capacity--120° Field of View, Information Display, 25-38 (July/Aug. 1968).

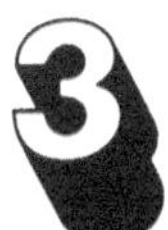

Physiology and Psychology of Depth Perception

In this chapter, the construction of human eyes is discussed first. It is stated that binocular vision using the two eyes is the most important source of depth perception. A pair of organs called the lateral geniculate body (LGB), located at the end of the optic nerves, is believed to play an important role in depth perception. In the latter half of this chapter, four physiological cues and six psychological cues for depth perception are described. Some recent studies on depth perception are also briefly discussed.

3.1 INTRODUCTION

Before dealing with technical subjects in the following chapters, we discuss in this chapter the sources of three-dimensional sensation, or in other words depth perception. It is not intended, however, to present a general and comprehensive treatise on depth perception. The descriptions will rather be restricted to what is required for later discussions. There are two reasons for this course. The first is that the author is not a specialist on the subject; a number of excellent review papers prepared by specialists in this field are now available. The second is that investigators in this field are now achieving great progress, taking advantage of new tools furnished by modern electronics. The present is not a good time, therefore, to survey this field. Several years ahead we shall have much better comprehension than at present of the mechanisms for depth perception.

3.2 HUMAN EYES

3.2.1 BINOCULAR VISION (1),(2)

The eyes of creatures can be classified into several classes according to their stage in the evolutionary process. Some eyes detect only the intensity of incident light, and would better be called detectors. There are two functions characteristic of highly advanced eyes such as those of humans: the imaging property and movement of the eyeball.

If we define an eye to be an optical device having the above two advanced functions, we can say that all creatures have two eyes.

There are some insects which apparently have a third eye at the center front of their heads; examples include locusts and cicadas; however, this organ is known to be merely an infrared detector. Another peculiar case is the "jumping spider" (Slticidae, Dendryphantinae). It has strongly interested investigators in the past several years because it has apparently eight eyes; however, investigations have revealed that only two of them are genuine eyes, whereas two are merely light detectors, and the other four are provided with the imaging property but not with mechanisms for movement. On the other hand, within present knowledge,

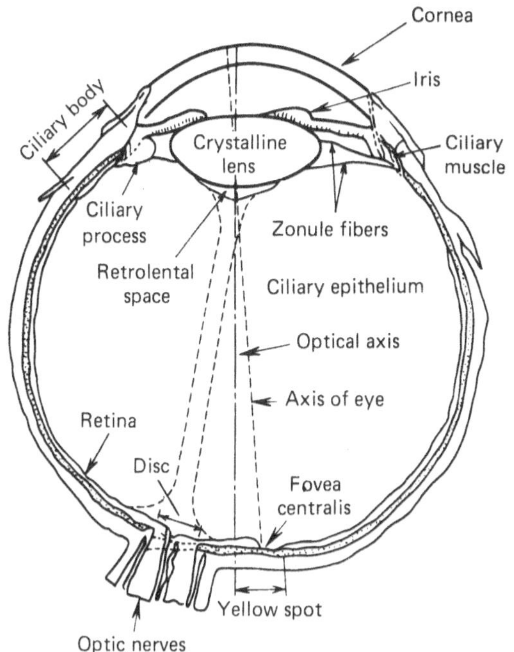

Fig.3.1 Construction of a human eye.

there is no creature that is provided with only one "genuine" eye. (The shellfish has one "eye," but it is merely a detector.)

It is accepted that binocular vision using a pair of eyes is the most important source of depth perception. The retina of an eye can collect only two-dimensional image information because it is of a spherical shape and hence, essentially two-dimensional. Therefore, cues of the third dimension (depth) can never be collected by the retina of a single eye.

Figure 3.1 shows the structure of the human eye.

3.2.2 OPTIC NERVES AND LATERAL GENICULATE BODY

Next we investigate how optic nerves connect the eyes and brain. The basic structure of the nervous

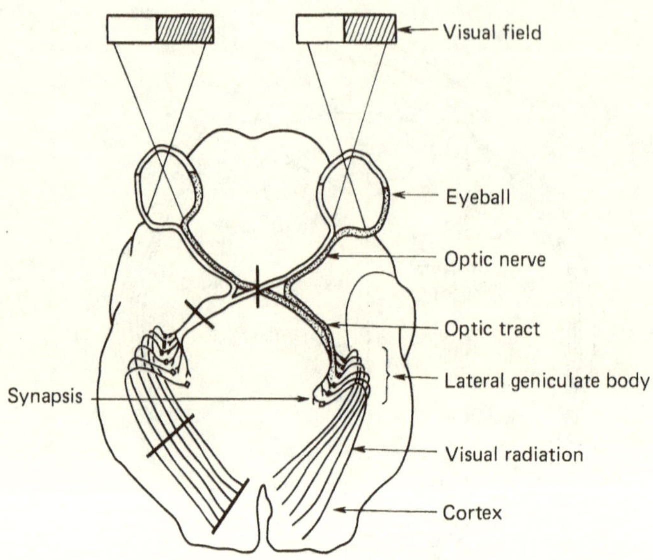

Fig.3.2 Information channels around the brain.

network is shown in Fig.3.2. The brain has two entrances where the optic nerves from retinas get into the cerebrum. At those entrances we find the pair of organs called the lateral geniculate body (LGB). An LGB has a diameter of a few centimeters and a thickness of approximately one centimeter.

An interesting fact is that the optic nerves coming from the two eyes intersect each other once before they enter the LGBs. Some part of the optic nerves transfer visual information from the right eye to the left LGB, and some part, from the left eye to the right LGB. We call these nerves the intersecting optic nerves. This fact, together with some experimental results obtained with animals, suggests that the LGBs have the function of preprocessing visual information from both eyes before it is forwarded to the cerebrum.

Study in this field has been energetic in the past decade. It has revealed so far that each LGB

TABLE 3.1

Layer	Right LGB	Left LGB
1	L	R
2	R	L
3	L	R
4	R	L
5	R	L
6	L	R

consists of six layers, which preprocess the right and left eye information separately as shown in Table 3.1, where the symbols R and L denote information from the right and left eyes, respectively.

We should note, however, that the function of depth perception cannot entirely be attributed to the LGBs. For example, it is known that when a channel-like organ called a callous body connecting both hemispheres of the cerebrum is cut off surgically, depth perception is lost. Thus it is proved that the LGBs carry out only a specific role in depth perception. The tension in the ciliary body for varying the focal length of the crystalline lens (accommodation; see the next section) and the relative directions of the optical axes of the two eyes (convergence; see the next section) also provide important information for depth perception. Obviously, information from these two sources has nothing to do with the LGBs because it is forwarded directly to the cerebrum.

3.2.3 PHYSIOLOGICAL PROPERTIES OF THE HUMAN EYE

Several important parameters and properties of the human eye will be mentioned in the following:

(1) *Average separation of the two eyes:* (pupil distance): 6.5 cm.

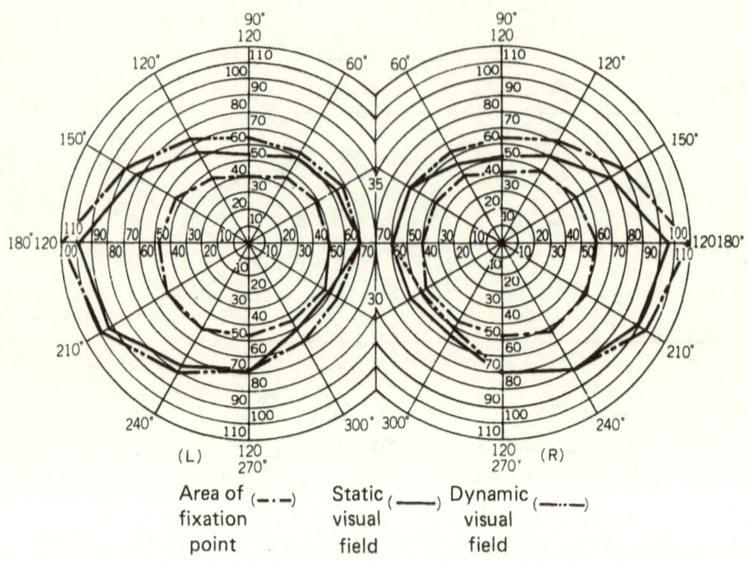

Fig.3.3 Measured range of visual fields of left eye (L) and right eye (R).

(2) *Diameter of the pupil:* 2-8 mm (depending upon brightness).

(3) *Maximum angular resolution:* Approximately 0.5" (1/120°).

(4) *Maximum transmission rate of information:* 4.3×10^6 bits/s for two eyes, and 5 bits/s for a single nerve. (The maximum transmission rate of sound information from the two ears is approximately 8000 bits/s. The human brain has a memory capacity of approximately 10^{10} bits.)

(5) *Visual field:* There are three principal subdivisions of the visual field, whose standard coverages are shown in Fig.3.3:

(a) Static visual field: The visual field obtained when the eyes are looking straight ahead.

(b) Movable area of the fixation point.

(c) Dynamic visual field: The visual field

obtained when the eyes are permitted to move while the head is held fixed. (This third term is not widely used.)

3.3 PHYSIOLOGICAL AND PSYCHOLOGICAL CUES OF DEPTH PERCEPTION

In current psychology it is accepted that ten cues are available for perceiving the depth of an image. These cues can be classified further into two major groups: physiological cues and cues obtained from the retinal image or psychological cues. The former four cues are believed to be more important than the latter six.

3.3.1 PHYSIOLOGICAL CUES

A. Accommodation

Accommodation is the muscular tension of the ciliary body for adjusting the focal length of the crystalline lens. This cue is available even when we see an object with a single eye; therefore, it is said to be a monocular depth cue. However, this cue is effective only when it is combined with other binocular cues, and for a viewing distance less than two meters.

B. Convergence

When we look at a certain point on an object with both eyes, the angle made by the two viewing axes is called the convergence angle. Obviously the muscular tension for rotating the eyeballs a little inside to look at a point gives a cue for depth perception. This binocular cue is called convergence.

Experiments show that the rotation of eyeballs consists of two components: a volitional (watching) component and a subconscious component, that is, a subconscious muscular tension to unify the retinal images of the two eyes. We call this latter action *pattern matching*.

Experiments also show that an interaction exists between convergence and accommodation. The information of convergence corresponding to a certain distance automatically brings about a certain degree of accommodation. On the other hand, information on accommodation influences convergence, though it is a weaker effect. The latter effect can be verified by a simple experiment in which we measure the movement of the eyeballs, shielding one eye and moving the object being watched by the other eye. For example, when the object distance is changed abruptly from infinity to 20 cm, it takes 0.2-0.3 s for convergence to respond to the distance information given by accommodation.

Let the distance between the eyes and object, the pupil distance, and the convergence angle be denoted by a, P_0, and α, respectively; then we obtain

$$\alpha = P_0/a \text{ [rad]} = 57.3P_0/a \text{ [deg]} \quad . \tag{3.1}$$

This equation gives $|d\alpha/da| = P_0/a^2$, which shows that the derivative is greater for shorter viewing distance and smaller for longer distance. For example, if we assume that $P_0 = 6.5$ cm, then the variation of the convergence angle α is one degree, for both of the following two cases:

when a varies from ∞ to 3.58 m and
when a varies from 25 to 23.4 cm.

In fact, it has been verified experimentally that convergence is an effective cue for short distances (especially when working with accommodation), but becomes ineffective beyond 10 m.

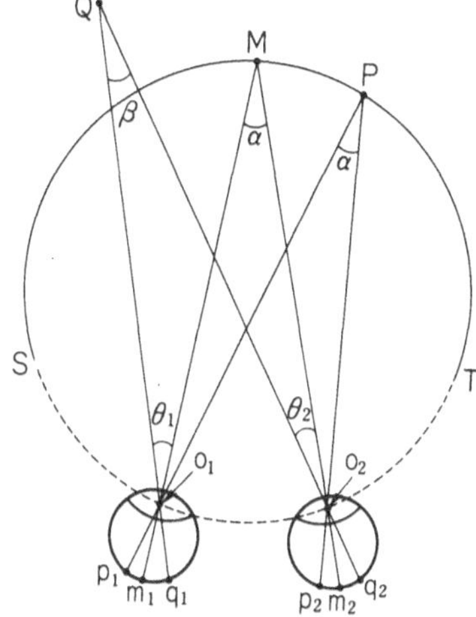

Fig.3.4 Binocular parallax.

C. Binocular Disparity (Binocular Parallax)

When an observer looks at a point on an object, the rays of light originating at that point focus upon the centers of the retinas (fovea centralis; see Fig. 3.1) in both eyes. Therefore, we may say that the two fovea centralis in the pair of eyes give "corresponding positions" on the retinas, with respect to which convergence is determined. However, the rays of light coming from points other than the watched point (point of fixation) do not always focus upon corresponding positions on the retinas. This effect is called binocular disparity or binocular parallax.

We first have to define what corresponding positions are in general. Suppose both eyes look at the point M shown in Fig.3.4. Let the centers of the crystalline lenses be O_1 and O_2. We draw straight lines connecting M and O_1, and M and O_2, and extend them to m_1 and m_2 on the retinas (fovea centralis). Now we define that two points p_1, p_2 on the retinas as shown in Fig.3.4 are at *corresponding positions* when

$$< p_1O_1m_1 = < p_2O_2m_2. \tag{3.2}$$

In reality, it is known experimentally that when the images of a point P are formed on such points p_1 and p_2 as defined above, binocular parallax disappears for the point P. Simple geometrical considerations will show that the ensemble of points P forms a circle passing through points O_1 and O_2 as illustrated in Fig.3.4. Such a circle may be said to give the *equidistance locus* in terms of binocular parallax, and is called the *holopter*.

It is easily seen that rays of light originating at a point off the holopter (like point Q shown in Fig.3.4) never focus at corresponding positions. Thus binocular parallax is perceived and the difference in the distance is recognized.

Let us consider the above statement in a little more quantitative manner. The discrepancy from the standard convergence angle is given as

$$\eta = \theta_1 - \theta_2 = \alpha - \beta. \tag{3.3}$$

Let the distances to points M and Q be denoted by a_M and a_Q, respectively, and define

$$a_Q - a_M = \delta. \tag{3.4}$$

Then we obtain, assuming $\delta \ll a_M$,

$$\eta = \alpha - \beta = P/a_M - P/a_Q \doteqdot P\delta/a_M{}^2. \tag{3.5}$$

For a healthy eye, the threshold of η for perceiving binocular parallax is 10" (approximately 5×10^{-5} rad). (It is interesting to note that the above threshold angle is smaller than the angular resolution of a healthy eye, which is about 30".) The minimum detectable difference in the viewing distance, therefore, is

$$\begin{aligned} \delta_{min} &= 0.8 \text{ mm} \quad \text{for} \quad a_M = 1 \text{ m}, \\ \delta_{min} &= 8 \text{ cm} \quad \text{for} \quad a_M = 10 \text{ m}, \\ \delta_{min} &= 8 \text{ m} \quad \text{for} \quad a_M = 100 \text{ m}. \end{aligned}$$

It is accepted that for medium viewing distances,

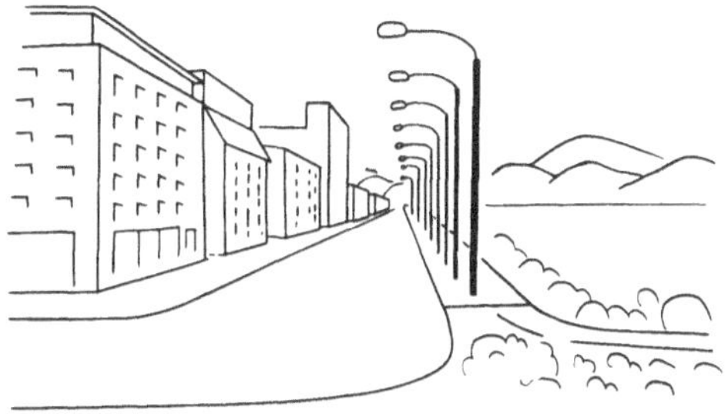

Fig.3.5 Linear and areal perspective.

binocular parallax information is the most important cue for depth perception.

D. Monocular Movement Parallax

Suppose we observe an object while shielding one eye. If the viewing position is held fixed, accommodation will be the only effective cue for depth perception. However, if displacement of the viewing position is permitted, we can make use of an effect like binocular parallax by observing the object from various directions. This effect is called monocular movement parallax. It is especially significant when the observer is moving rather rapidly, for example aboard an airplane or a very fast train.

3.3.2 CUES OBTAINED FROM RETINAL IMAGE (PSYCHOLOGICAL CUES)

The above four cues appeal directly to visual sensation and are believed to be the most important cues for depth perception. In reality, however, we always get a certain degree of depth sensation from the planar image reproduced on the retina being assisted by experience and imagination. Such effects can be sorted into six categories.

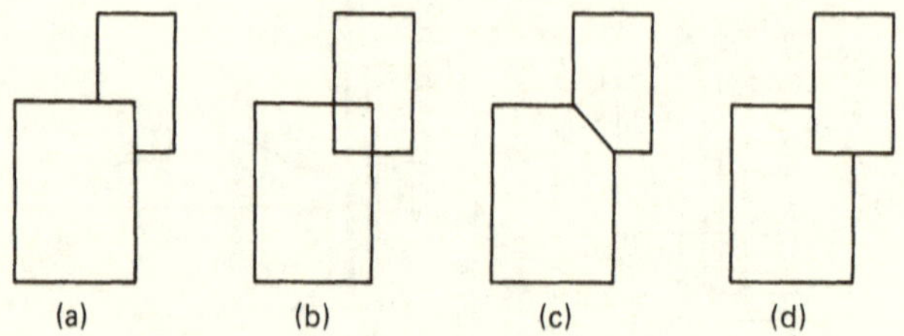

Fig.3.6 Depth perception based upon overlapping.

A. Retinal Image Size

We know to a certain degree the actual size of many objects. Therefore, the size of an image on the retina tells us the distance to the object. Sometimes tricks based upon this principle are employed in movies or in still photography.

B. Linear Perspective

For example, let us suppose that we are looking at a lakeside scene as shown in Fig.3.5. In such a case, an important cue for depth sensation will be the perspective of the buildings, that is, the gradual reduction of the image size in the distant scene. This effect is called linear perspective.

C. Areal Perspective

In the same figure (Fig.3.5), if the distant mountains and houses look more or less hazy, depth sensation will be emphasized. The reason for this effect is that distant landscapes often look hazy due to the scattering of light by small particles in the air (dust, fog, smog, mist, etc.), and that we are accustomed to take this fact into account subconsciously. This effect is called areal perspective.

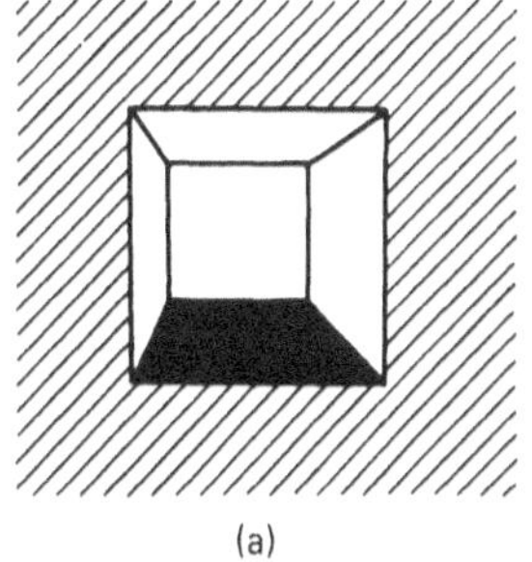
(a)

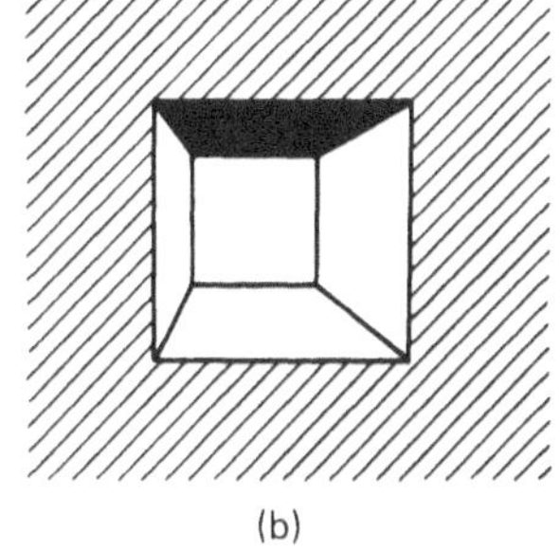
(b)

Fig.3.7 Depth perception based upon the shades and shadows.

D. Overlapping

As shown in Figs.3.6a and 3.6d, the overlapping relation of the outlines of objects also offers a cue for depth perception. In Figs.3.6b and 3.6c, no cue is present for depth perception. Generally, an object whose outline pattern looks relatively continuous is felt to lie closer, and vice versa.

E. Shades and Shadows

Shades and shadows are also important psychological cues for depth perception. Figures 3.7a and 3.7b both show pictures of a square convexity made on a wall. An interesting fact is that Fig.3.7a really looks like a convexity but Fig.3.7b looks like a concavity on a wall or a window in a jail. Moreover, once we believe so, it seems more and more natural to believe so. This is a well-known psychological trick; the origin of such an illusion is the fact that most illumination (either natural or artificial) comes downward. If the reader looks at Fig.3.7 upside down, he will find that Fig.3.7a looks like a concavity, whereas Fig.3.7b looks like a convexity.

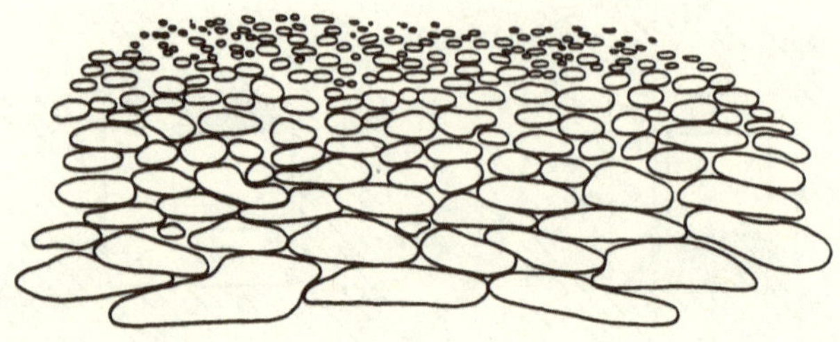

Fig.3.8 Texture gradient.

F. Texture Gradient

Texture gradient might be said to be a kind of linear perspective. When we look at a uniform texture gradient like brick pavement or a gravel road as shown in Fig.3.8, the gradient in its roughness offers a cue for depth perception.

3.4 SOME RECENT STUDIES ON DEPTH PERCEPTION

Before closing this chapter, some of the recent studies on depth perception will be described.

Julesz (3),(4) took advantage of recent progress in computer technology to produce random patterns for use in psychological studies of depth perception and performed a series of experiments using these random patterns. An example of a pair of the random patterns used in the psychological test is shown in Fig.3.9, which has been produced by T. Hatada of Nippon Hoso Kyokai (Japan Broadcasting Corporation) in a manner like Julesz's. The subject holds these random patterns in an upright position, looks at the right pattern with the right eye and the left pattern with the left eye, and tries to achieve pattern matching between the right and left images with respect to the topmost portions of the patterns. He will then find a three-

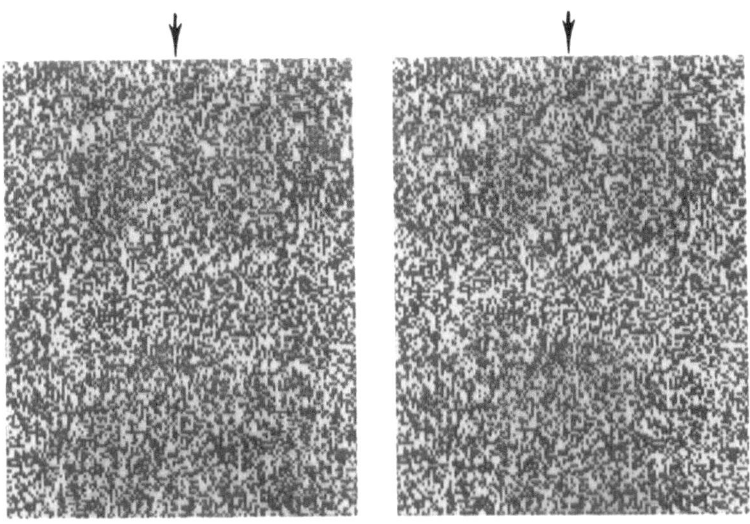

Fig.3.9 A random pattern (courtesy of Mr. T. Hatada, Nippon Hoso Kyokai).

dimensional object (in this case a square dent) in the random pattern.

The characteristic of depth-perception experiments using random patterns is as follows. When we use a random pattern, accommodation does not contribute at all to depth sensation, although it might rather give a negative effect by telling the observer the true distance to the pattern. All of the psychological cues also lose their sense because a random pattern does not at all elicit any memory or knowledge from the observer. Thus, eliminating all the cues appealing to a posteriori knowledge about the object, the random pattern offers a tool for systematic investigation of the process of pattern matching. Julesz took advantage of the above feature to study the validities of three assumptions about depth perception:

(1) Pattern recognitions are first performed for each set of monocular visual information, and the

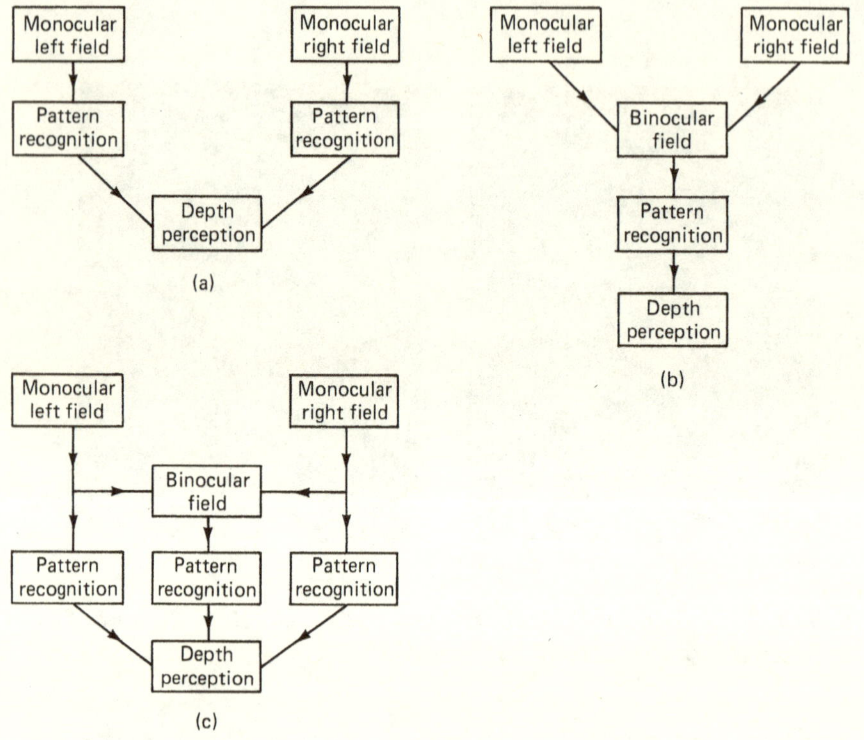

Fig.3.10 Three assumptions of Julesz.

results of these pattern recognitions are then processed to achieve depth perception (Fig.3.10a).

(2) Two pieces of monocular visual information are first compared to produce binocular information, from which the binocular disparity is later extracted to achieve depth perception (Fig.3.10b).

(3) The information processing actually performed is a combination of the above two mechanisms (Fig.3.10c).

Julesz's conclusion was that assumption 3 was the most reasonable. The establishment of the means for depth perception studies eliminating the a posteriori factors seems very significant, although physiological proofs of the assumption are still lacking.

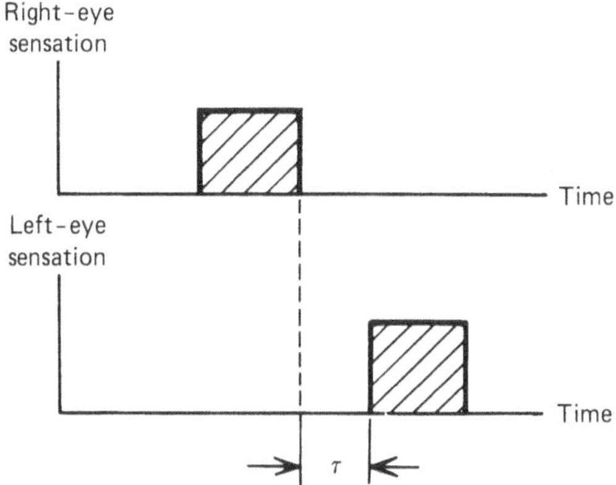

Fig.3.11 Ogle's experiment [after Ogle (5)].

Ogle studied depth perception from various aspects. An interesting example of his experiments is the measurement of the delay-time effect on binocular disparity (5). He gave the subject right-eye and left-eye information with a short time interval delay as shown in Fig.3.11, and proved that depth perception is possible if τ is less than 100 ms. This fact suggests that short-term memory must be operative for binocular-parallax information processing.

References

(1) Yamada, Y., Oyama, T., and Imai, S., "Handbook of Psychology of Perception." Seishin Shobo, Tokyo, 1969.

(2) Graham, C. H., "Vision and Visual Perception." Wiley, New York, 1965.

(3) Julesz, B., Binocular Depth Perception of Computer-Generated Patterns, *Bell Sys. Tech. J. 39*, No. 5, 1125-1162 (Sept. 1960).

(4) Julesz, B., "Foundations of Cyclopean Perception." Univ. of Chicago Press, Chicago, Illinois, 1971.

(5) Ogle, K. N., Some Aspects of Stereoscopic Depth Perception, *J. Opt. Soc. Amer. 57*, No. 9, 1073-1081 (Sept. 1967).

Lens-Sheet Three-Dimensional Pictures

As described in the introduction, three principal three-dimensional imaging techniques are lens-sheet three-dimensional pictures, projection-type three-dimensional displays, and holography. Among these lens-sheet three-dimensional pictures are the most suitable example for considering the optimum design problem of three-dimensional pictures. In this chapter, the optimum design of lens-sheet pictures will mainly be discussed. Also described are picture recording techniques and special integral photography techniques. In the appendix, an analysis of aberration of the lens sheet is presented.

4.1 INTRODUCTION

To begin, the reader's attention is drawn to the title of this chapter. The terminology used, "lens-sheet three-dimensional picture," is new; it is used generically to include integral photography (IP) and lenticular-sheet three-dimensional imaging. As will be seen later, these two three-dimensional imaging techniques can be treated in a unified manner as far as their picture resolution and optimum design are concerned. Furthermore, as shown in Chapter 5, specific types of lens-sheet screens used in projection-type three-dimensional imaging also permit a similar treatment. To avoid needless confusion, however, the design problem of the projection screen will not be considered here and will be discussed separately in the next chapter.

The history and state of the art of integral photography and lenticular-sheet three-dimensional pictures have been described in Chapter 2.

Integral photography is a basic three-dimensional imaging technique; its technical importance is widely accepted; however, fabrication of a high quality fly's-eye lens sheet requires extremely refined techniques. Due to this fact integral photography at present is far from being a practical technique.

On the other hand, the art of lenticular-sheet three-dimensional pictures has made great progress in the past decade. It became practical late in the 1960s. Lenticular-sheet pictures are often seen in everyday life. Such a big difference in the developments of integral photography and lenticular-sheet pictures results from two factors. First, a lenticular sheet is much easier to fabricate than a fly's-eye lens sheet. Second, the picture-taking process is much simpler for lenticular-sheet pictures than for IP because in the former vertical parallax information is abandoned.

In the next section the recording process for lens-sheet pictures will be discussed.

TABLE 4.1 Features of Various Methods of LS Picture Recording

Methods			Picture obtained	Picture quality	Size of the LS	Suitable object	Practicability
			Technical features				
(1) Continuous methods	(a) Moving camera methods	(1a-1) Keystone-free movement	A single picture to be fixed on the rear surface of LS sheet	Mainly determined by the preciseness of mechanism	Predetermined	Large still object	No
		(1a-2) Linear movement					Yes
		(1a-3) Arc movement					Yes
	(b) Moving object methods	(1b-1) Keystone-free camera				Small still object	No
		(1b-2) Rotating camera					No
		(1b-3) Rotating table					Yes
		(1b-4) Segmented linear movement					No

(2) Discontinuous methods	(a) Single-camera methods	(2a-1) Keystone-free camera	Plural number of ordinary pictures	Mainly determined by the number of exposures or cameras	Freely enlargeable	Entirely immobile object	No
		(2a-2) Linear movement					No
		(2a-3) Arc movement					Yes
		(2a-4) Moving lens					No
	(b) Multi-camera methods	(2b-1) Linearly arrayed multi-camera					Yes
		(2b-2) Circularly arrayed multi-camera				Mobile object	No

4.2 RECORDING OF LENS-SHEET THREE-DIMENSIONAL PICTURES

4.2.1 RECORDING OF INTEGRAL PHOTOGRAPHS

This has been described in Section 2.4.2. (Also refer to Sections 4.4.3 and 4.5.)

4.2.2 RECORDING OF LENTICULAR-SHEET THREE-DIMENSIONAL PICTURES

The basic principle of recording a lenticular-sheet three-dimensional picture is to collect picture information from the object observed from various directions, usually using a special camera. This principle is identical to that of recording a parallax panoramagram described earlier (see Figs.2.6 and 2.8). The purpose of the discussion in this section is to show and compare various recording methods in more detail and to clarify their advantages and disadvantages.

In Ref.(1), Koseki *et al.* showed a table in which various recording methods for lenticular-sheet three-dimensional pictures are sorted into 10 categories, and compared the technological features of these methods. In this section, a table prepared by the present author summarizing their discussion will be shown. However, their table has been modified a little to include some new versions; the number of methods has been increased from 10 to 13.

A. Classification of Recording Methods

The proposed classification is shown in Table 4.1. According to Koseki *et al.*, the methods of recording a lenticular-sheet three-dimensional picture are first divided into two major groups: continuous methods and discontinuous methods. Then these are further

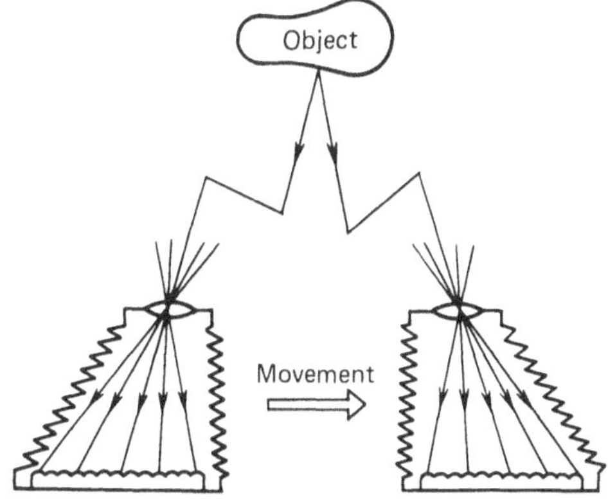

Fig.4.1 Keystone-free movement (1a-1).

sorted into finer groups. The 13 methods mentioned in the table will be described in the order of their occurrence there.

(1) Continuous methods

In continuous methods, a lenticular sheet backed by a photographic plate (or combination of a photographic plate and a slit sheet as used in a parallax panoramagram) is installed in the camera. In taking a three-dimensional picture, images of the object observed from various directions are recorded directly and continuously on the photographic plate. After development, the stripelike picture obtained on the photographic plate is used as the original picture for reproduction.

(a) Moving camera methods

(1a-1) *Keystone-free movement.* This method features the use of a camera having relatively small lens aperture and a flexible body permitting gradual change in lens position relative to the photographic plate,

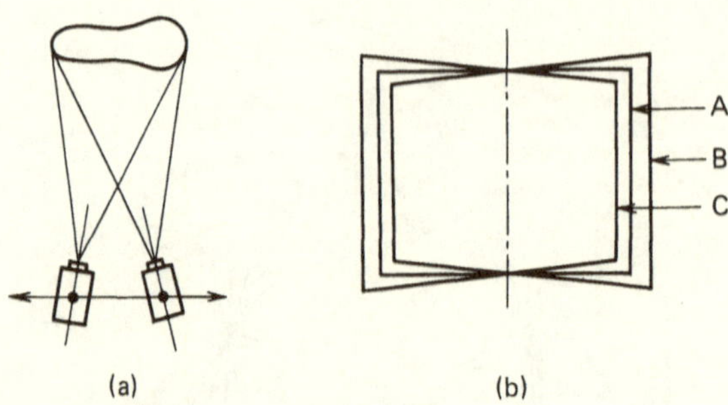

Fig.4.2(a) Linear movement (la-2). (b) Illustration of keystone distortion.

as shown in Fig.4.1. The principle of this method is entirely identical to that for a parallax panoramagram as shown in Fig.2.6. As to the origin of the term *keystone-free*, refer to the next item.

(la-2) *Linear movement*. This is a simplified version of the above method (la-1). A camera consisting of a small-aperture lens and a rigid body moves along a linear guide, rotating gradually as it moves so that it always fixes on the object as shown in Fig.4.2a. A lenticular sheet backed by a photographic plate is installed in the camera, and it also rotates as the camera moves, so that images of the object seen from different positions are recorded on the lenticular/photographic plate from different directions. [Instead of rotating the lenticular/photographic plate, in some cameras the lenticular sheet (or the slit sheet) is shifted minutely in the horizontal direction so that images of the object seen from different directions are recorded on different positions behind each lenticule or slit.]

It should be noted that this method generates more or less of the so-called keystone distortion which is illustrated in Fig.4.2b for the case when a square, two-dimensional object is recorded. In such a case, if the object is seen from the right, the right-hand side is observed as larger than the left-

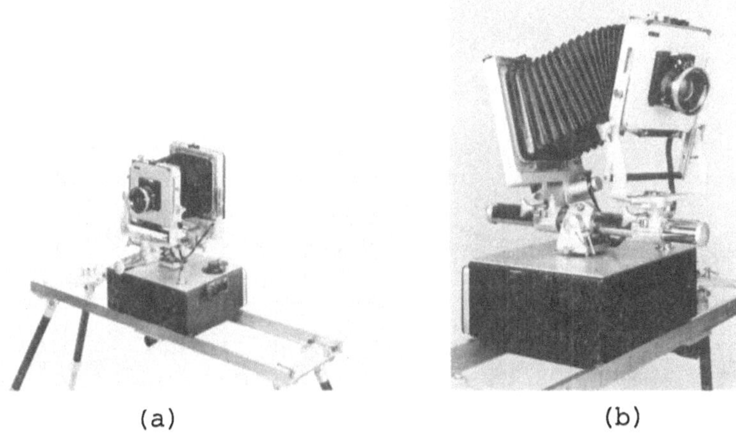

(a) (b)

Fig.4.3 An example of a camera used in the linear movement method. (a) Guiding mechanism. (b) Camera body (courtesy of Mr. M. Horiuchi, Dai-Nippon Printing Company).

hand side, and vice versa. This is called *keystone distortion*. This distortion is not generated in the keystone-free movement described above; this fact is the very origin of the name of that method.

Figures 4.3a and 4.3b show the photographs of a camera used in the linear movement method.

(1a-3) *Arc movement*. This is also a simplified version of method (1a-1), and is illustrated in Fig. 4.4. Keystone distortion is also generated with this method. The lenticular/photographic plate rotates in the camera as it moves, as in method (1a-2). When a relatively small object is to be recorded, this method is superior to method (1a-2) since the mechanism is usually simpler.

(b) Moving object methods

(1b-1) *Keystone-free camera*. A modified version of method (1a-1), in which the object moves instead of the camera. See Fig.4.5a.

(1b-2) *Rotating camera*. A version of method

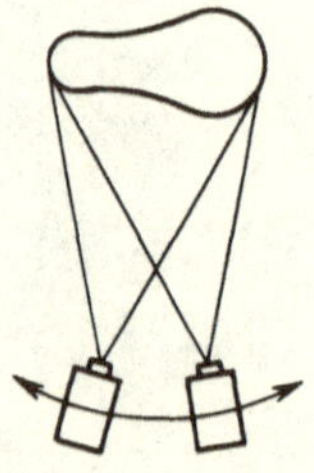

Fig.4.4 Arc movement (1a-3).

(1a-2), in which the object moves linearly instead of the camera. Rotation of the camera and rotation or displacement of the lenticular (or slit) sheet still must occur. See Fig.4.5b.

(1b-3) *Rotating table.* A version of method (1a-3) in which the object table rotates instead of the camera. See Fig.4.6a. Rotation or displacement of the lenticular sheet must also occur.

(1b-4) *Segmented linear movement.* A version of method (1b-1) or (1b-2) in which the movement of the objects is segmented according to the depth of the image to be presented to the observer. This method is usually used in producing synthesized three-dimensional pictures. See Fig.4.6b.

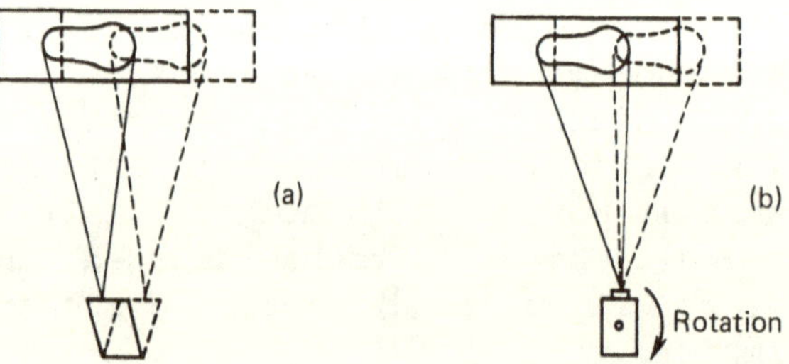

Fig.4.5(a) Moving object, keystone-free camera method (1b-1). (b) Moving object, rotating camera method (1b-2).

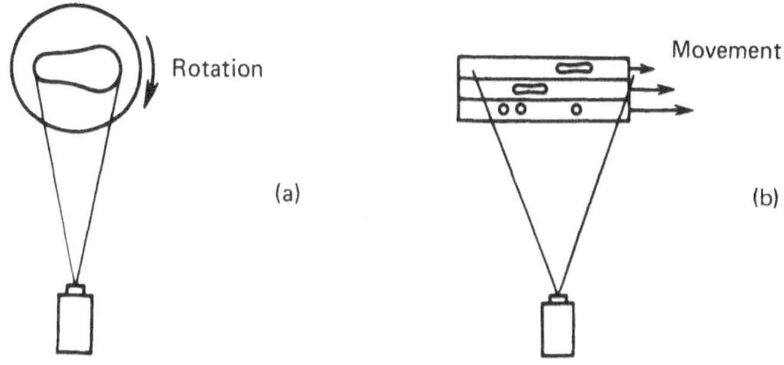

Fig.4.6(a) Moving object, rotating table method (1b-3). (b) Moving object, segmented linear movement method (1b-4).

(2) Discontinuous methods

In discontinuous methods, a plane photographic plate or a plane sheet film is installed in a camera or cameras. A number of pictures are obtained first, and they are synthesized by using a special printing device to give a lenticular picture.

(a) Single-camera methods

(2a-1) *Keystone-free camera.*
(2a-2) *Linear movement.*
(2a-3) *Arc movement.*

The above three methods correspond to the continuous methods (1a-1), (1a-2), and (1a-3) described

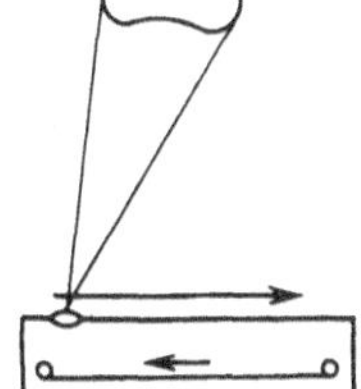

Fig.4.7. Signal camera, moving lens method (2a-4).

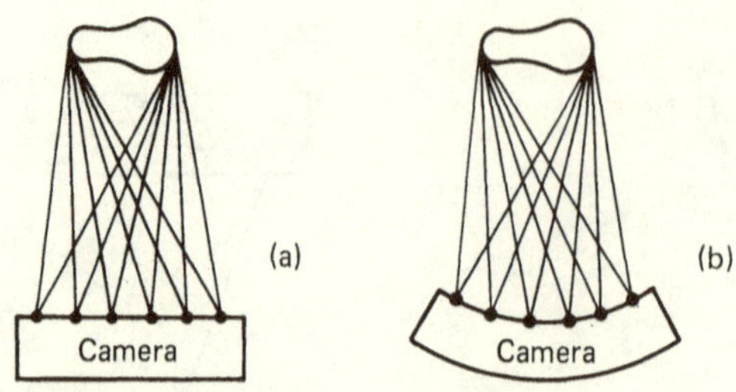

Fig.4.8(a) Linearly arrayed multicamera method (2b-1). (b) Circularly arrayed multicamera method (2b-2).

in the preceding subsection. In these discontinuous methods, however, the camera moves step-by-step to take N different, ordinary photographs of the object observed from N different directions. As to the advantages and disadvantages of the three methods, refer to the previous discussion on the corresponding three methods.

(2a-4) *Moving lens.* This method features the use of a special camera, in which the lens moves laterally to take N photographs from N directions. See Fig.4.7.

(b) Multicamera methods

(2b-1) *Linearly arrayed multicamera.* A number of photographs are taken from N different directions by using an array of N cameras (Fig.4.8a). An example of this type of camera is shown in Fig.4.9.

(2b-2) *Circularly arrayed multicamera.* This is a version of method (2b-1) in which the cameras are arrayed circularly. See Fig.4.8b.

B. Synthesis of a Three-Dimensional Picture from Many Photographs

In discontinuous methods, a picture to be lined upon the rear surface of the lenticular sheet must be

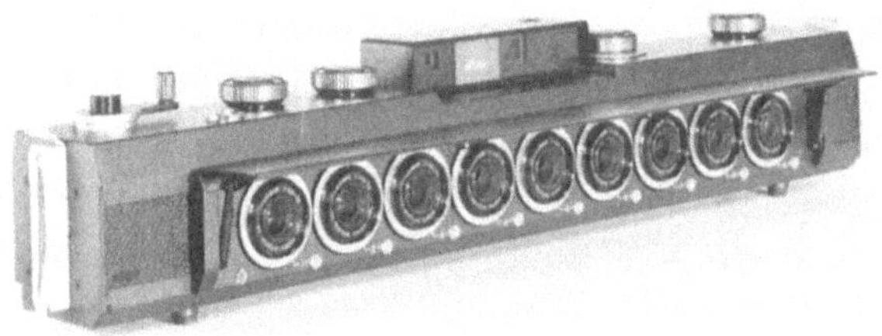

Fig.4.9 An example of the camera used in the linearly arrayed multicamera method.

synthesized from N ordinary photographs taken from N different directions. This synthesis is performed using an arrangement as shown in Fig.4.10.

The N original, ordinary photographs are installed in the projectors shown in the lower part of Fig.4.10. The N images are projected onto a lenticular sheet lined by photographic emulsion (or backed by a photographic plate). If we assume, for simplicity, that the films installed in the projectors have been obtained by N cameras with the same focal length and geometrical arrangement as the projectors, the N projectors will produce a three-dimensional real image just at the same place as the object existed, and the three-dimensional image will be recorded on the photographic material behind the lenticular sheet.

In practice, the reconstructed image does not always have the original size. Furthermore, the distances in recording and in projection, the focal lengths of the cameras and projectors, and the separations between cameras and between projectors are usually not equal to each other. These differences bring about the "scaling problem" which is common to both projection-type and lens-sheet three-dimensional pictures. This will be discussed in Section 4.3.3.A.

C. *Summary of Advantages and Disadvantages*

The advantages and disadvantages of various recording methods are summarized in Table 4.1. From

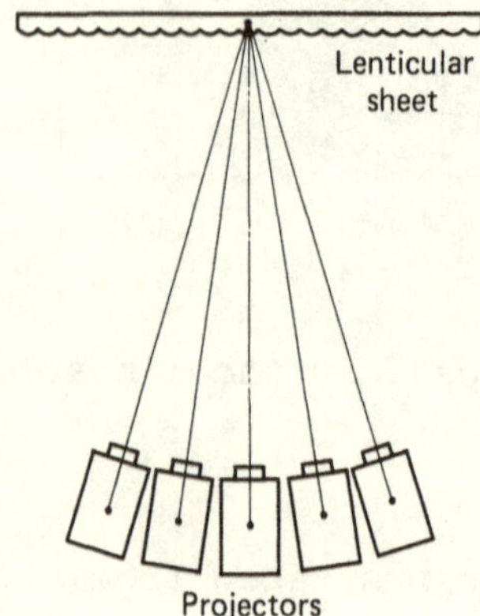

Fig.4.10 Method of producing a final lenticular-sheet three-dimensional picture from original "slide pictures" taken with discontinuous methods.

this table we may conclude that:

(1) Practically recommendable ones are (1a-2), (1a-3), (1b-3), (2a-3), and (2b-1).

(2) Technically interesting ones (but not without practical difficulty) are (1a-1) and (2b-2).

4.3 OPTIMUM DESIGN OF LENTICULAR-SHEET PICTURES

4.3.1 INTRODUCTION

The term *optimum design* in the title of this section denotes the optimum design of the whole picture-making process, i.e., including both the picture-taking process and lens-sheet fabrication.

In the past decade there has been great progress in lenticular-sheet three-dimensional pictures; however, within the author's knowledge, little theoretical consideration has been reported concerning its optimum design. Lenticular-sheet three-dimensional pictures have technical similarities with IP. As to IP, Burckhardt presented a design theory to determine the optimum diameter of the fly's-eye lenslet (2);

however, as will be discussed in Section 4.4.3, the assumptions in his theory are not realistic for ordinary LS pictures and most integral photography.

In this section, the optimum design theory developed by the author will be described, taking into account aberration, diffraction, and sampling effects of the lens sheet (3). It features the use of depth resolution as a measure of three-dimensional image quality. Optimum parameters in the picture-recording process, optimum lens pitch, and depth-resolution limitation will be considered. The results obtained are also applicable to specific types of IP and projection-type three-dimensional displays including projection-type holography. These applications will be described in other corresponding sections and chapters.

In Section 4.3.2, the process to be considered is described briefly. In Sections 4.3.3 and 4.3.4, optimum design parameters and depth-resolution limitations are discussed. Section 4.3.5 describes some experimental data of presently obtainable lenticular sheets.

4.3.2 RECORDING OF A LENTICULAR-SHEET PICTURE

As an example of various recording methods, the simplest LS picture-taking process will be considered in the following. No loss of generality will be brought about by this simplification.

A. *Primary Picture Taking*

We consider a process as shown in Fig.4.11a. Photographs (negatives) of an object (having a depth b) are taken from N directions using N cameras. The distance between the cameras and the object, focal length of the cameras, separation between cameras, and the lens apertures are denoted by a_1, f_1, d_1, and D_1, respectively.

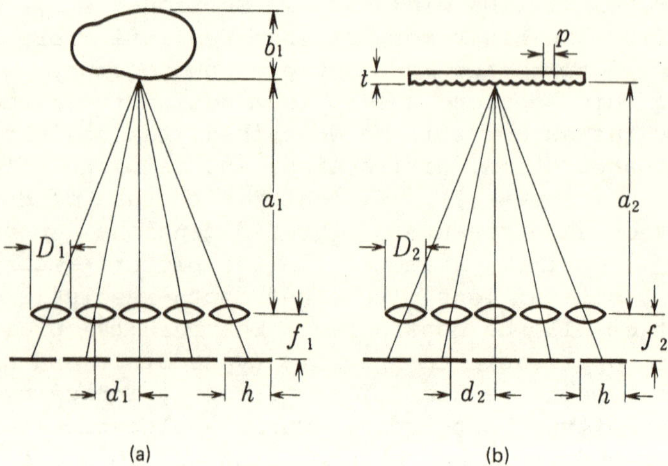

Fig.4.11(a) Primary picture taking using N cameras. (b) Secondary picture taking using N projectors.

B. *Secondary Picture Taking*

The N photographs are projected onto a lenticular sheet backed by a photographic emulsion using a device similar to a slide projector (see Fig.4.11b). The emulsion is then developed, and an LS picture is completed. The distance between projectors and the lens sheet, the focal length of the projectors, separation between projectors, and the lens aperture are denoted by a_2, f_2, d_2, and D_2, respectively. The pitch of the lens sheet (width of the cylindrical lens) and its thickness are denoted by p and t, respectively.

C. *Assumptions for the Analysis*

(1) The camera lens aperture D_1 can be chosen independently from d_1, but cannot exceed d_1.

(2) The camera focuses upon the front of the object.

(3) The projector lens aperture D_2 can be chosen

arbitrarily, but cannot exceed d_2.

(4) In the secondary picture-taking process, the N pictures are projected so that their focus is upon the lenticular sheet.

(5) The width of the film h is smaller than either d_1 or d_2.

(6) The thickness of the lenticular sheet is chosen so that a parallel incident beam focuses upon the rear surface, that is, upon the emulsion.

4.3.3 OPTIMUM PARAMETERS IN PRIMARY PROCESS

A. Scaling Rule

We first consider the scaling rule between the primary and secondary picture taking. In other words, we consider the conditions for obtaining a proper relation between lateral magnification and depth magnification. This rule is necessary in the following discussion on the optimum projector separation.

In the preparation of LS pictures, complete identities in the primary and secondary picture takings, $d_1 = d_2$, $f_1 = f_2$, and $a_1 = a_2$, are not necessarily required. If we define scaling parameters

$$\alpha = d_2/d_1, \tag{4.1}$$

$$\beta = f_2/f_1, \tag{4.2}$$

$$\gamma = a_2/a_1, \tag{4.3}$$

a scaling rule is found between these parameters as follows.

To facilitate the analysis, the arrays of cameras and projectors are shown back-to-back in Fig.4.12. Let the width and depth of the object be w_1 and b_1. If the visual width and depth of the image are denoted by w_2 and b_2 as shown in Fig.4.12, a simple geometrical calculation leads to

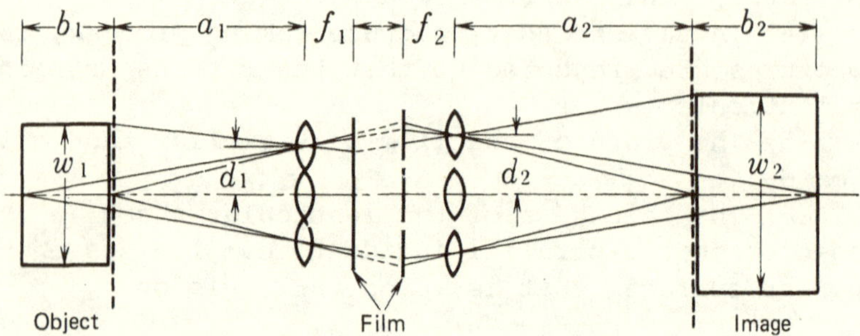

Fig.4.12 Symbols used in the derivation of the scaling rule.

$$\frac{b_2}{a_2} = \frac{b_1}{a_1} \cdot \frac{1}{(b_1/a_1)(K-1)+K}, \tag{4.4}$$

$$\frac{w_2}{a_2} = \frac{w_1}{a_1} \cdot \frac{1}{\beta}, \tag{4.5}$$

therefore

$$\frac{b_2}{w_2} = \frac{b_1}{w_1} \cdot \frac{\beta}{(b_1/a_1)(K-1)+K}, \tag{4.6}$$

where

$$K = \alpha\beta/\gamma. \tag{4.7}$$

These equations tell us that the present scaling rule should be stated in the following three steps.

First, suppose we want to conserve only the depth-to-distance ratio, that is, to make $b_2/a_2 = b_1/a_1$. In this case, from Eq.(4.4), only $K = 1$ is required. The relation between b_2/a_2 and b_1/a_1 is shown in Fig. 4.13. When $K \neq 1$, the relation between b_1 and b_2 cannot be linear. As illustrated in Fig.4.13:

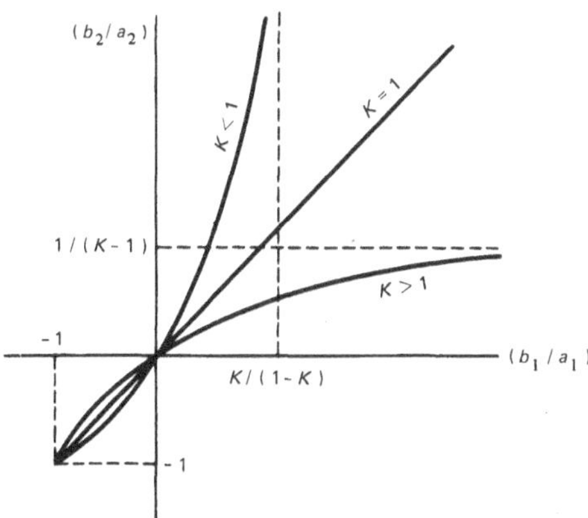

Fig.4.13 Expansion and compression of the image depth.

(1) when $K > 1$, the depth is compressed $[b_2/a_2 = 1/(K-1)$ for $b_1 = \infty]$, and

(2) when $K < 1$, the depth is expanded [the region $b_1 > a_1K/(1-K)$ cannot be reproduced].

Second, suppose we want to conserve, in addition to the above quantity, the depth-to-width ratio (b/w). Then $\beta = 1$ will be required in addition to $K = 1$ [see Eq.(4.6)]. Therefore $\alpha/\gamma = 1$ must hold.

Third, complete identity is defined as $\alpha = \beta = \gamma = 1$.

In actual LS pictures, sometimes K is chosen as $K > 1$ intentionally to prevent eye fatigue, and sometimes $K < 1$ to emphasize depth sensation. In the following analysis, we shall not always assume that $\alpha = \beta = \gamma = 1$, but will assume $\beta = 1$ and $K = 1$ (therefore $\alpha = \gamma$) throughout.

B. Optimum Projector Separation

The most important step in the design for primary picture taking is to determine the optimum projector separation.

For a spot reproduced deep inside the lenticular sheet, a "flipping" will be observed due to the finite number of projectors. In other words, when the observer moves his head, the observed spot will "flip," or will be "switched" from one image reproduced by one projector to another image reproduced by an adjacent projector. Since the magnitude of the flipping is obviously proportional to d_2, a smaller d_2 is more desirable to minimize the flipping; however, since d_2 also gives the maximum size of the primary pictures to be projected, reducing d_2 results in reduced resolution due to the resolution limitation of the emulsion (turbidity) of the primary slide pictures. If we further reduce d_2, d_1 is also reduced; but d_1 gives the maximum camera lens size D_1, which finally brings about resolution reduction due to diffraction.

On the other hand, to make the flipping less visible, we may take advantage of defocus for spots away from the lenticular sheet by properly increasing the lens size of the camera D_1.

The above quantities are given with respect to the innermost spot as follows:

(1) *Flipping*. Figure 4.14a immediately shows that the magnitude of the flipping measured upon the lenticular sheet is given as

$$F = d_2 b_2/(b_2 + a_2). \tag{4.8}$$

(2) *Defocus*. As stated in Section 4.3.2.C, the cameras focused upon the front of the object. Therefore, in primary picture taking, the focus of the innermost spot will be formed between the film and lens (at f_1) as illustrated in Fig.4.14b. From this illustration and the lens formula, we obtain

$$\Delta f = f_1 - f_1' \doteqdot f_1^2 b_1/a_1(a_1+b_1),$$

therefore

$$\delta = D_1(\Delta f/f_1'). \tag{4.9}$$

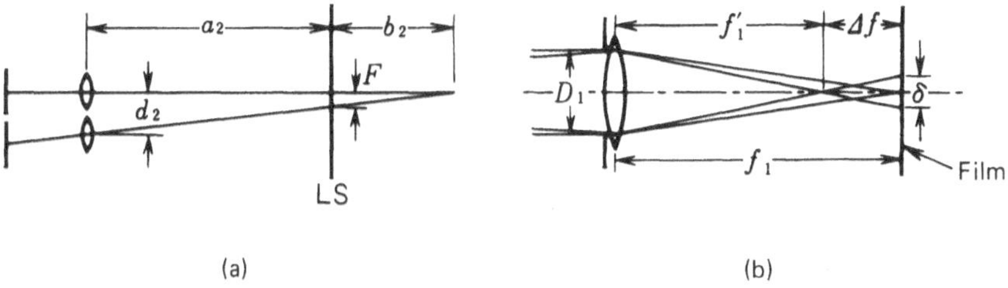

Fig.4.14(a) The "flipping" observed for the spots reproduced inside the lenticular sheet. (b) The defocus for the spots to be reproduced behind the lenticular sheet.

Next, we consider the secondary process. We define the ratios of the lens aperture and separation as

$$k_1 \overset{\Delta}{=} D_1/d_1, \quad k_2 \overset{\Delta}{=} D_2/d_2. \tag{4.10}$$

When the primary picture is projected, the width of the blurring due to defocus measured upon the lenticular sheet is given as

$$\begin{aligned} S &= (a_2/f_2)\delta = (a_2/f_2)\ (D_1/f_1')\Delta f \\ &\doteqdot k_1 d_2 b_2/(a_2+b_2) = k_1 F. \end{aligned} \tag{4.11}$$

In obtaining the above equation, $\beta = 1$ and $\alpha = \gamma$ (therefore, $f_2 = f_1$ and $d_2/d_1 = a_1/a_2 = b_2/b_1$) are assumed. The final form in Eq.(4.11) suggests that when $k_1 = 1$, $S = F$ holds; that is, the flipping is just erased by defocus for any depth. This is an important fact practically.

(3) *Resolution limitation (turbidity) of film.* If the resolvable distance upon the film is denoted by s, it is enlarged upon the lenticular sheet to

$$G = sa_2/f_2 \doteqdot sw_2/h = sw_2/md_2 \ \left[\text{where } m \overset{\Delta}{=} h/d_2\ (<1)\right]. \tag{4.12}$$

(4) *Resolution limitation due to diffraction.* As will be seen in the following paragraphs, this limitation is not predominant in primary picture taking. Hence, only the results of computation will be presented. (For more detail, refer to Section 4.4.2.B.) When the camera lens aperture is D_1, the resolution limit due to diffraction measured upon the film (δ_D) is given as

$$\delta_D = 1.22 f_1 \lambda / D_1, \tag{4.13}$$

where λ denotes the wavelength. Therefore, the resolution limit upon the lenticular sheet is given as

$$R = \sqrt{2} a_2 \delta_D / f_2 = 1.22\sqrt{2}\lambda a_2 f_1 / k_1 d_1 f_2, \tag{4.14}$$

where $\sqrt{2}$ represents the resolution reduction in the secondary process (see Section 4.3.4). Again assuming $\beta = 1$ and $\alpha = \gamma$, we obtain

$$R = 1.22\sqrt{2} a_2{}^2 / a_1 k_1 d_2. \tag{4.15}$$

The above four quantities, F, S, G, and R, are shown in Fig.4.15 as functions of the projector separation d_2. Since $k_1 < 1$, Eq.(4.11) leads to $F > S$. Therefore, to eliminate flipping by taking advantage of defocus, we should make $k_1 = 1$. In other words, if the lens aperture (diameter) is equal to the camera separation, we may obtain a flip-free picture even using discontinuous methods.

On the other hand, $G \gg R$ holds in most practical cases: for example, when $s = 0.01$ mm, $w_2 = 60$ cm, $m = 0.5$, $a_1 = a_2 = 100$ cm, $k_1 = 1$, and $\lambda = 500$ nm, we obtain $R/G = 0.07$. Hence, the overall resolution for the innermost spot $(S^2 + G^2 + R^2)$ will be minimized, approximately, for $S = G$, which gives

$$d_{2\,opt} = \left[s w_2 (a_2 + b_2) / m b_2 \right]^{1/2} \tag{4.16}$$

and

$$\left(S^2 + G^2\right)^{1/2}_{min} = \left[2 s w_2 b_2 / m (a_2 + b_2) \right]^{1/2}. \tag{4.17}$$

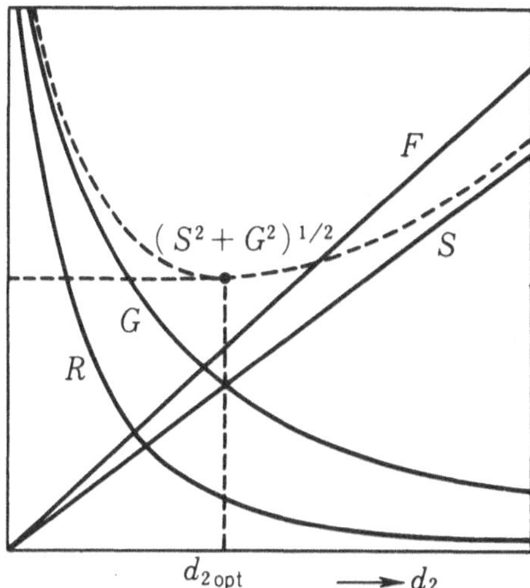

Fig.4.15 Factors affecting the lateral resolution upon the lenticular sheet shown as functions of d_2. The optimum value of d_2 and the obtainable resolution are also illustrated.

For example, when s = 0.01 mm, w_2 = 60 cm, m = 1.0, a_2 = 100 cm, and b_2 = 30 cm, then $d_{2\,opt}$ = 5 mm. When s = 0.02 mm, w_2 = 100 cm, m = 0.5, a_2 = 200 cm, and b_2 = 50 cm, then $d_{2\,opt}$ = 14 mm. For each case, $(S^2 + G^2)^{1/2}_{min}$ is 1.7 and 4 mm, respectively.

4.3.4 OPTIMUM PARAMETERS IN THE SECONDARY PROCESS

If we investigate carefully commercially available LS pictures, we find that the pitch of the cylindrical lens is usually fine in small LS pictures and relatively large in large LS pictures. The most important design parameter in the secondary process is the pitch of the lenticular sheet. To determine the optimum lens pitch, several preliminary discussions are needed.

A. *Measure for Three-Dimensional Picture Quality*

To determine the "optimum" lens pitch, we have to decide beforehand what a good three-dimensional picture is, or in other words, what is the measure for three-dimensional picture quality.

The quality of a three-dimensional picture can be defined in various ways; however, the most reasonable measure may be depth resolution, that is, the distance between two resolvable spots in the depth direction, or the resolvable spot number in that direction.

An objection might arise that the lateral resolution can be equally as important; however, as will be seen in the following section, this can be included in depth resolution. In any three-dimensional imaging using stereopsis, good lateral resolution is a premise for good depth resolution.

B. *Sampling Effect*

The thickness of the lenticular sheet has been assumed to be chosen so that the focus of a parallel incident beam is formed upon the emulsion (assumption 6). In such a case, in observation, a light beam from a spot on the emulsion will form, at best, a parallel bundle as illustrated in Fig.4.16. Therefore, an observer can never distinguish two lateral spots inside a single cylindrical lens in any given direction. In other words, the lens width gives the minimum resolvable width, or the lateral picture element.

This effect will hereafter be called the sampling effect. As shown in Section D, this effect limits the depth resolution of the three-dimensional image.

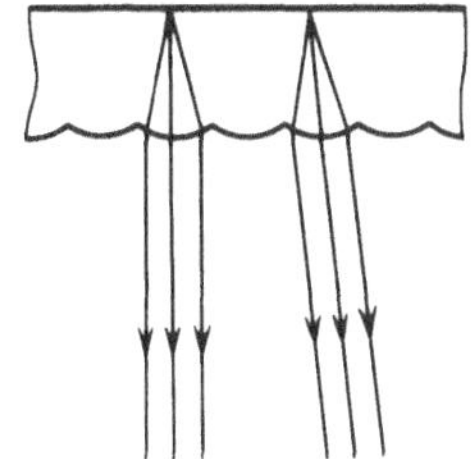

Fig.4.16 Light bundle generated at a spot upon the emulsion.

C. *Directivity of the Lens Sheet*

Another factor limiting depth resolution is directivity of the lens sheet; this is defined as follows. Suppose that a parallel light beam is incident upon the lens sheet. The emulsion is exposed to the beam, developed, and then illuminated from the rear to reconstruct the original parallel beam (see Fig.4.16). In practice, however, the beam is no longer parallel but has an angular spread, which we define as the directivity of the lens sheet. Hence it carries units of radians (or degrees).

There are four factors affecting directivity:

(1) *Diffraction.* If we consider only the sampling effect, a finer lens pitch is more desirable to get a finer picture; however, when the pitch is small, directivity is deteriorated due to several factors, among which diffraction is the most essential.

The diffraction effect was discussed rigorously by Burckhardt in his analysis of IP (2). In the present analysis, we shall use a rather crude approximation, which permits easier access to the problem.

Diffraction causes a spread of the focus upon the emulsion in the secondary picture-taking process. Let the lateral displacement upon the emulsion be denoted by x as shown in Fig.4.17a. We use the approximation that the corresponding angular spread θ ($\doteqdot x/t$) is given as the Fraunhofer diffraction of a slit having width p (see Fig.4.17b). When the incident light has wavelength λ' in the lens, the amplitude distribution

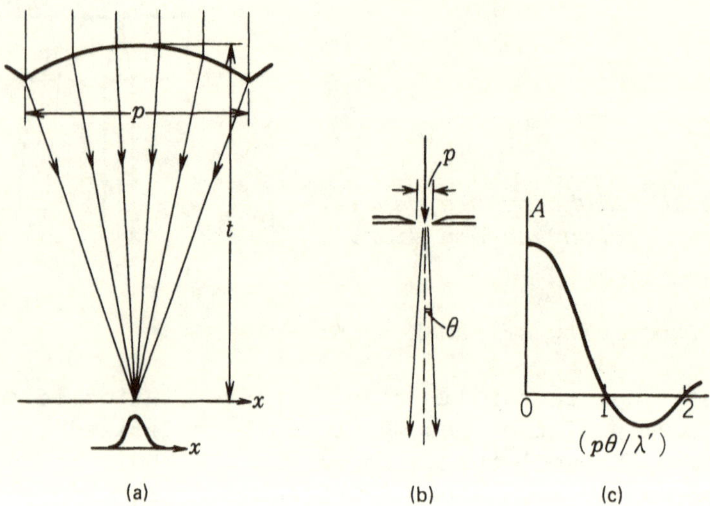

Fig.4.17 Symbols used in the analysis of the diffraction.

along x is then given as

$$A(x) \doteq A_1[\sin(\pi p\theta/\lambda')/(\pi p\theta/\lambda')]$$
$$= A_1[\sin(\pi px/t\lambda')/(\pi px/t\lambda')], \qquad (4.18)$$

where A_1 is a constant. The relation of Eq.(4.18) is shown in Fig.4.17c. Following Burckhardt [see Eq.(10) in Ref.(2)], we shall approximate this formula as

$$A(x) = A_1 \exp\left[-(2x/w_D)^2\right], \qquad (4.19)$$

where

$$w_D = 2t\lambda'/(2.04)^{1/2}\, p. \qquad (4.20)$$

Next we consider the reconstruction (observation) process. If other conditions are ideal, the angular spread of the reconstructed light beam will be given as (see Fig.4.18)

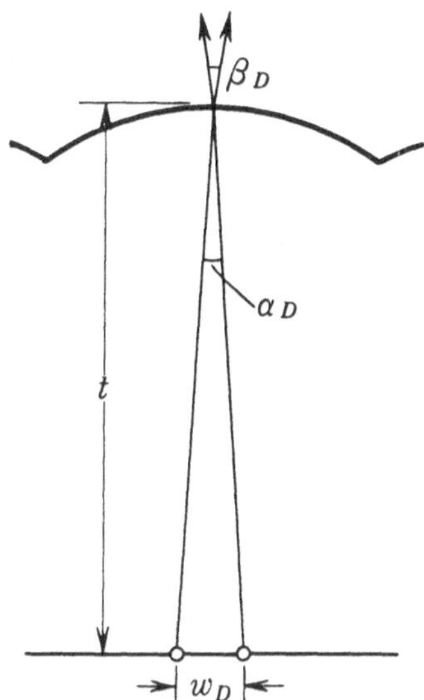

Fig.4.18 Angular spread due to diffraction.

$$\beta_D = \sqrt{2}w_D n/t = \left[2\sqrt{2}/(2.04)^{1/2}\right](\lambda/p) \doteqdot 2.0\lambda/p, \quad (4.21)$$

where n denotes the refractive index of the lens and $\lambda = n\lambda'$ is the wavelength in free space. The factor $\sqrt{2}$ represents the fact that the reconstruction process is also accompanied by the same amount of diffraction. Thus the overall spread due to the two diffractions is given by the autoconvolution of the spread given by the picture-taking process, and the factor $\sqrt{2}$ is required. Note that β_D is inversely proportional to p but independent of n and t.

(2) *Turbidity (resolution of the emulsion) and surface roughness.* The resolution of the emulsion also limits directivity. The angular spread due to this limitation is given as

$$\beta_E = nw_E/t = nw_E M/p \quad (M \triangleq p/t), \quad (4.22)$$

where w_E denotes the spread measured upon the emulsion and M is the pitch-to-thickness ratio, which ranges from 0.3 to 0.1 in most cases.

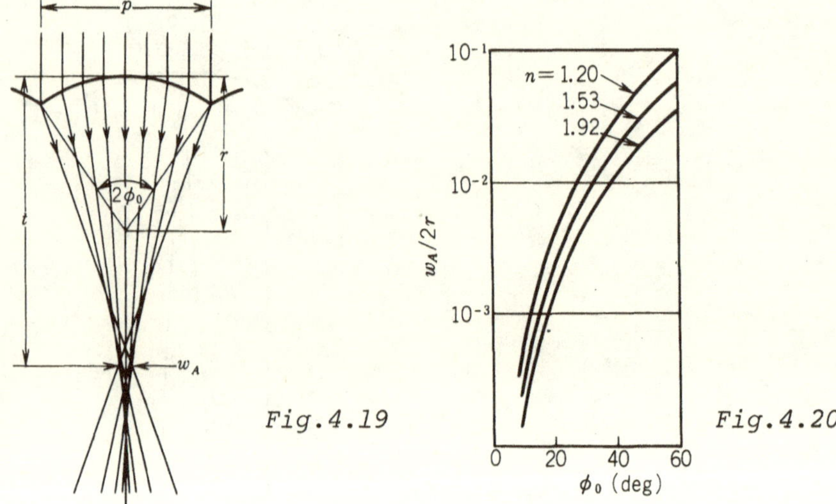

Fig.4.19 Aberration in a cylindrical/spherical lenslet.
Fig.4.20 Spread of the focus due to aberration.

The surface roughness of the lenticular sheet (both of the front and rear surfaces) also gives a limitation to directivity, which is proportional to p^{-1}. Therefore, when necessary, we may take it that w_E and β_E also include this effect.

(3) *Aberration*. This effect is difficult to evaluate but very important practically.

In LS pictures, the use of a noncylindrical lenticular sheet is possible; on the other hand, in IP it is fairly difficult to use nonspherical lenses because the fly's-eye lens sheet is usually produced from a copper mold originally pressed by using many tiny steel balls.

Theoretically, aberration can be reduced without limit by the use of a noncylindrical/nonspherical lens sheet; however, for light beams with skew incidence, the improvement is less, and sometimes even deterioration occurs. Therefore, in the present analysis, to

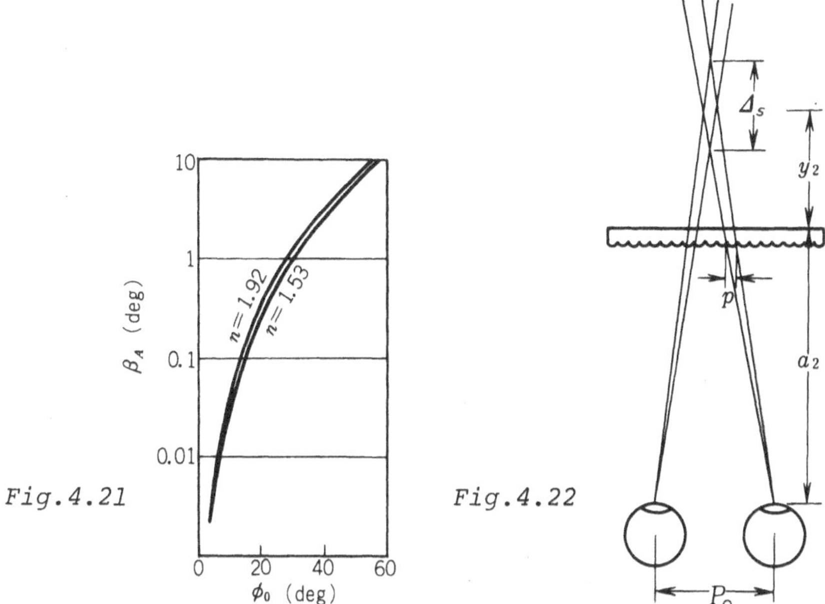

Fig.4.21 Angular spread due to diffraction.
Fig.4.22 Depth-resolution limitation due to a finite lens pitch.

give a standard measure of aberration, a cylindrical or spherical lens will be considered.

Let the radius of curvature of the lens be r and the angle giving the relative aperture be ϕ_0 (see Fig.4.19). We now assume that the thickness t is chosen so that the back surface is upon the minimum spread of the light bundle for a parallel incident beam. (Therefore t is a function of ϕ_0.) This minimum spread w_A can be determined with a computer, and the result is shown in Fig.4.20 as a function of ϕ_0 for three values of the refractive index n (3).

The spread w_A results in an angular spread of the beam when the picture is observed. This spread β_A was also determined with a computer, and the result is shown in Fig.4.21. It is noteworthy that the

TABLE 4.2 Factors Affecting Directivity and Their Dependence on Lens Pitch p[a]

Factors affecting directivity	Angular spread	Dependence on lens pitch p
Diffraction β_D	$\beta_D = 2.0\lambda/p$	p^{-1}
Resolution of emulsion β_E	$\beta_E = nw_E M/p$	p^{-1}
Aberration β_A	(see Fig.4.21)	p^0 (if p/t = const)
Projector-lens aperture β_L	$\beta_L = k_2 d_2/a_2$	p^0

[a]From Ref.(3).

angular spread is almost independent of the refractive index. It is about 0.035° for ϕ_0 = 10°, 0.3° for ϕ_0 = 20°, 1.1° for ϕ_0 = 30°, and 3° for ϕ_0 = 40°.

For details of the computation giving the results shown in Figs.4.20 and 4.21, refer to the appendix of this chapter.

(4) *Limitation due to projector-lens aperture.* When the lens diameter of the projector is finite (not zero), in secondary picture taking, each of the N pictures are projected from a certain width of angle (= D_2/a_2) onto the emulsion on the rear surface of the lenticular sheet. Therefore, it will give a spread to the image upon the emulsion. This degrades the resultant directivity. The angular directivity due to this effect is given, from simple geometrical considerations, as

$$\beta_L = D_2/a_2 = k_2 d_2/a_2. \qquad (4.23)$$

The only way to minimize this effect is to use as small a projector lens as possible (but not so excessively small as to cause a diffraction problem) in

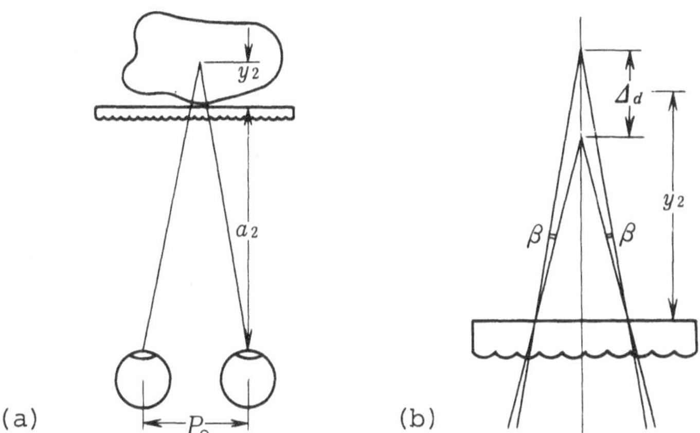

Fig.4.23 Depth-resolution limitation due to a finite directivity of the lens sheet.

the secondary picture-taking process.

This discussion on the four factors is summarized in Table 4.2.

D. Depth-Resolution Limit

On the basis of the analysis performed in the preceding Section C, a depth-resolution limit will be obtained.

(1) *Limitation due to sampling effect.* Depth resolution is limited partly by the sampling effect. Let the distance between two human eyes be P_0 ($\doteqdot$ 6.25 cm). The depth-resolution limit (distance between two resolvable spots) due to the sampling effect at a depth y_2 is computed directly from Fig.4.22 as

$$\Delta_s \doteqdot 2p(a_2 + y_2)^2/a_2P_0. \qquad (4.24)$$

(2) *Limitation due to finite directivity.* Depth resolution is limited also by the finiteness of the

lens directivity, as illustrated in Fig.4.23. Directly from the figure, this limitation is obtained as

$$\Delta_d \doteqdot 2\beta y_2(a_2+y_2)/P_0, \tag{4.25}$$

where β denotes the resultant directivity, including β_D, β_E, β_A, and β_L.

(3) *Overall depth-resolution limit.* If we define the overall depth-resolution limit as $(\Delta_s+\Delta_d)$,* it is given as

$$\Delta = \Delta_s+\Delta_d = (2pa_2/P_0)(1+Y_2)\{1+Y_2[1+(\beta a_2/p)]\}, \tag{4.26}$$

where $Y_2 = y_2/a_2$. Then the number of resolvable spots (resolution) in the depth direction between $y_2 = 0$ and $y_2 = b_2$ is given as

$$N_D = \int_0^{b_2}(1/\Delta)dy_2 = \int_0^{B_2}(a_2/\Delta)dY_2$$
$$= (P_0/2a_2\beta)\log\{1+(a_2\beta/p)[B_2/(1+B_2)]\}, \tag{4.27}$$

where $B_2 = b_2/a_2$.

Four variables, β, p, B_2, and a_2, appear in Eq. (4.27). However, it is found in Eq.(4.27) that a variation of a_2 can be replaced by that of β, and a variation of B_2 can be replaced by that of p.

Let us consider this problem in a little more detail. Suppose there are two sets of p and B_2 and that they are represented by p, B_2 and p', B_2'. From Eq.(4.27), it is seen as far as the relation

---

* An alternative, probably more reasonable definition would be $\Delta' = (\Delta^2+\Delta^2)^{1/2}$. The definition in the text (algebraic sum) is used here only for simplicity. The difference from the second definition only makes the estimation of the overall resolution a little more pessimistic. It is easy to show that the ratio Δ/Δ' is between 1 and $\sqrt{2}$.

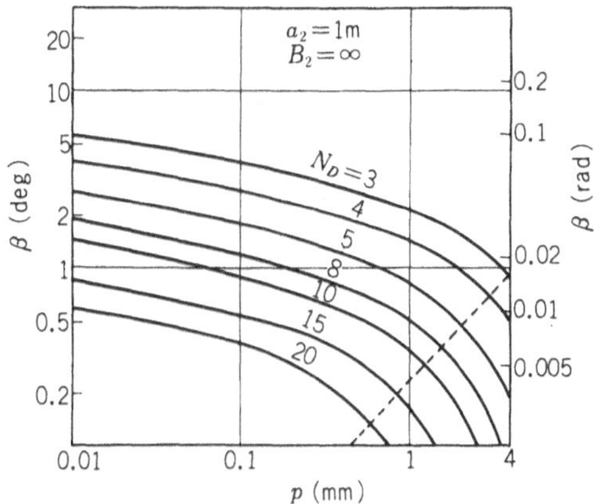

Fig.4.24 Overall depth resolution N_D as a function of the lens pitch and directivity for $a_2 = 1$ m and $B_2 = \infty$. The broken line is the symmetry axis. The contours of N_D for other values of a_2 and B_2 can be obtained by translations.

$$B_2/p(1+B_2) = B_2'/p'(1+B_2') \tag{4.28}$$

holds, the resolution N_D is identical for both cases. On the other hand, if there are two sets of a_2 and β and they are represented by a_2, β and a_2', β', and the relation

$$a_2\beta = a_2'\beta' \tag{4.29}$$

holds, N_D is also identical for the two cases.

Therefore, a single figure like Fig.4.24 can be used universally. In this figure, the contours of N_D are shown as a function of p and β for $a_2 = 1$ m and $B_2 = \infty$ as an example. For different values of a_2 and B_2, contours of N_D can easily be obtained by translations of the original contours in the vertical and horizontal directions, respectively. The amount of translation is determined by Eqs.(4.28) and (4.29).

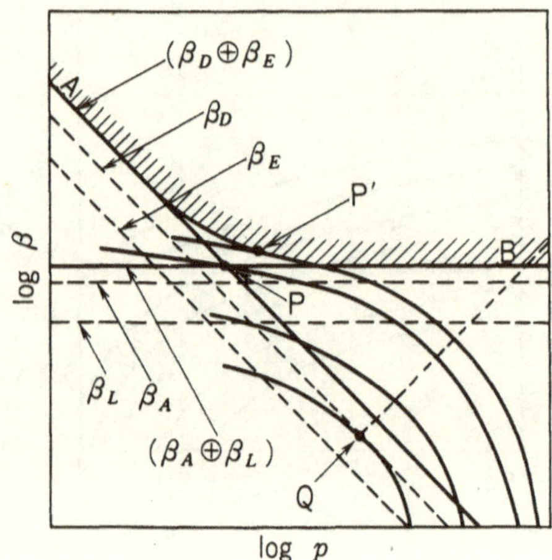

Fig.4.25 Realizable region in the p-β diagram. Point Q gives the optimum design in the case $\beta = \beta_D$*. The symbol ⊕ denotes a convolution.*

E. Optimum Pitch of Lens Sheet

Figure 4.24 gives a guide to the discussion of the optimum lens pitch. We first note that among the four factors affecting directivity, diffraction and turbidity give rise to angular spreads inversely proportional to pitch, whereas the spreads caused by aberration and the projector aperture are primarily independent of pitch (see Table 4.2);

$$\beta_D \propto p^{-1}, \quad \beta_E \propto p^{-1}, \quad \beta_A \propto p^0, \quad \text{and } \beta_L \propto p^0. \qquad (4.30)$$

Therefore, the realizable region in the p-β plane will be something like the hatched region in Fig.4.25. This figure suggests that an optimum lens pitch is present maximizing the resolvable spot number. In Fig.4.25 it is given by point P', or approximately by the cross point P of lines A and B. For example, when

$$\beta_D \oplus \beta_E = 3\beta_D = 6.0\lambda/p = 0.003 \text{ mm}/p \qquad (\lambda = 500 \text{ nm}),$$

$\beta_A \oplus \beta_L = 0.5°$, ($\oplus$ denotes convolution)

$a_2 = 1$ m, $B_2 = 0.25$,

and we use the approximation that the point P gives the optimum condition, we obtain

$p_{opt} = 0.35$ mm and $N_{D\ max} = 6.5$.

As stated earlier, among the four factors affecting directivity, diffraction is the most essential. In an idealized case where the directivity is limited only by diffraction, the optimum lens pitch and depth resolution limit (corresponding to point Q in Fig.4.25) can be obtained analytically.' Putting

$$\beta = \beta_D = k\lambda/p \quad (k = \text{constant}) \tag{4.31}$$

into Eq.(4.27) and differentiating with respect to p, we find that depth resolution is maximized at

$$p_{opt} = \left\{(k\lambda a_2/3.92)\left[B_2/(1+B_2)\right]\right\}^{1/2}. \tag{4.32}$$

The constant 3.92 appearing in Eq.(4.32) is the root of the equation $\log(1+x)-2x/(1+x) = 0$.

Then the maximum resolution is given, from Eqs. (4.27), (4.31), and (4.32), as (all lengths in meters)

$$\begin{aligned} N_{D\ max} &= \log 4.92(P_0 p_{opt}/2k\lambda a_2) \\ &= 0.0251\left[B_2/k\lambda a_2(1+B_2)\right]^{1/2}. \end{aligned} \tag{4.33}$$

For a lenticular sheet, k = 2.0 as shown in Eq. (4.21). The values of p_{opt} and $N_{D\ max}$ are tabulated in Table 4.3 for $B_2 = 0.25$ and $\lambda = 500$ nm.

4.3.5 MEASURED DIRECTIVITY OF LENTICULAR SHEET

To demonstrate how the directivity of a lens sheet is degraded by the essential limitation (diffraction),

TABLE 4.3 Optimum Lens Pitch and Maximum Resolution of Lenticular Sheet[a]

Lenticular Sheet		
a_2 (m)	p_{opt} (mm)	$N_{D\,max}$
0.25	0.11	22.4
0.5	0.16	15.4
1.0	0.23	11.2
2.0	0.32	7.9

[a]Only the diffraction and the sampling effect are considered [after Okoshi (1971)].

measured directivities of lenticular sheets will be described briefly. Details of the measuring setup and method will be described in Chapter 5.

To facilitate measurement, four samples of commercially available LS sheets were backed with white diffusing material (vinyl paint), and the reflection directivities of such LS screens were measured. If we suppose that the spread of a spot image due to the diffusing material is equal to the turbidity of the emulsion, the measured reflection directivity should be equal to the directivity of the LS picture.

The measured data are tabulated in Table 4.4 and are also plotted in Fig.4.26. This figure shows that the directivity is at least ten times worse than the diffraction limitation in those samples. (The "directivity" shown in Table 4.4 and Fig.4.26 denotes the angular spread of the reflected beam where the light intensity is decreased to one-half the central intensity.)

4.3.6 DISCUSSIONS

The important features of the results of the above considerations can be summarized as follows:

TABLE 4.4 Measured Directivity (Half-Intensity Spread) of Lenticular Sheets[a]

Sample number	Dimensions (mm) r	p	t	ϕ_0 (deg)	Directivity (deg)
1	0.52	0.438	1.30	24.9	4.8
2	0.64	0.292	1.10	13.2	8.0
3	0.52	0.625	1.30	36.9	6.8
4	0.24	0.418	0.525	60.5	10.4

[a]For the symbols used, refer to Fig.4.19.

(1) Theoretically, the optimum projector separation is obtained as ranging between several millimeters and 20 mm. These values seem smaller than the conventional separation determined empirically. The reason for this discrepancy is probably ignorance of the mechanical inaccuracy in the present theory.

(2) Theoretically, the optimum lens pitch is obtained as ranging between 0.1 and 0.5 mm. These values are in good agreement with the empirical optimum design.

(3) The computed depth resolution N_D ranges ordinarily between 3 and 10; this is surprisingly small as compared with what we feel when we observe LS pictures. This fact tells that our depth perception is very ambiguous and is easily deceived. In other words, we recognize a three-dimensional image mostly through psychological processes, with the aid of a small number of depth cues but a lot of experiences and imagination.

4.4 OPTIMUM DESIGN OF INTEGRAL PHOTOGRAPHY

4.4.1 INTRODUCTION

In this section we consider two-step integral photography as devised by H. E. Ives (4). The term

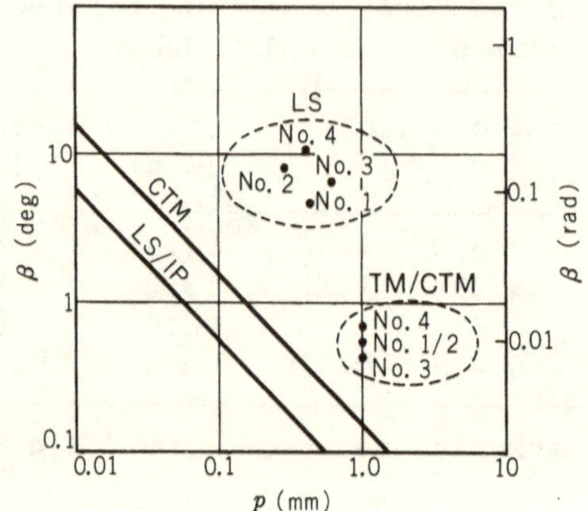

Fig.4.26 Diffraction limitations of the directivities of LS, IP, TM (triple mirror) screen, and CTM (curved triple mirror) screen. The measured directivities are also shown for comparison. For the theory and data of TM and CTM screens, refer to Section 5.7.

"optimum design" again refers to the design of the whole picture-making process, that is, the primary taking of an integral photograph and its inversion to the secondary integral photograph.

As stated in Chapter 2, until the late 1960s it had been almost impossible to make a practical integral photograph. The first scientific approach to the design for IP was made by C. B. Burckhardt in 1968 (2), who presented the first optimum design theory and showed the resolution limitation.

As will be described later, however, it seems to the author that one of the assumptions made by Burckhardt does not always fit the setup for the fabrication of an integral photograph. This is the assumption that in making the secondary integral photograph, the secondary photograph is placed just in front of the primary photograph. This assumption is valid in some cases as described by DeMontebello (5), but not realistic in many practical cases. For the latter, it leads to the unrealistic conclusion that the optimum pitches of the primary and secondary integral photographs are equal to each other.

In this section, the optimum design theory of LS pictures described in the preceding section will first be modified for application to IP. Next, Burckhardt's theory will be introduced and compared with the "modified" theory mentioned above.

4.4.2 *EXTENSION OF THE THEORY ON LS PICTURES*

As stated in Chapter 2, the only essential difference between IP and LS pictures is that the former give vertical parallax information, whereas the latter lack it. Therefore, it is evident that the discussion on LS pictures described in Section 4.3 is also valid for specific types of integral photographs. We should, however, interpret some figures in Section 4.3 in somewhat different ways; that is:

(1) The cameras and projectors shown in Fig.4.11 should be arranged in two-dimensional arrays;
(2) Figures 4.17a and 4.18 should represent cross sections of fly's-eye lens sheets.

Another modification required is the evaluation of diffraction. In the preceding section, the diffraction given by a lenticular sheet was approximated by the Fraunhofer diffraction of a slit having a width p. In IP, a fly's-eye lens sheet is used, for which a similar, but slightly different analysis is required.

A. *Case of Square Lenslets*

We first consider a fly's-eye lens sheet consisting of many square lenslets, having an area $p \times p$ and a thickness t. In such a case, the spread of light (amplitude distribution) corresponding to Eq.(4.18) is given as

$$A(x,y) = A_1 \frac{\sin(\pi p\theta/\lambda')}{(\pi p\theta/\lambda')} \cdot \frac{\sin(\pi p\theta'/\lambda')}{(\pi p\theta'/\lambda')} . \tag{4.34}$$

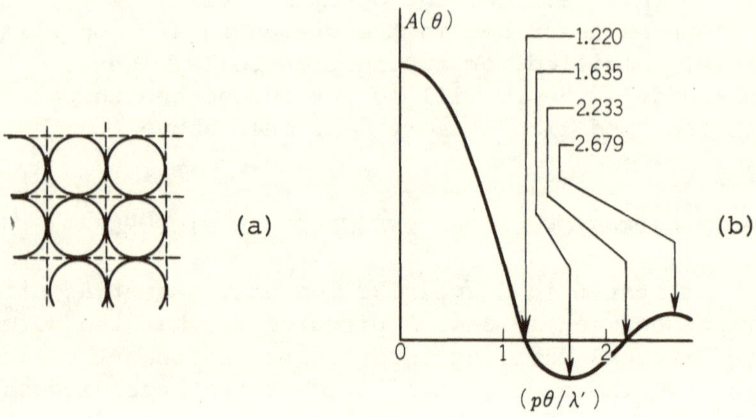

Fig.4.27(a) A fly's-eye lens sheet having circular apertures. (b) The Airy function.

In this equation, A_1 is a constant, λ' is the wavelength in the lens material, and θ, θ' are the spread angles with respect to the lateral directions ($\theta = x/t$, $\theta' = y/t$). Equation (4.34) tells that Eqs.(4.19)-(4.21) for an LS picture are also valid for IP as far as the beam spread due to diffraction is concerned.

B. *Case of Circular Lens Apertures*

To improve the picture quality of an integral photograph, a fly's-eye lens sheet having circular masks as shown in Fig.4.27a is sometimes used. Such masks remove the light from each corner of the lenslets, where the aberration is nonnegligible.

In the above case, the Fraunhofer diffraction of a circular aperture may be used as a measure of diffraction given by the lenslets. The amplitude distribution of the Fraunhofer diffraction of a circular aperture is given by the Airy function:

$$A(\theta) = \frac{2J_1(p\theta/\lambda')}{(p\theta/\lambda')}, \tag{4.35}$$

where p denotes the diameter of the aperture, which in the present case is equal to the pitch of the lens sheet. The relation of Eq.(4.35) is shown in Fig. 4.27b. Noting that the angle giving the first zero ($\theta = 1.22\lambda'/p$) is a little greater than the corresponding angle for a square aperture ($\theta = \lambda'/p$), instead of Eq.(4.21) we may write

$$w_D = 1.22 \times 2.0\lambda/p = 2.44\lambda/p. \tag{4.36}$$

Therefore, for an IP with circular lens apertures, we have to use k = 2.44 in Eqs.(4.32) and (4.33).

4.4.3 BURCKHARDT'S THEORY

Burckhardt considered a process for making an integral photograph in which the primary image is recorded with an ordinary arrangement as shown in Fig.4.28a; whereas the three-dimensional image reconstructed from the primary photograph is then transferred to the secondary photograph placed just in front of the primary. The virtual image is observed deep behind the secondary photograph as shown in Fig. 4.28b. His theory will be introduced in the following, while the symbols used in the analysis are modified so as to be consistent with the preceding discussions.

A. Principle of the Analysis

Burckhardt considered that the picture recorded in IP is deteriorated by two causes: diffraction and defocus. The "flipping" discussed in the previous section is not considered. However, since the lens aperture-to-separation ratio is unity [$k_1 = k_2 = 1$ in Eq.(4.10)] in the primary picture-taking process under consideration, the width of flipping is equal to the blurring due to defocus. Therefore, Burckhardt might have considered that the estimation of the defocus effect automatically includes the flipping effect.

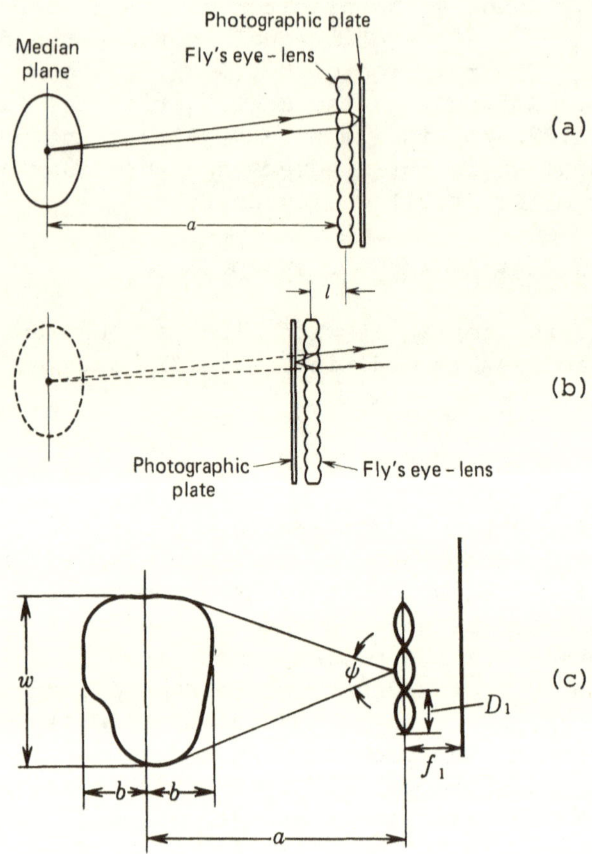

Fig.4.28(a) Recording of the primary integral photograph. (b) Recording of the secondary photograph in Burckhardt's process. (c) Symbols used in the analysis (w and ψ will be used in Chapter 7) [after Burckhardt (2)].

As in the analysis in the preceding section, therefore, the blurring due to defocus will be excessive when the lens pitch is too large, whereas image deterioration due to diffraction will be present when the lens pitch is too small. The optimum lens pitch can be given as a compromise between the above two effects. The following assumptions are made in the analysis.

B. *Assumptions in the Analysis*

(1) The primary photograph is focused at distance a. (As shown in Fig.4.28a, we assume that the lens sheet and the photographic plate are separated and the separation is adjustable.)

(2) The lenslets are square, and are tightly arrayed ($k_1 = k_2 = 1$).

(3) $a \gg b$ in Fig.4.28c.

(4) In recording the secondary photograph, the secondary photograph is placed just in front of the primary. Or in other words, if the distance between the primary and secondary integral photographs is denoted by ℓ, $\ell \ll a$ holds. (This assumption is not described explicitly in his paper, but is evident from the analysis.)

(5) The lenslets in the secondary photograph focus upon the median plane of the reconstructed image.

C. *Analysis of the Primary Photograph*

We first compute the light intensity distribution upon the emulsion of the first photograph for distances a+b or a-b for which blurring due to defocus is largest. Since the results obtained are rather easy to understand, only the results of the computation will be shown.

The distribution I(x) along the lateral displacement x is expressed approximately as

$$I(x) = \exp\left(-4x^2/w_1{}^2\right) \tag{4.37}$$

where w_1 denotes the spread of the spot image and is given as

$$w_1 = \left(0.415 \frac{b^2 f_1{}^2}{a^4} D_1{}^2 + 0.989 \frac{\lambda^2 f_1{}^2}{D_1{}^2}\right)^{1/2} . \tag{4.38}$$

In this equation, f_1 and D_1 denote the focal length and the lens diameter of the integral photograph, respectively, and λ denotes the wavelength. Note that the first term in Eq.(4.38) is proportional to $b^2 D_1{}^2$, while the second term is proportional to $\lambda^2 D^{-2}$. These facts suggest that the first term represents the blurring due to defocus, and the second term represents the blurring due to diffraction.

It is obvious from Eq.(4.38) that the value of D_1 making w_1 minimum is given as

$$D_{1\ opt} = 1.24a\sqrt{\lambda/b}. \tag{4.39}$$

For example, when a = 50 cm, b = 5 cm, and λ = 500 nm, we obtain $D_{1\ opt}$ = 1.96 mm. For the optimum D_1, w_1 is given as

$$w_{1\ min} = 1.13\left(f_1/a\right)\sqrt{b\lambda}. \tag{4.40}$$

D. Analysis of the Secondary Photograph

The primary integral photograph reconstructs the real image of the object, whereas the secondary integral photograph reconstructs its virtual image. This is the only essential difference between the two stages. Therefore, if we assume $\ell \ll a$ in Fig.4.28 [assumption (4) stated previously], then $D_{1\ opt} \doteqdot D_{2\ opt}$ is obtained; that is, the optimum parameter of the secondary photograph is approximately equal to that of the primary. (Burckhardt might have taken this fact as self-explanatory and assumed that the primary and the secondary photographs have identical structure.) In the following discussions, therefore, the subscripts 1, 2 will be omitted.

Let the spread of the real image reproduced by the primary photograph be denoted by w_{total}. Since the spread is given as the autoconvolution of the first spread given by Eqs.(4.37) and (4.38), and the whole image is then further enlarged by (a/f), w_{total} is given as

$$w_{total} = \sqrt{2}\,(a/f)w_1. \tag{4.41}$$

When this real image is recorded upon the emulsion of the secondary photograph, the spread of the image upon the emulsion is given as

$$w_2 = \sqrt{3}\,(a/f) \times (f/a)w_1 = \sqrt{3}\,w_1, \tag{4.42}$$

where the factor $\sqrt{3}$ represents the triple autoconvolution.

Finally, the lateral spread of a spot in the image reconstructed from the secondary photograph is given as

$$w_{final} = \sqrt{4}\,(a/f)w_1 = \left(1.66\,\frac{b^2}{a^2}\,D^2 + 3.93\,\frac{\lambda^2 a^2}{D^2}\right)^{1/2}. \tag{4.43}$$

E. Comparison of the Author's and Burckhardt's Theory

Burckhardt's theory was presented solely for IP; however, as described before, there ought not to be any essential difference between IP and LS pictures as far as the optimum design parameters are concerned. Hence, the difference between the obtained optimum parameters:

(1) in the present author's theory: $D_{opt} \simeq 10$ mm for the primary stage (see Section 4.3.3), and $p_{opt} = 0.1 \sim 0.5$ mm for the secondary stage (see Section 4.3.6), whereas

(2) in Burckhardt's theory: $D_{opt} \simeq 2$ mm for both stages

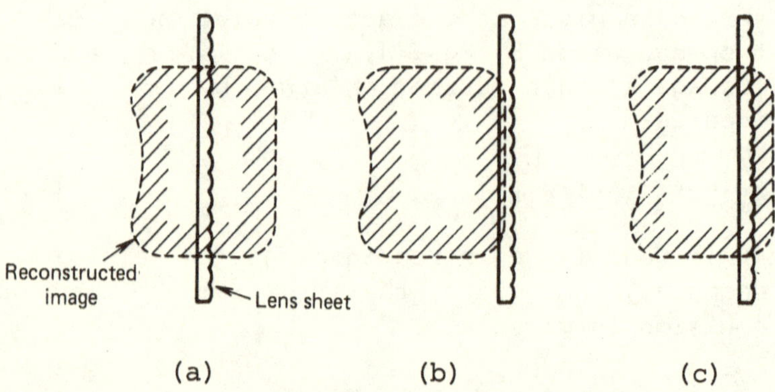

Fig.4.29 Relative position of the reconstructed image with respect to the lens sheet.

should not be attributed to the difference between IP and LS pictures, but to the difference in the assumptions employed in the two analyses.

It seems that Burckhardt's assumption (4) (see Section B above) is not always realistic. In actual LS pictures, such an arrangement is never employed. It is also rather rare, within the author's knowledge, in IP. Ordinary LS pictures or secondary integral photographs are made in an arrangement as shown in Fig.4.29; that is, they are located:

(1) upon the median plane of the reconstructed three-dimensional image (Fig.4.29a), or

(2) upon the nearest end of the reconstructed image (Fig.4.29b), or

(3) between the nearest end and the median plane (Fig.4.29c).

It is not without reason that the above arrangements are employed. Method (1) is theoretically the best because the blurring due to defocus can be minimized; in this case the maximum defocus is equal to one-half of that corresponding to the total depth of the object. However, if we really construct an IP or an LS picture satisfying condition (1), there will be an unnatural appearance to the nearer half of the reconstructed image. This is because the finiteness of the picture size inevitably makes us conscious of the presence of the frame, which gives an unnatural impression for the nearer part of the reconstructed image floating in front of the frame. Sometimes we even feel that this part of the image is "pulled back" along its edges toward the frame. This is due to a psychological rule that the presence of a floating image in front of a truly existing frame can hardly be accepted. Such an effect is called, in Ref.(6), a "frame canceling" effect.

Method (2) described above is sometimes employed because it is free from frame canceling, although the blurring due to defocus increases.

Method (3) is nothing but a compromise between (1) and (2), and is the most commonly employed method in LS pictures. In most LS pictures, 70-90% of the image is reconstructed behind the lens sheet, whereas the remaining 10-30% is reconstructed in front of it. As to IP, it is difficult to talk about "common practice" because there are only a few examples. The IP produced by MDH Products Corporation described in Chapter 2 employs method (2), and the beaded plate IP of Bell Telephone Laboratories, to be discussed in the next section, employs method (1).

In most of the above three cases, Burckhardt's assumption (4) cannot hold, and $D_{1\ opt} = D_{2\ opt}$ is no longer valid. The only possible exception is the case when we use method (2) together with the first integral photograph placed very close to the object.

Furthermore, assumption (5), that the secondary photograph is focused at a specific depth of the image, is hardly feasible. This is because, as illustrated in Fig.4.30, such an adjustment is possible when the image is reconstructed far beyond the second photograph (Fig.4.30a), but is impossible when it is reconstructed around the integral photograph. The validity of the above statement will easily be seen if we consider an extreme case in which point P to be focused approaches point Q (Fig.4.30b). In such a case the

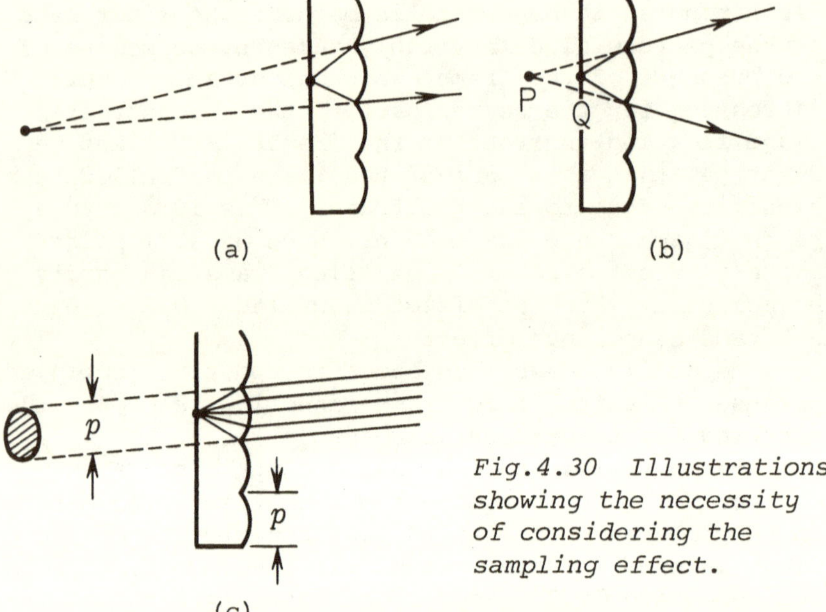

Fig.4.30 Illustrations showing the necessity of considering the sampling effect.

focusing function of the lens should not be present.

Therefore, when the second photograph lies across or just in front of or very close to the reconstructed image, the only way of optimizing the integral photograph thickness is to adjust it so that the reconstructed light forms a parallel beam, as shown in Fig. 4.30c. Therefore, the light beam inevitably has a certain width. This is the reason that the sampling effect is considered in the author's theory.

The author calls a reconstruction such as that shown in Fig.4.28b "remote imaging," and those shown in Fig.4.29 "close imagings." Burckhardt considered remote imaging in his theory on IP, but has also dealt with the case of close imaging in his theory on the fly's-eye lens autocollimating screen (7). This will be described in the next section.

4.5 BEADED-PLATE INTEGRAL PHOTOGRAPHY

When the development of IP techniques began early in the 1960s, the greatest difficulty encountered was how to fabricate a usable fly's-eye lens sheet. It was, and still is, a technologically very difficult task. To avoid this difficulty of fabricating a lens sheet, Burckhardt and Doherty (8) devised another method in which a large number of tiny glass balls (typically with a diameter of 50 μm) are densely arrayed upon the emulsion of a photographic plate. They called such a plate a *beaded plate*. In this section, their beaded-plate IP technology will be described.

4.5.1 PRINCIPLE

In the appendix to this chapter where the aberration of a lens sheet is considered, a simple paraxial analysis is first presented. Equation (A.2) shows the optimum thickness of the lens sheet under the paraxial approximation. Setting n (the refractive index) = 2 in Eq.(A.2), we obtain $t = 2a$, that is, the optimum thickness is equal to twice the radius of the lens curvature. Therefore, within the paraxial approximation, the rear surface of a glass ball with $n = 2$ forms a back focal plane. If a large number of tiny glass balls with $n = 2$ are arrayed upon the emulsion of a photographic plate as shown in Fig.4.31a, the array of balls is equivalent to a fly's-eye lens sheet, and such a "beaded plate" can be used to take an integral photograph.

4.5.2 EXPERIMENT

Burckhardt and Doherty used the above method in the second stage of integral photography. The beaded plate is prepared in the following way. A Kodak 649GH plate is immersed in Kodak Photo-Flo 200 for 5 min at

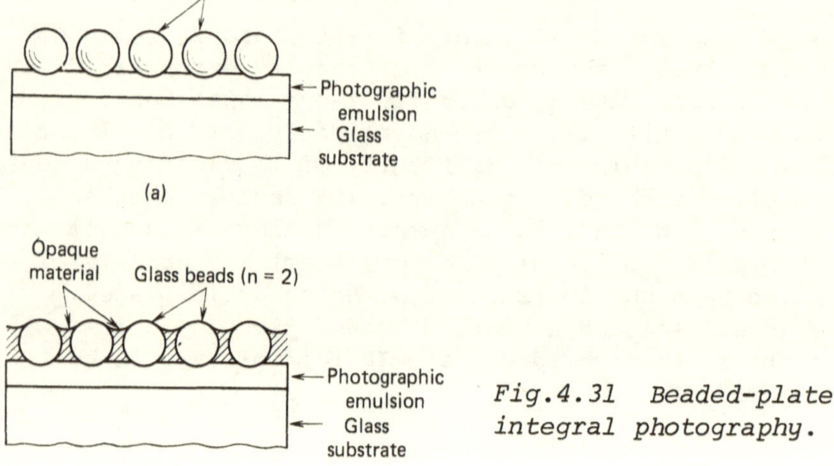

Fig.4.31 Beaded-plate integral photography.

room temperature. The plate is then taken out and held upright to remove excess liquid on the surface. Before the surface becomes quite dry, a large number of glass balls having diameter of 53-62 μm and n = 1.92 are scattered on the plate. The plate is then laid horizontally for 1 hr. Next, it is again immersed in water with Photo-Flo, and shaken gently. During this process the excess glass balls flow away, and a single layer of balls remains upon the emulsion. The plate is dried for one night and a beaded plate is completed. Note that the entire process must be carried out in a darkroom.

In Burckhardt's experiment, the first stage of IP used a fly's-eye lens sheet made of Lucite, having dimensions of 20 × 25 cm, thickness of 4.7 mm, and lens diameter 2.8 mm. A beaded plate integral photograph was employed for the second stage. Since the first stage was a "positive" picture, a complex conversion process was performed in developing the second photograph to make it again positive.

In Ref.(8), two photographs of a three-dimensional bust of Beethoven produced in this fashion, taken from two different directions, are shown.

4.5.3 OPTIMUM LENS DIAMETER

The optimum design theory of such special IP can be constructed almost in the same manner as those for LS pictures or ordinary IP; however, a beaded glass integral photograph is equivalent to a fly's-eye lens sheet (more exactly, a cat's-eye lens sheet; see Section 5.3.2) having a very large "aperture angle" ϕ_0 shown in Fig.4.19. Therefore, the aberration is usually large and cannot be neglected.

In discussing the optimum lens diameter for beaded-plate IP (8), Burckhardt used the optimum diameter formula

$$D_{opt} = 1.24\sqrt{\lambda b}, \qquad (4.44)$$

which he originally obtained in Ref.(7) for a fly's-eye lens sheet autocollimating screen (see Section 5.3). He states in Ref.(8) that direct application of the above equation leads to $D_{opt} = 200\ \mu m$, whereas actually $D_{opt} = 50\ \mu m$ was obtained experimentally. This discrepancy between the simple theory and experiment is probably due to the large aberration. (It was shown experimentally that when the ball diameter was reduced to 20 μm, the diffraction became comparable to the aberration.)

In the above paper, Burckhardt and Doherty also proposed a modification of beaded-plate IP as shown in Fig.4.31b, in which the lens aperture is reduced intentionally by filling the gaps between balls with opaque material. In reality, however, this modification would not be easy; one of the difficulties is that the developing chemical must penetrate the opaque material freely. The optimum design theory for such "masked" beaded-plate IP has not been developed. Such a mask will reduce both aberration and diffraction, but will leave the sampling effect unchanged.

4.6 FUTURE ENGINEERING TASKS RELATED TO LENS-SHEET THREE-DIMENSIONAL PICTURES

In this section, future tasks will be reviewed briefly. In either LS pictures or IP, the biggest technical problem is how to make a good lens sheet. The optical feature of the lenslets constituting these lens sheets is that they usually must have very small F numbers. Therefore, spherical aberration cannot be neglected; the resolution limitation is given by the aberration and surface roughness rather than by the diffraction in most practical cases. Moreover, the focusing property is greatly degraded for skew incidences.

The complete design theory of such lens sheets, taking into account deterioration of the performance for skew-incident light has not been established. In the present state of the art, most lens sheets are designed on a spherical/cylindrical basis, the aberration effect being neglected. Furthermore, most practical lens sheets are made of a single material, in contrast to ordinary camera lens systems in which multilayer structures using various optical materials is common practice. In the near future, to improve performance of the lens sheet toward its theoretical limit, the use of nonspherical/noncylindrical curvature and/or the use of curved rear surface and/or the use of nonuniform lens materials will become important technical tasks. The aberration and performance for skew-incident light can be improved only through such innovations.

The theoretical bases for the above innovations will be established sooner or later by taking advantage of recent progress in computers, and finally the use of nonuniform (or multilayer) lens sheets will become common practice. At that stage, a series of plastic lens materials having various refractive indexes (if possible, almost continuously), and also having good chemical and physical affinities with each other, will be demanded. We should note that for traditional glass lens systems, a series of optical glasses having different refractive indexes has already been developed and is ready for practical use.

Traditional optics has principally dealt with discontinuous optical media. In the near future, it will be extended to continuous optical media in which both the refractive index and absorption are given as functions of position. Earlier examples of such continuous media are the so-called gas lens developed by Bell Laboratories and the collimating optical fiber developed by Nippon Electric Company.

As to fabrication technology, many interesting engineering tasks still remain for IP, whereas the technology for LS pictures seems fairly matured. Generally speaking, LS technology has been well developed in France and Japan, whereas IP technology has been advanced in the United States.

4.7 SUMMARY

In this chapter, LS pictures and IP have been dealt with in a unified manner; the optimum design theory and theoretical limitations of the picture performance have been discussed. It should be noted that some of the discussions described in this chapter are also applicable to the projection-type imaging to be discussed in Chapter 5.

APPENDIX. ABERRATION IN SPHERICAL/CYLINDRICAL LENS SYSTEMS

A.1. SIMPLE ANALYSIS OF LENS SHEETS

As a first-order approximation of the problem, a paraxial ray solution for the optimum lens thickness will first be derived. From the so-called lens formula, a simple relation

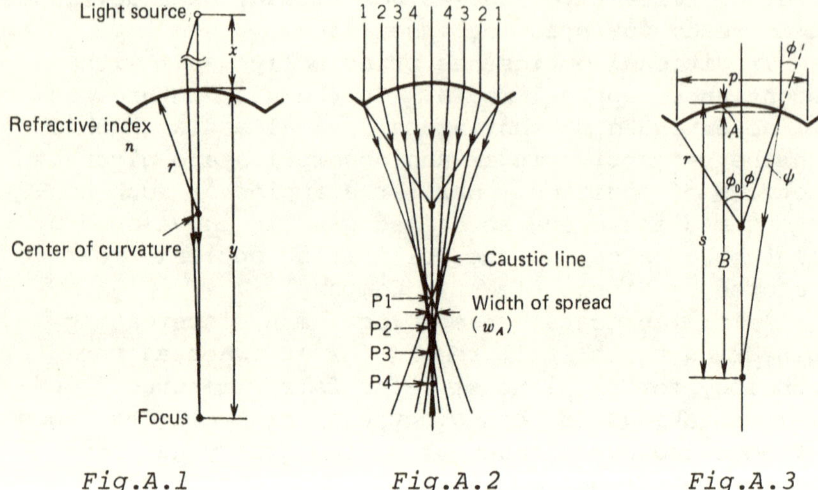

Fig.A.1 *Fig.A.2* *Fig.A.3*

Fig.A.1 Paraxial approximation.
Fig.A.2 Caustic lines in a thick lens.
Fig.A.3 Symbols used in the analysis.

$$(1/x)+(n/y) = (n-1)/r \tag{A.1}$$

is obtained for the quantities illustrated in Fig.A.1. The symbol n denotes the refractive index of the lens material. If the lens thickness t is chosen so that $y = t$ for $x \to \infty$,

$$t = [n/(n-1)]r. \tag{A.2}$$

For example, when $n = 1.5$, $t = 3r$. Obviously, Eq.(A.2) holds both for fly's-eye lenses and for lenticular sheets.

Equation (A.2) gives a first-order approximation of the optimum lens thickness. However, since the relation (lens pitch) $\ll$ t does not hold for ordinary fly's-eye lenses or lenticular sheets, the above formula based upon the paraxial approximation is not practical. A more exact computation taking into account aberration is required.

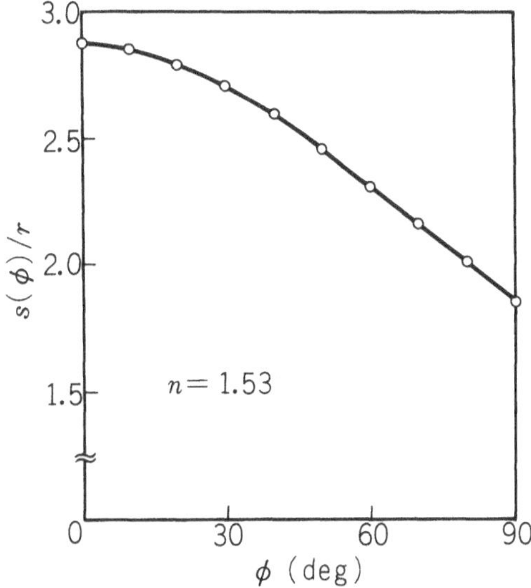

Fig.A.4 Relation between s(ϕ) and ϕ.

A.2. ANALYSIS OF SPHERICAL/CYLINDRICAL ABERRATION (9)

If we draw precisely each optical path for various incident positions as shown in Fig.A.2, we immediately see that the cross points of the paths and the axis move along the axis inward, like P_1, P_2, P_3, P_4. Thus, the light flux is collimated along an envelope (caustic line), and is never focused to a pinpoint but only to a finite size. Therefore, the best choice of the lens thickness is to place the rear surface at the "waist" position where the spread of the light bundle is smallest. The waist position and the size of this minimum spread will be analyzed in the following.

As shown in Fig.A.3, the lens radius of curvature, lens pitch, and the angle determining the lens aperture are denoted by r, p, and $2\phi_0$, respectively. The position of the cross point s is computed first as a function of the incident position ϕ:

$$s = A+B = r(1-\cos\phi)+r\sin\phi\cot(\phi-\psi). \qquad (A.3)$$

On the other hand, the law of refraction is expressed as

$$\psi = \sin^{-1}(\sin\phi/n). \tag{A.4}$$

From the above equations,* we obtain

$$s(\phi)/r = (1-\cos\phi)+\sin\phi\cot\{\phi-\sin^{-1}(\sin\phi/n)\} \tag{A.5}$$

$$= 1+\left[\left(\sqrt{n^2-\sin^2\phi}+\cos\phi\right)/(n^2-1)\right]. \tag{A.6}$$

The relation of Eq.(A.6) is shown in Fig.A.4 for n = 1.53 as an example. When ϕ tends to zero, Eq.(A.6) approaches n/(n-1) given by Eq.(A.2).

Next, we compute the minimum spread w_A. For this purpose we have to know first the position of the waist point of the caustic line. As seen in Fig.A.2, the caustic line is not smooth, but intersects the optical path for $\phi = \phi_0$ at a finite angle, and this intersecting point gives the point where the spread is of minimum size.

Therefore, by referring to Fig.A.5, the waist position can be obtained as follows:

---

* Derivation of Eq.(A.6):

$$\text{Eq.(A.5)} = 1-\cos\phi+\sin\phi\,\frac{\cos\{\phi-\sin^{-1}(\sin\phi/n)\}}{\sin\{\phi-\sin^{-1}(\sin\phi/n)\}}$$

$$= 1+\frac{\sin[\phi-\{\phi-\sin^{-1}(\sin\phi/n)\}]}{\sin\{\phi-\sin^{-1}(\sin\phi/n)\}}$$

$$= 1+\frac{1}{n}\,\frac{1}{\sqrt{1-\sin^2\phi/n^2}-\cos\phi/n}$$

$$= \text{Eq.(A.6)}$$

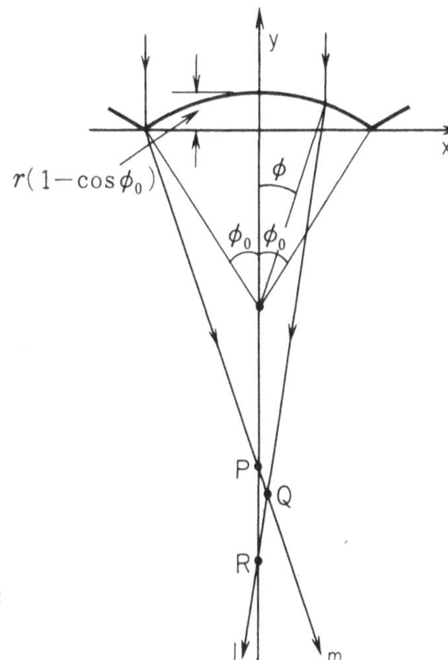

Fig.A.5 Illustration showing the computation of the minimum spread.

(i) Compute the coordinates (x,y) of point Q where the optical path for angle ϕ and the path m for $\phi = -\phi_0$ intersect, as functions of ϕ.

(ii) Compute the ϕ that makes $x(\phi)$ maximum.

(iii) The maximum value of x thus obtained gives half the minimum spread ($w_A/2$).

Noting that:

coordinates of P: $(0,\ r(1-\cos\phi_0)-s(\phi_0))$;
coordinates of Q: $(x(\phi),\ y(\phi))$;
coordinates of R: $(0,\ r(1-\cos\phi)-s(\phi))$;

we obtain equations expressing lines m and ℓ as

$$y = \frac{r(1-\cos\phi_0)-s(\phi_0)}{r\sin\phi_0}\,x+\{r(1-\cos\phi_0)-s(\phi_0)\}, \qquad (A.7)$$

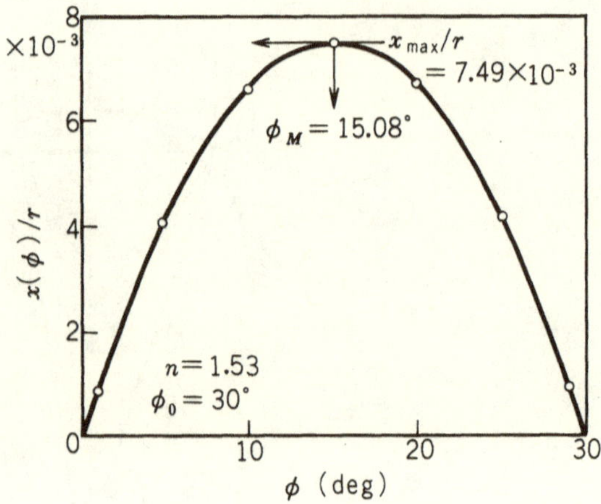

Fig.A.6 Relation between x and ϕ.

$$y = -\frac{r(1-\cos\phi)-s(\phi)}{r\sin\phi}x+\{r(1-\cos\phi)-s(\phi)\}, \qquad (A.8)$$

respectively. After equating the above two expressions and performing some algebraic computations, we obtain

$$x(\phi) = r\frac{f(\phi)-f(\phi_0)}{g(\phi)+g(\phi_0)}, \qquad (A.9)$$

where

$$f(\phi) = \sqrt{n^2-\sin^2\phi}+\cos\phi,$$

$$g(\phi) = \frac{\sqrt{n^2-\sin^2\phi}+n^2\cos\phi}{\sin\phi}.$$

The value of ϕ maximizing $x(\phi)$ cannot be computed analytically; numerical analysis is required. As an example, the x versus ϕ relation for $n = 1.53$, $\phi_0 = 30$ is shown in Fig.A.6. It is seen that the curve is

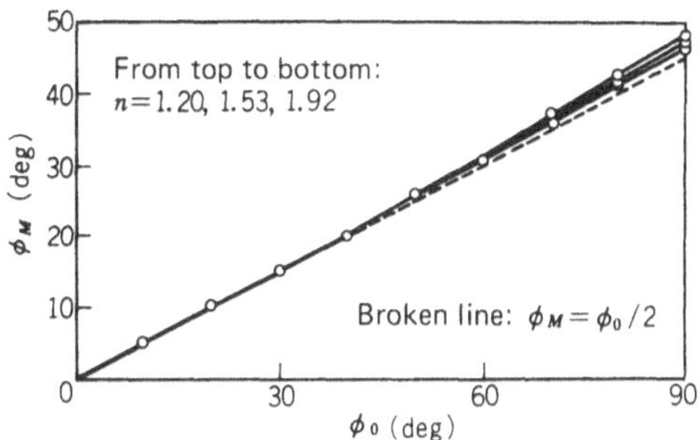

Fig.A.7 Relation between ϕ_M and ϕ_0. Broken line: $\phi_M = \phi_0/2$.

nearly parabolic, and that x'_{max} is obtained for $\phi \doteqdot \phi_0/2$. The angle ϕ giving x_{max} will hereafter be denoted by ϕ_M. The relation between ϕ_0 and ϕ_M is shown in Fig.A.7 for n = 1.20, 1.53, and 1.92. Figures A.6 and A.7 show that ϕ_M is not critical in giving x_{max}, and within $\phi_0 < 30°$,

$$\phi_M \doteqdot \phi_0/2 \tag{A.10}$$

holds with good accuracy.

Figure 4.20 in the text shows the minimum spread as a function of ϕ_0, the spread being obtained by putting the exact values of ϕ_M (not the approximate value $\phi_0/2$) into Eq.(A.9). It is seen that if $\phi_0 < 30°$, we may consider that $(w_A/2r) \lesssim 10^{-2}$.

A.3. OPTIMUM LENS THICKNESS

The most important design problem discussed in the text is the determination of the optimum lens pitch. In this section, another optimum design problem, determination of the optimum lens thickness, will be considered.

Putting Eq.(A.9) into Eq.(A.7), we obtain the y coordinate of point Q as

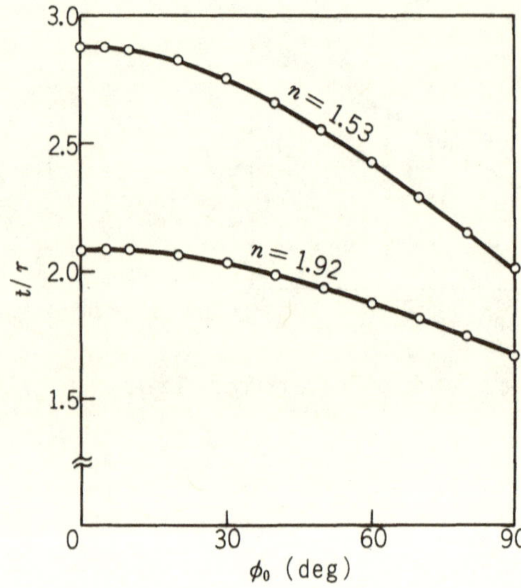

Fig.A.8 Relation between the optimum lens thickness t and ϕ_0.

$$y(\phi) = -\left[g(\phi_0)/(n^2-1)\right]\{x(\phi)+r\sin\phi_0\}. \qquad \text{(A.11)}$$

Hence, the optimum lens thickness is given as

$$t = \left[g(\phi_0)/(n^2-1)\right]\{x(\phi_M)+r\sin\phi_0\}+r(1-\cos\phi_0). \qquad \text{(A.12)}$$

The relation of Eq.(A.12) is shown in Fig.A.8.

A.4. LIMITATION OF DIRECTIVITY DUE TO ABERRATION

So far we have considered the picture-taking process of an integral photograph or an LS picture assuming parallel incident light, and obtained the width of the blurring w_A due to aberration.

The next step is to compute the angle spread of the light flux in the reconstruction process. We first consider ten beams as shown in Fig.A.9, five of them being emitted at point U and the other five at V; the two points are separated by w_A. The ten beams pass the lens surface at points characterized

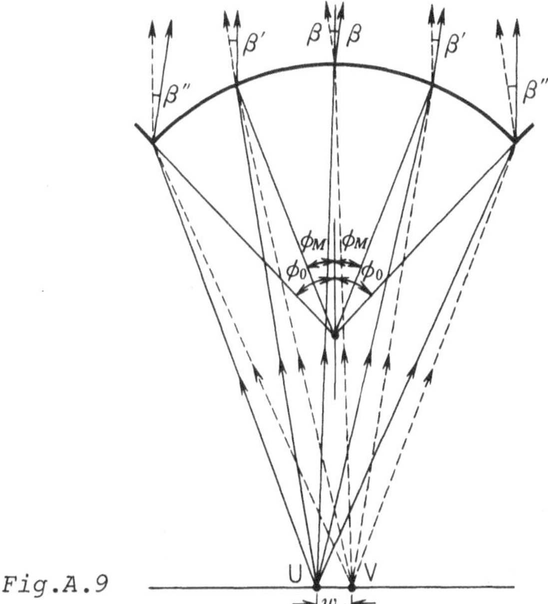

Fig.A.9

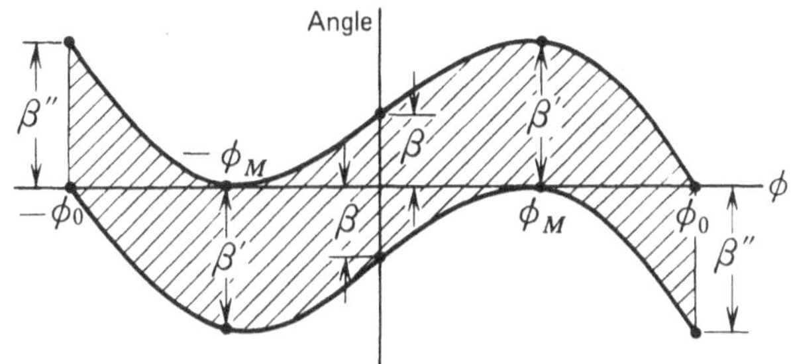

Fig.A.10

Fig.A.9 Spread of the light bundle in the reconstruction process, I.

Fig.A.10 Spread of the light bundle in the reconstruction process, II.

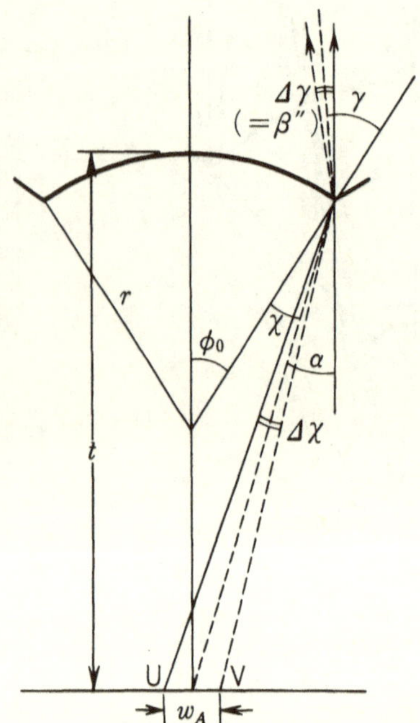

Fig.A.11 Symbols used in computing β".

by $\phi = 0$, $\phi = \pm\phi_M$, and $\phi = \pm\phi_0$. If we consider only the right-hand half of Fig.A.9, the light paths for each beam will be given as follows. (We should remember how angle ϕ_M has been defined, and how points U and V have been determined in the preceding section.)

(i) For $\phi = \phi_0$, after refraction the beam from U proceeds normal to the lens sheet, whereas the beam from V proceeds inward. (The angle between these paths is denoted by β''.)

(ii) For $\phi = \phi_M$, the beam from V proceeds normal, whereas the beam from U proceeds outward. (The angle between these paths is denoted by β'.)

(iii) For $\phi = 0$, both beams from U and V spread outward. (The angle between these paths is denoted by β.)

Hence, to summarize, the whole light bundle in the

reconstruction process will show a spread as illustrated in Fig.A.10, where the angle deviation and ϕ are shown on the ordinate and abscissa, respectively.

If we assume a paraxial model, that is $\phi_0 \ll \pi$, we may consider that $2\beta = \beta' = \beta''$. Therefore, to compute the maximum angle spread due to aberration, we should first compute β' and β'', and then take the bigger angle (see Fig.A.10); however, since the computation of β' is rather cumbersome, we shall temporarily assume that $\beta' \doteqdot \beta''$ and compute β''. (Practically, the difference between β' and β'' is not very important.)

We define several parameters as shown in Fig.A.11, and first compute $\Delta\gamma$ as a function of $\Delta\chi$.* Then, we

---

* Derivation of Eq.(A.13): We first note that

$$(t-r+r\cos\phi_0)\tan\alpha = r\sin\phi_0.$$

On the other hand, since

$$\left[(t-r+r\cos\phi_0)/\cos\alpha \ \Delta\chi\right] = w_A\cos\alpha$$

holds approximately, we obtain

$$\Delta\chi = w_A\cos^2\alpha/(t-r+r\cos\phi_0).$$

Next, we use $\sin\gamma = n\sin\chi$ to relate $\Delta\gamma$ and $\Delta\chi$; that is,

$$d\gamma = (n\cos\chi/\cos\gamma)d\chi = \left(n\sqrt{1-\sin^2\phi_0/n^2}/\cos\phi_0\right)d\chi.$$

Therefore, β'' $(= \Delta\gamma)$ is given as

$$\Delta\gamma = \frac{n\sqrt{1-\left(\sin^2\phi_0/n^2\right)}}{\cos\phi_0} \cdot \frac{w_A\cos^2\alpha}{t-r+r\cos\phi_0}$$

Eliminating $\cos^2\alpha$ from the first and last equations, we obtain Eq.(A.13) in the text.

obtain the overall angle spread β_A due to aberration as

$$\beta_A = 2\beta'' = 2\Delta\gamma = \frac{4Td\sqrt{n^2-\sin^2\phi_0}}{\cos\phi_0\left(T^2+\sin^2\phi_0\right)} \tag{A.13}$$

where $T = t-r(1-\cos\phi_0)$.

The values of β_A have been computed for two values of the refractive index n, and are shown in Fig.4.21 in the text. This figure shows that, regardless of the value of n,

$\beta_A \doteqdot 0.035°$ for $\phi_0 = 10°$,
$\beta_A \doteqdot 0.3°$ for $\phi_0 = 20°$,
$\beta_A \doteqdot 1.1°$ for $\phi_0 = 30°$,
$\beta_A \doteqdot 3°$ for $\phi_0 = 40°$.

For $\phi_0 > 40°$, the spread due to aberration becomes intolerable.

References

(1) Koseki, K., Horiuchi, M., and Suzuki, T., Studies on Three-Dimensional Effect by Lenticular Screen Method, TAGA (Tech. Assoc. of the Graphic Arts) Meeting, May 1969, Washington, D.C.
(2) Burckhardt, C. B., Optimum Parameters and Resolution Limitation of Integral Photography, *J. Opt. Soc. Amer. 58*, No. 1, 71-76 (Jan. 1968).
(3) Okoshi, T., Optimum Parameters and Depth Resolution of Lens-Sheet and Projection-Type Three-Dimensional Displays, *Appl. Opt. 10*, No. 10, 2284-2291 (Oct. 1971).
(4) Ives, H. E., Optical Properties of a Lippmann Lenticulated Sheet, *J. Opt. Soc. Amer. 21*, 171-176 (March 1931).
(5) DeMontebello, R. L., U.S. Patent, 3,503,315 (1970).

(6) Valyus, N. A., "Stereoscopy." The Focal Press, London, 1962.

(7) Burckhardt, C. B., Collier, R. J., and Doherty, E. T., Formation and Inversion of Pseudoscopic Images, *Appl. Opt. 7*, No. 3, 627-631 (March 1968).

(8) Burckhardt, C. B., and Doherty, E. T., Beaded Plate Recording of Integral Photographs, *Appl. Opt. 8*, No. 11, 2329-2331 (Nov. 1969).

(9) Okoshi, T., and Ando, K., Characteristics of Stereoscreens for Projection-Type Three-Dimensional Imaging, National Convention of the Institute of Television Engineers of Japan, Paper No. 10-15 (Oct. 17, 1969).

Projection-Type Three-Dimensional Displays

A screen that always reflects an incident ray of light into the original direction is called an autocollimating screen. Well-known examples of such screens are a fly's-eye lens sheet backed by white diffusing material or a screen consisting of many tiny triple mirrors. Many versions of such screens have been devised. A projection-type three-dimensional display system usually consists of such a special screen and many projectors, from which pictures of the object originally taken from various directions are projected onto that special screen.

In this chapter, the principle of the projection-type display and various screens devised for that purpose will be described first. In the latter half of this chapter, two most promising screens, the diffuser-backed lenticular screen and the curved triple-mirror screen, will be described in detail.

5.1 INTRODUCTION

A good lenticular-sheet (LS) picture may give satisfactory three-dimensional sensation to an observer viewing it from an arbitrary direction, using only presently available, relatively simple techniques. However, two technical drawbacks seem to limit its versatility in future three-dimensional imaging technology.

First, when we make a large (for example, 1 × 1 m) LS picture, the lenticular sheet is very expensive; price is nearly proportional to the picture area. We may generally say that LS picture techniques are not well suited for large imaging. The above difficulty might be overcome if we could make an LS picture in which the "rear picture" is not fixed but detachable from the lens sheet; however, such a modification would be accompanied by new technical difficulties: registration and reproducibility.

The second drawback will be found when such a technique is used in the display of animated images. For example, if we intend in the future to use an LS picture technique in three-dimensional television, we have to animate the "rear picture," which is usually very fine, having resolution of typically 10 μm. Such a technique might not be so difficult as one for animating a hologram, but is nevertheless not presently available.

The projection-type three-dimensional display technique to be discussed in this chapter overcomes both of the above problems. On the other hand, it is accompanied by some other drawbacks; for example, a projection-type display cannot be observed in a bright place. However, undoubtedly the projection-type display will become a powerful tool in future three-dimensional imaging techniques.

The discussion in this chapter is limited to multiple-photography-type projection techniques. There is, however, another category in projection-type imaging: projection-type holography, in which a hologram is used as the projector (1). Although very few experiments have been reported on this technique,* it too seems an interesting idea because it overcomes some of the technical difficulties inherent in conventional holography. Projection-type holography will be described in Chapter 7 in connection with information reduction in holograms.

There is another aspect to the combination of projection techniques and holography: the fabrication of a special projection screen (direction-selective screen) using holographic methods; this will also be an important technique in the future. Some proposals and experiments on holographic screens will be described in Chapter 8.

5.2 PRINCIPLE OF PROJECTION-TYPE THREE-DIMENSIONAL DISPLAYS

A projection-type three-dimensional display is defined as a display that gives three-dimensional sensation to an observer by projecting many pictures of an object originally taken from various directions (or taken from continuously varying directions) onto a direction-selective or an autocollimating screen. The original idea is very old; it was proposed in the early 1930s as described in Chapter 2. Since then many versions of projection-type displays have been proposed and experimented with; these include, for example, those using conically or cylindrically curved screens or projection-type holography as mentioned above.

---

* This situation has drastically changed. In 1972-1975 many papers appeared concerning various types of projection-type holography: sampled-hologram versions, enlarged viewing-zone type, white-light-reconstruction type, and projection-type image holography. Detailed descriptions are given in a review article by Okoshi (2).

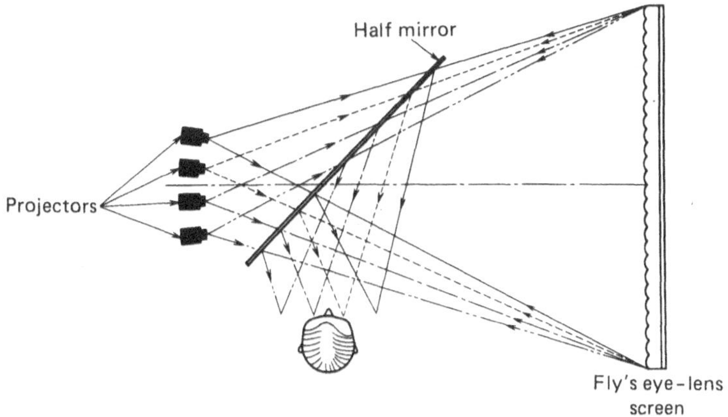

Fig.5.1 Projection-type three-dimensional display using a regular direction-selective screen and a half mirror.

To facilitate understanding, however, a simple, multiple-photography-type projection display using a flat screen will first be considered.

5.2.1 PROJECTION-TYPE THREE-DIMENSIONAL DISPLAY USING REGULAR DIRECTION-SELECTIVE SCREEN

We consider a fictitious "slide-projection" system having $m \times n$ slide projectors, m in the vertical direction and n in the horizontal, as shown in Fig.5.1. To prepare the "slides," $m \times n$ cameras arranged in a similar array with separations of, for example, 5 cm are used. After exposure, the pictures are reversed, and the positive pictures obtained are put into the projectors and projected onto a direction-selective screen.

A direction-selective screen collimates a ray of light emitted from a point source toward that original source point, or toward some other point. A well-known example of such a screen is the fly's-eye lens sheet backed by a white diffusing plane. The diffusing

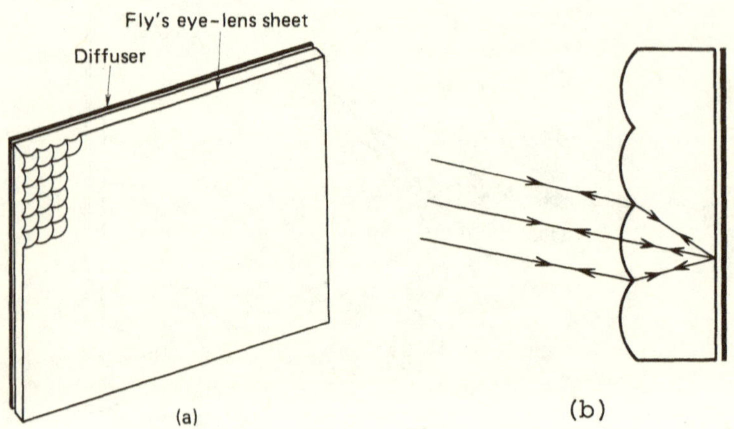

Fig.5.2 A fly's-eye screen.

plane may either be a little apart from the lens sheet to facilitate adjustment of focus, or be made simply by painting white the rear surface of the lens sheet. The latter method is usually more practical. In any case when the diffusing plane is placed just upon the focal plane of the lens sheet, such a screen always reflects the incident ray of light into its original direction as shown in Fig.5.2b; such a property is called an *autocollimating* or *retrodirective* property. (Note that the autocollimating property is a concept narrower than the direction-selective property. In other words, the latter includes the former. Both of them can be used in projection-type imaging.)

Therefore, when the projected images are observed from the projector side, three-dimensional sensation will be obtained, provided that m and n are satisfactorily large. However, such a system has two obvious drawbacks.

First, since the optimum viewing positions coincide with the projector positions, a half mirror must be used in observing the projected image as shown in Fig.5.1.

Second, in the system as described above having, for example, a viewing zone of 50 cm (width) × 20 cm

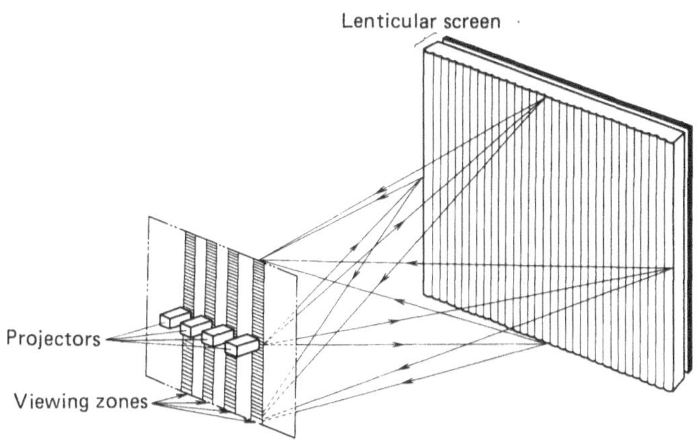

Fig.5.3 Projection-type three-dimensional display using a unidirectionally direction-selective screen.

(height) and projector separation of 5 cm, the required projector number is m × n = 10 × 4 = 40. This is not a practical number.

5.2.2 PROJECTION-TYPE THREE-DIMENSIONAL DISPLAY USING UNIDIRECTIONALLY DIRECTION-SELECTIVE SCREEN

The above two difficulties will readily be overcome if we use a lenticular sheet in the screen instead of a fly's-eye lens sheet. A screen consisting of a lenticular sheet and a diffusing plane has the autocollimating property in the horizontal direction but a diffusing property in the vertical direction (3). Such a screen is called a *unidirectionally direction-selective* (UDS) screen (4). We may set up a projection-type three-dimensional imaging system as shown in Fig.5.3, using a UDS screen and a single array of projectors; the viewing zone is expanded in the vertical direction by virtue of the vertically diffusing property of the UDS screen.

The fly's-eye lens screen and the lenticular-sheet

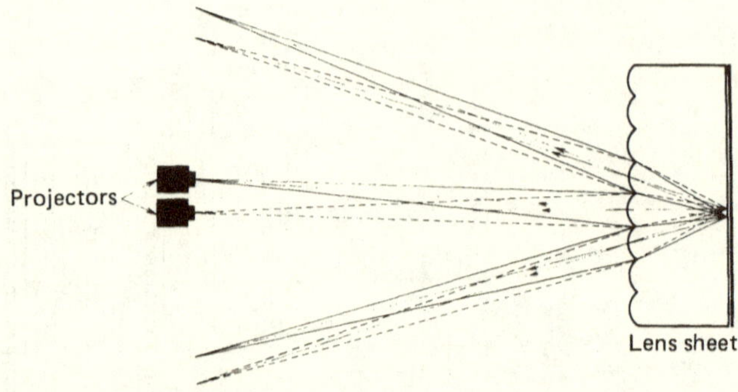

Fig.5.4 Generation of side lobes in a lens-sheet screen.

screen, which are called generically "lens-sheet screens," have a common property of generating "side lobes." This property is attributed to the fact that some of the light energy scattered upon the diffusing plane is collimated by the neighboring (or next to the neighboring) cylindrical lenslet as shown in Fig. 5.4. This "side-lobe" effect is often utilized in expanding the viewing zone of projection-type displays. It is merely to simplify the illustration that only two projectors are shown in Fig.5.4. However, in reality the side-lobe effect is particularly useful when only two projectors are used as in Fig.5.4, that is, in binocular three-dimensional display systems.

In addition to the diffuser-backed lenticular screen described above, various unidirectionally direction-selective screens have been devised. An example of one of these consists of a large number of tiny triple mirrors having a slight curvature on one of the three mirror surfaces; this is called a curved triple-mirror screen. Various UDS screens will be described in Section 5.3.

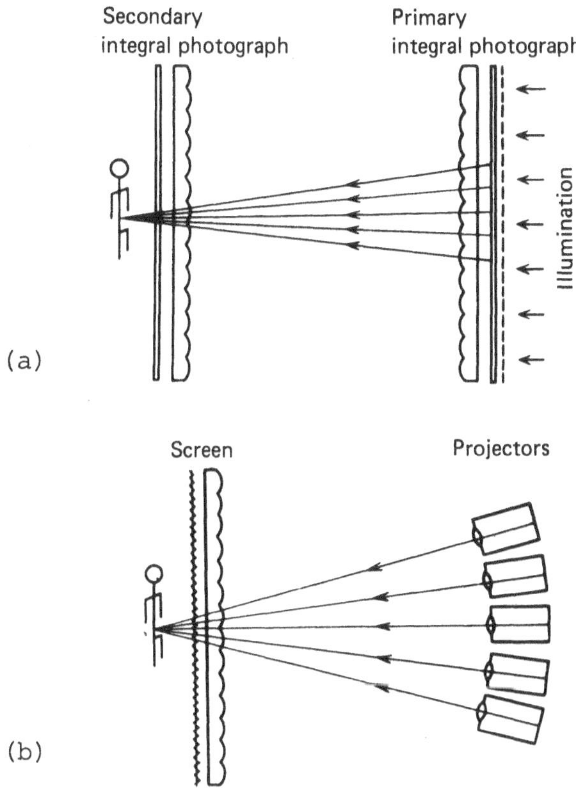

Fig.5.5 Comparison of (a) two-step integral photography, and (b) projection-type three-dimensional display.

5.2.3 RELATION BETWEEN LENS-SHEET THREE-DIMENSIONAL PICTURES AND PROJECTION-TYPE THREE-DIMENSIONAL DISPLAYS

As has been described in Chapter 2, Lippmann proposed for the first time a method of complete three-dimensional imaging: integral photography. Later H. E. Ives proposed two-step integral photography in 1931 to overcome the difficulty of pseudoscopic images. The projection-type three-dimensional imaging described in this chapter has a close similarity with two-step integral photography.

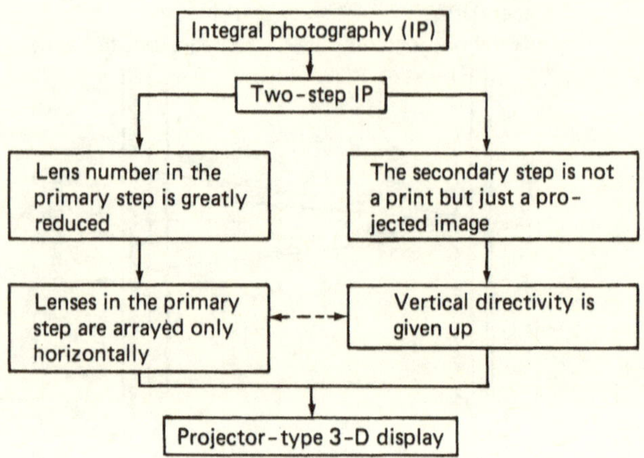

Fig.5.6 Relation between two-step integral photography and multiple-picture projection-type three-dimensional display.

We consider for example a projection-type three-dimensional imaging system in which pictures are projected from N projectors onto a diffuser-backed lenticular screen. The similarity between IP and such a system is illustrated in Fig.5.5. It is seen that when

(1) the number of lenses in the primary step is greatly reduced,

(2) the secondary integral photograph is not printed but just observed as an image projected onto the screen,

(3) the vertical directivity of the screen is given up, and hence

(4) the lenses of the primary step are arrayed unidirectionally (only horizontally),

integral photography is transformed into a multiple-picture projection-type display. The above transformations are illustrated in Fig.5.6.

Transformation (3) fits the arrangement of human eyes. Items (1) and (4) enable us to set up the system more economically. Item (2) gives a better versatility especially when animated images are to be reconstructed. These are the reasons projection-type imaging is particularly important as a basic three-dimensional imaging technique.

A comment should be made concerning item (1). When two-step IP was described in Chapter 3, the pitches of the primary and secondary integral photographs were assumed equal to each other. On the contrary, in projection-type displays, the primary pictures are usually much larger (and fewer in number) than the lens size (pitch) of the screen. (The dimensions in Figs.5.3 and 5.4 are of course not realistic.) It is important to know which choice is more reasonable; on this point we should recall the discussion on optimum design of lens-sheet pictures (Section 4.4.3), and further note the similarity between lens-sheet pictures and projection-type imaging. As was concluded in Section 4.4.3, in both IP and LS pictures, the lens size in the first step should be relatively large, whereas that in the secondary step relatively small, as far as the final image is "closely" reconstructed. The same statement holds for projection-type imaging.

Thus, in most practical lens-sheet pictures or projection-type displays, the optimum lens number in the primary step ranges between 10 and 20, sometimes even less than 10. Such a scheme is sometimes called *multiple photography*. This term will hereafter be used generically for both lens-sheet and projection-type "multiple picture" schemes.

5.3 CLASSIFICATION OF DIRECTION-SELECTIVE SCREENS

5.3.1 INTRODUCTION

In addition to the fly's-eye and lenticular screens described above, various types of screens have been proposed; some of these have been put into

TABLE 5.1 Various Screens Used in the Projection-Type Display

Category		Typical examples	Type of screen[a]	Availability of side lobes	Figure	Reference[b]
(1) Regular direction-selective screens (RDS screens)	(1-A)	Fly's-eye screen	R	Yes	5.2	-
	(1-B)	Cat's-eye screen	R	No	5.8	-
	(1-C)	Triple-mirror screen	R	No	5.9	-
	(1-D)	Transmission-type double fly's-eye lens screen	T	Yes	5.10	(9)
	(1-E)	Large concave mirror	R	No	5.11	(5)
	(1-F)	Large convex lens	T	No	5.12	(5)
(2) Unidirectionally direction-selective (UDS) screens, I (vertically diffusing screens)	(2-A)	Diffuser-backed parallel lenticular screen	R	Yes	5.3	(3)
	(2-B)	Curved triple-mirror screen	R	No	5.14	(6)
	(2-C)	Combination of a right-angle double-mirror screen and a parallel lenticular sheet	R	No	5.14	(4)[c]
	(2-D)	Combination of a TM screen and a parallel lenticular sheet	R	No	5.15	(4)

	(2-E) Transmission-type double-lenticular screen	T	Yes	5.16	(9)
	(2-F) Combination of a large convex lens and a parallel lenticular sheet	T	No	5.17	(4)
(3) Unidirectionally direction-selective (UDS) screens, II (radially diffusing screens)	(3-A) Diffuser-backed radial lenticular screen	R	Yes	5.7	(5)
	(3-B) Combination of a radial right-angle double-mirror screen and a concentric lenticular sheet	R	No	5.18	(4)
	(3-C) Combination of a TM screen and a concentric lenticular sheet	R	No	5.19	(4)
	(3-D) Transmission-type double radial-lenticular screen	T	Yes	5.20	(4)
	(3-E) Combination of a large convex lens and a concentric lenticular sheet	T	No	5.21	(4)

[a] T: transmission; R: reflection.

[b] The reference indicates the paper in which a screen was proposed for the first time.

[c] A screen of similar type was proposed in Ref.(10). See Section 2.7.2.

practical use. In this section these various screens are classified and their features compared (4).

The proposed classification is shown in Table 5.1. Some particular screens made with holographic methods are omitted; they will be described in Section 8.7.

The first category in Table 5.1 includes regular direction-selective screens. The fly's-eye screen is the typical example. As described earlier, a half-mirror is needed to observe a reconstructed image projected onto this kind of screen. An interesting fact is that a large concave mirror or a large convex lens can also furnish a similar (but essentially different; see Section 5.4) function (5).

The second category is characterized by a direction-selective property in the horizontal direction but a diffusing property in the vertical direction. As described in Section 5.2.2, the half-mirror is not needed and the information is reduced in three-dimensional display systems using this kind of screen.

Screens in the third category have a radially diffusing property and a direction-selective action in the azimuthal direction; these are sometimes called *radial raster* screens. A radial raster screen is a version of the *parallel raster* screen for furnishing a wider viewing zone. Its principle will be described in the following, using the radial lenticular screen as a typical example. (The inventor of the radial raster screen is not definitely known; according to Valyus (5), who succeeded first in constructing one, it was S. P. Ivanov of the Soviet Union.)

We consider a diffuser-backed lenticular sheet having radially arranged cylindrical lenses as shown in Fig.5.7. We assume that all the radial lines in the screen pattern meet at a common point O called the center of convergence of the pattern, and that the lens-sheet plane intersects the diffusing plane along a straight line AB which includes point O. Many projectors are placed along line S_0S_1. However, in Fig. 5.7, we tentatively assume that only two projectors are used and they are placed at S_0 and S_1.

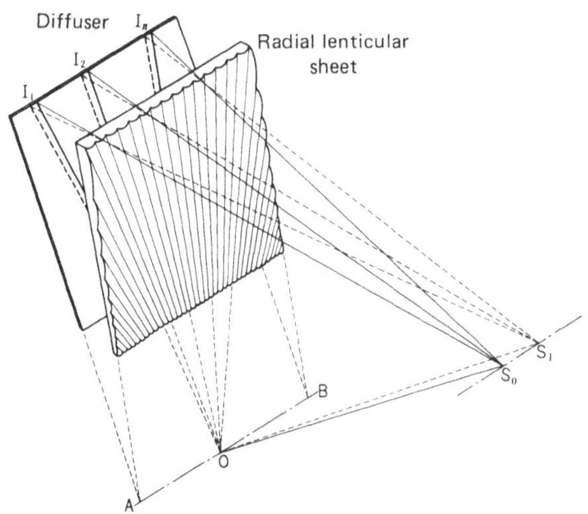

Fig.5.7 A radial lenticular screen.

In the above arrangement, rays from a light source at S_0 pass through the cylindrical lenses and make light strips I_1, I_2, ... on the diffusing plane, which are also radially directed, their extensions meeting together at point O. Therefore, all the planes including point S_0 and those cylindrical lenses and light strips will meet together along the straight line OS_0. Similarly, all the planes including S_1 and the cylindrical lenses will meet along OS_1.

We now consider an observer whose left eye is placed on OS_0 and right eye on OS_1. His eyes will observe flipping-free pictures over the entire screen. However, since the binocular separation is fixed (about 6.5 cm average), the optimum viewing distance is predetermined and only a single observer can occupy the optimum position.

The above statement is valid so long as only two projectors are used. However, if we prepare a large number of projectors along line S_0S_1 with shorter separations, 2 cm for example, then any observers whose eyes are located at any place upon the plane including

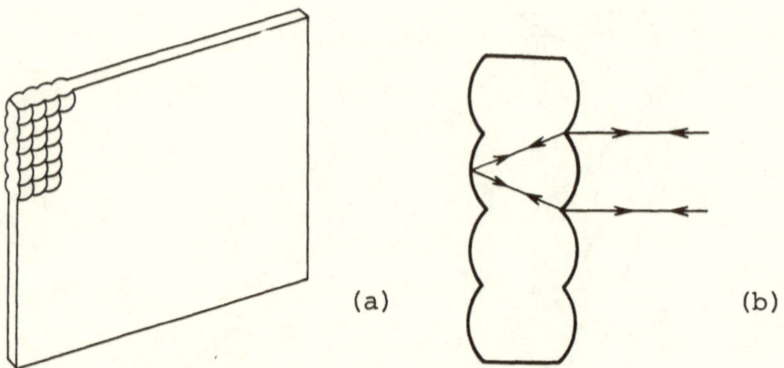

Fig.5.8 A cat's-eye screen.

points O, S_0, and S_1 will be able to observe flipping-free pictures over the entire screen. Thus the radial raster screen enables us to prepare a very wide viewing zone. The side-lobe effect is also usable if desired.

Note that the above function could not be furnished by a parallel raster screen having the vertical diffusing property. This is because such screens correspond to an extreme case of the radial raster screen in which point O is located far below, at an effectively infinite distance. Therefore, the plane OS_0S_1 giving optimum viewing positions is a vertical plane; more than one observer could not occupy the optimum viewing positions unless they were stacked vertically.

The above feature makes the radial raster screen suitable for large-size, theater-form three-dimensional displays. Engineers in the Soviet Union have been eager to develop this kind of screen. The three-dimensional movie theater described in Section 2.8 employed a radial lenticular screen and succeeded in furnishing binocular pictures to several hundred people by taking advantage of the side-lobe effect.

We now go back to Table 5.1 and investigate the various screens in the order of the table.

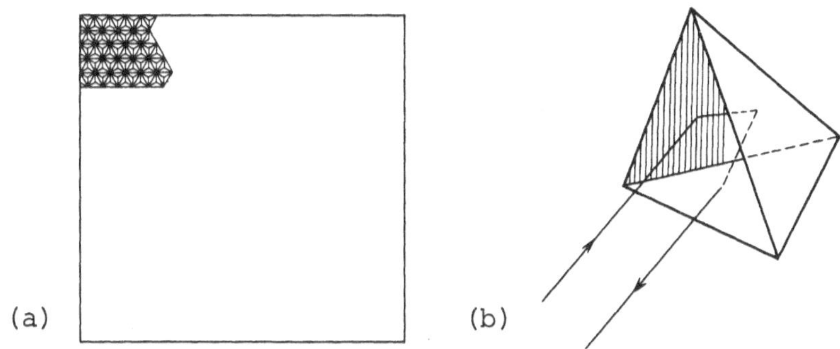

Fig.5.9 Triple-mirror screen (TM screen).

5.3.2 REGULAR DIRECTION-SELECTIVE SCREENS

(1-A) *Fly's-eye screen.* Described in Section 5.2.1.

(1-B) *Cat's-eye screen.* A cat's-eye screen is composed of many convex lenslets; each has two spherical surfaces and the center of curvature of the rear surface coincides with the back nodal point (Fig.5.8). The rear surface may either be a diffusing surface or a specular surface. In either case this screen exhibits an autocollimating property, but the reflection efficiency is much improved when the rear surface is sputtered with metallic material to form a specular film.

As discussed in Section 4.5 concerning beaded-plate IP, when the refractive index of the lens material is equal to 2.0, the optimum curvature of the rear surface is equal to that of the front lens surface and both surfaces have a common center of curvature. Therefore, the easiest way to make a cat's-eye screen is to arrange many tiny glass balls (having n = 2) on a soft specular (or diffusing) plane. This kind of screen, which may be called a "beaded-plate" screen, has been widely used in traffic signs.

(1-C) *Triple-mirror screen.* A triple-mirror screen consists of a large number of tiny triple mirrors (cubic

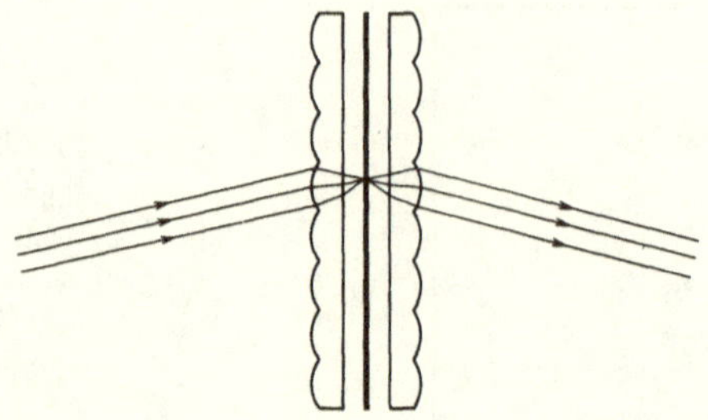

Fig.5.10 Transmission-type double fly's-eye lens screen.

corners) arranged in a panel as shown in Fig.5.9. This will hereafter be called the TM screen for short. The triple mirror has three adjacent, normally intersecting reflecting surfaces, and has a well-known "retrodirective" property as shown in Fig.5.9b. This property obviously results in an autocollimating property when a large number of triple mirrors are assembled to form a TM screen. The major advantages of this screen are high reflection efficiency and absence of the spurious surface reflection often seen in lens-type screens.

(1-D) *Transmission-type double fly's-eye lens screen.* A transmission-type screen consisting of a thin translucent sheet sandwiched by two well-aligned fly's-eye lens sheets has a direction-selective property (9). Such a screen "folds" the incident ray of light in a symmetrical way with respect to the screen plane as shown in Fig.5.10. In practice, however, the alignment (registration) of the lenticular sheets is rather difficult; within the author's knowledge no experiment has ever been reported.

(1-E) *Large concave mirror.* The imaging property of a large concave mirror can also be used in a three-dimensional imaging system as shown in Fig.5.11. Such a system features a very high reflection efficiency. In Fig.5.11 only two projectors are shown, but more can be used. However, the imaging property of a concave mirror is essentially different from that of (1-A),

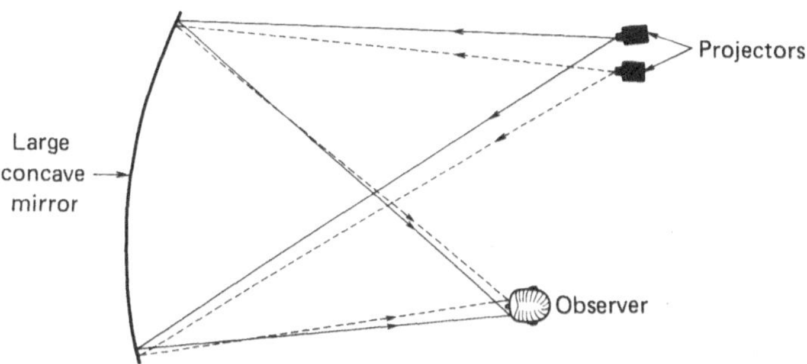

Fig.5.11 Observation of a three-dimensional image by using a large concave mirror.

(1-B), and (1-C) described above. Screens (1-A), (1-B), and (1-C) exhibit the autocollimating property, whereas the large concave mirror exhibits just ordinary lens properties. As a result, the concave mirror does not have the pseudoscopic-to-orthoscopic or orthoscopic-to-pseudoscopic inversion property (to be discussed in Section 5.4) which is inherent in autocollimating screens and plays an important role in three-dimensional imaging systems.

(1-F) *Large convex lens.* A large convex lens is essentially a transmission-type version of the large concave mirror as shown in Fig.5.12. It also exhibits only lens properties and lacks the inversion property to be discussed in Section 5.4. The relation of (1-A) to (1-E) is similar to that of (1-D) to (1-F).

A thick lens, as shown in Fig.5.12, is not practical because of the weight and cost. Fresnel-lens sheets are usually used since they are lighter and less expensive. For experimental use in the laboratory, cheap Fresnel-lens sheets for overhead projectors are often employed.

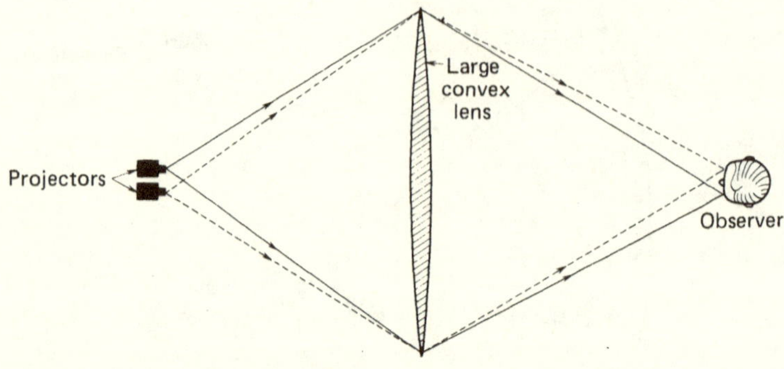

Fig.5.12 Observation of a three-dimensional image by using a large convex lens.

5.3.3 UNIDIRECTIONALLY DIRECTION-SELECTIVE SCREENS, I (VERTICALLY DIFFUSING SCREENS)

(2-A) *Diffuser-backed parallel lenticular screen.* This screen has been described in Section 5.2.2 and also in Section 2.7.1. Some optical properties will be discussed in more detail in Section 5.6.

(2-B) *Curved triple-mirror screen.* See Fig.5.13. This is a version of the TM screen. It is composed of a number of tiny triple mirrors, one reflecting surface of each of which is curved slightly to furnish a diffusing property in the vertical direction. This screen, sometimes called a CTM screen for short, was proposed by the author in 1968 and first fabricated by Yano, *et al.* (6). The horizontal directivity (0.5° half-intensity spread) and the reflection efficiency (approximately 30%) are both the best among all the UDS screens. A drawback is expensiveness. The optical properties will be discussed in detail in Section 5.7.

(2-C) *Combination of a right-angle double-mirror screen and a parallel lenticular sheet.* See Fig.5.14. A right-angle double-mirror gives autocollimation in

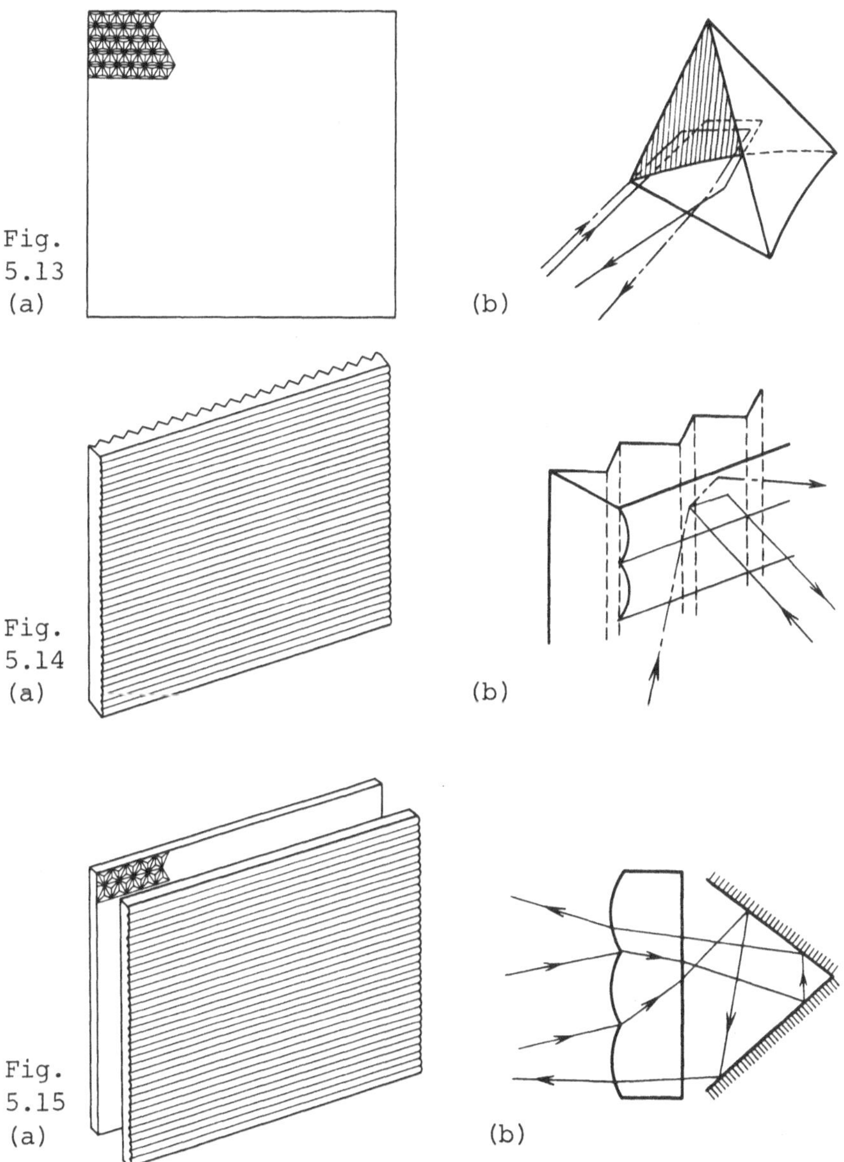

Fig.5.13 Curved triple-mirror (CTM) screen.
Fig.5.14 Combination of a right-angle double-mirror screen and a parallel lenticular sheet.
Fig.5.15 Combination of a triple-mirror screen and a parallel lenticular sheet.

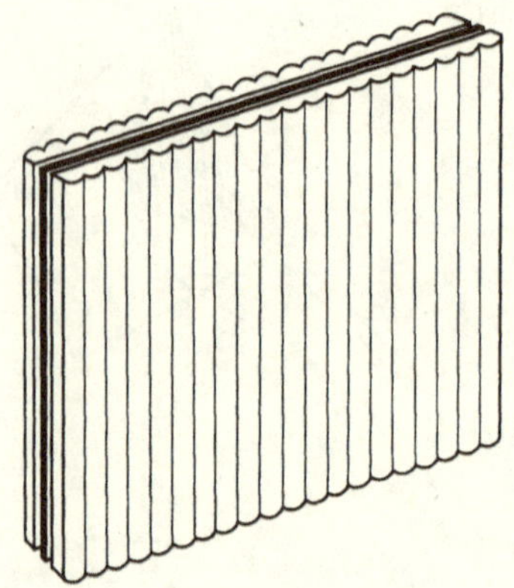

Fig.5.16 Transmission-type double-lenticular screen.

the horizontal direction, whereas the horizontally arrayed cylindrical lenses of the parallel lenticular sheet provide a vertical diffusing property. This is a practical version of the screen proposed by H. E. Ives in 1931, which has been described in Section 2.7.

(2-D) *Combination of a TM screen and a parallel lenticular sheet.* See Fig.5.15. The TM screen gives autocollimation, whereas the lenticular sheet provides vertical diffusion.

(2-E) *Transmission-type double-lenticular screen.* See Fig.5.16. This is a vertically diffusing version of the transmission-type double fly's-eye lens screen (1-D).

(2-F) *Combination of a large convex lens and a parallel lenticular sheet.* See Fig.5.17. The convex lens gives the imaging property, whereas the lenticular sheet provides vertical diffusion. However, in practice the use of a thick lens as shown in Fig.5.17 is not favorable because of weight and cost; a combination of a Fresnel-lens sheet and a lenticular sheet is usually employed. In the latter case, the two lens sheets are separated intentionally by a small distance (typically 1 cm) to avoid a moiré pattern generated by the interference between the two lens patterns.

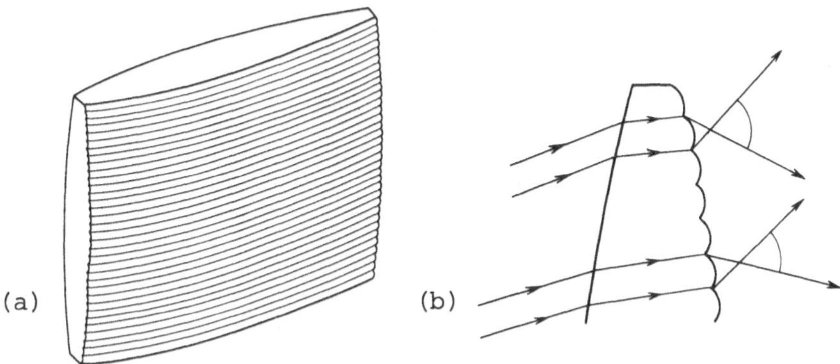

Fig.5.17 Combination of a large convex lens and a parallel lenticular sheet.

5.3.4 *UNIDIRECTIONALLY DIRECTION-SELECTIVE SCREENS, II (RADIALLY DIFFUSING SCREENS)*

(3-A) *Diffuser-backed radial lenticular screen.* The principle is described in Section 5.3.1. Very few practical uses have ever been reported. One of them is the big screen used in the three-dimensional movie theater in the Soviet Pavilion at Expo-70 in Osaka, mentioned in Section 2.8. According to Ref.(35) of Chapter 2, the dimensions of the screen are:

(1) Screen size: 4 m (width) × 3 m (height),
(2) Lens pitch: 2.3 mm (at the bottom) ∿ 3.6 mm (at the top),
(3) Separation between lens sheet and diffusing plane: 180 mm (at the bottom) ∿ 280 mm (at the top).

(3-B) *Combination of a radial right-angle double-mirror screen and a concentric lenticular sheet.* See Fig.5.18. This is a radial-diffusion version of (2-C).

(3-C) *Combination of a TM screen and a concentric lenticular sheet.* See Fig.5.19. This is a radial-diffusion version of (2-D).

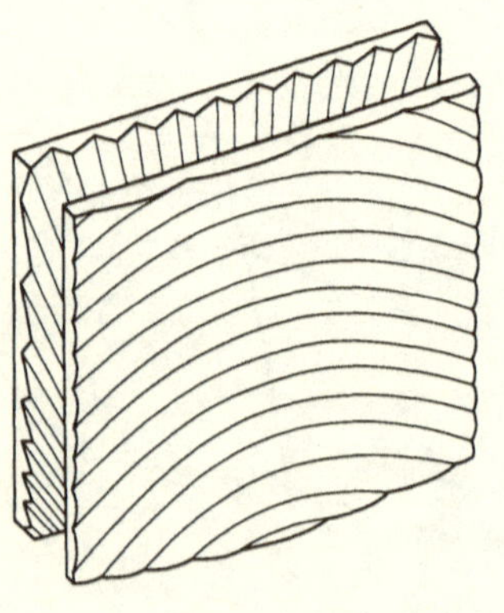

Fig.5.18

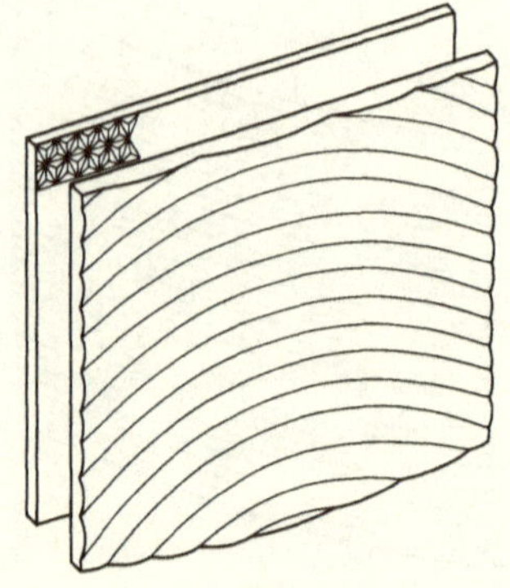

Fig.5.19

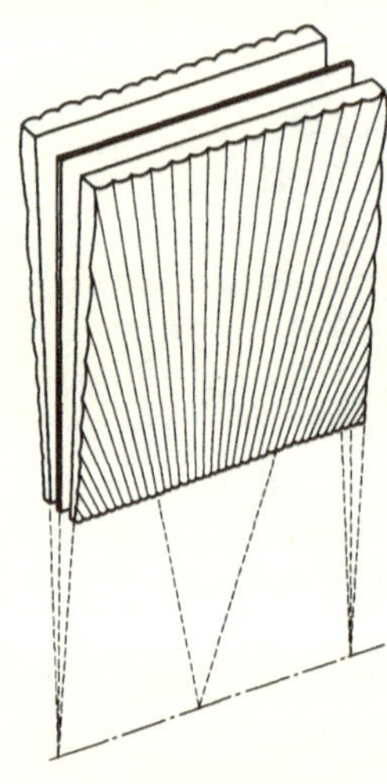

Fig.5.20

Fig.5.18 Combination of a radial right-angle double-mirror screen and a concentric lenticular sheet.
Fig.5.19 Combination of a TM screen and a concentric lenticular sheet.
Fig.5.20 Transmission-type double-radial-lenticular screen.

(3-D) *Transmission-type double-radial-lenticular screen.* See Fig.5.20. This is a radial-diffusion version of (2-E).

(3-E) *Combination of a large convex lens and a concentric lenticular sheet.* See Fig.5.21. This is a radial-diffusion version of (2-F). In practice the convex lens should be replaced by a Fresnel lens.

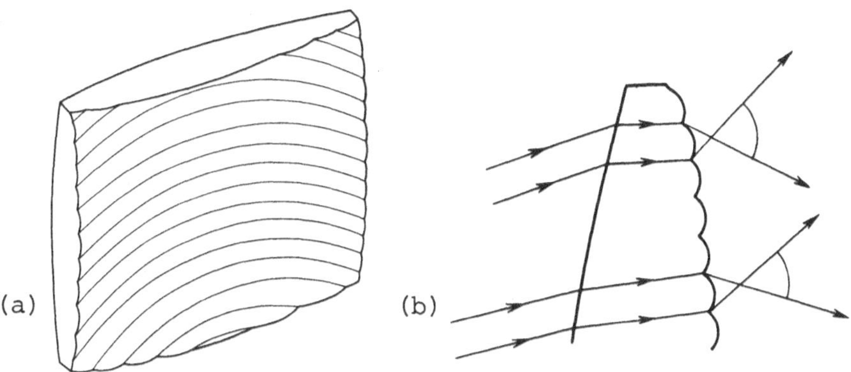

Fig.5.21 Combination of a large lens and a concentric lenticular sheet.

5.3.5 DISCUSSIONS

The screens mentioned above are merely typical examples of the direction-selective screens to be used in projection-type three-dimensional displays. Obviously many other combinations are possible.

At present the screen most commonly employed in practical display systems is type (2-A). Types (1-A), (1-C), (2-B), (2-F), and (3-A) are investigated and used in laboratories. Type (2-B) gives the best performance except for cost. In Sections 5.6 and 5.7, types (2-A) and (2-B) will again be discussed in more technical detail.

5.4 INVERSION BETWEEN ORTHOSCOPIC AND PSEUDOSCOPIC IMAGES USING AN AUTOCOLLIMATING SCREEN

Before we discuss technological matters of the screen, we should note an essential property of the autocollimating screen: orthoscopic-to-pseudoscopic or pseudoscopic-to-orthoscopic inversion.

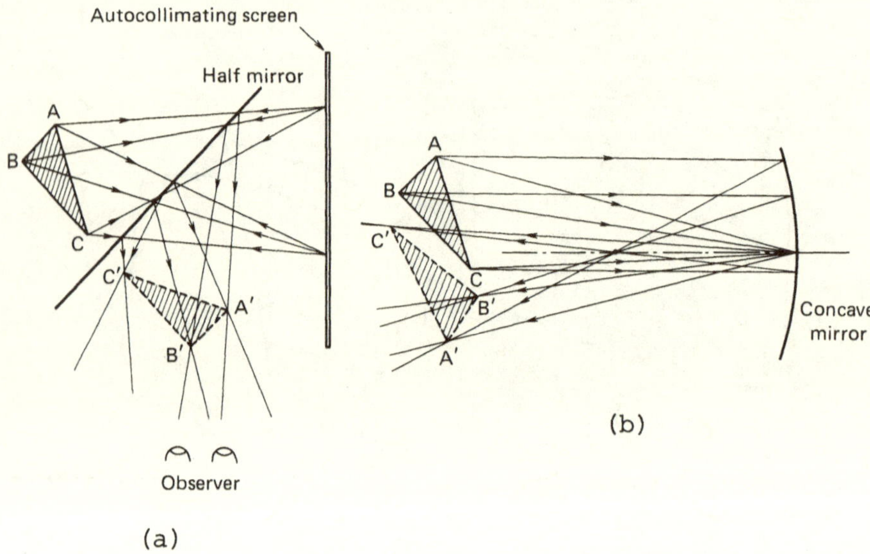

Fig.5.22(a) Inversion of a three-dimensional image by an autocollimating screen. (b) Imaging property of a concave mirror.

We consider a TM screen as an example and assume that each triple mirror is so small that we never recognize individual triple mirrors. Suppose we use such a TM screen like an ordinary mirror to observe an object (or an optical image) as illustrated in Fig.5.22a. Then we shall see an image in which depth is inverted as shown in Fig.5.22a; that is, a real object or an orthoscopic image will be observed as a pseudoscopic image, whereas a pseudoscopic image will become orthoscopic. Note that a half mirror is used in Fig.5.22a because the image reconstructed by an autocollimating screen is produced at the same position as the original object (or the original image) position; hence the separation is needed.

The inversion property described above might have long been noticed, but its technical significance was first pointed out by Burckhardt *et al.* (7). It should be noted that such an inversion is never obtained using a concave mirror (see Fig.5.22b).

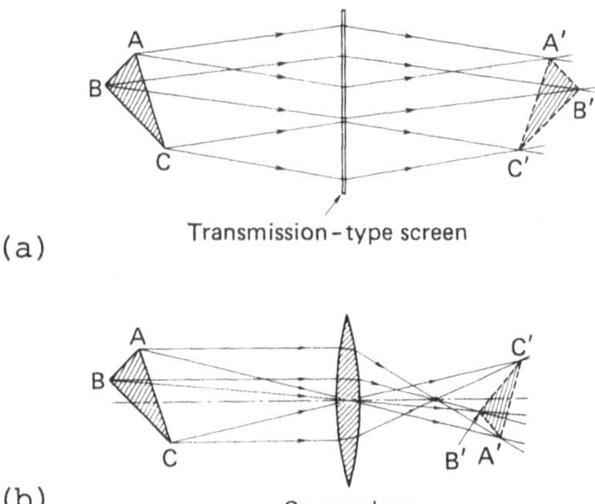

Fig.5.23(a) Inversion of a three-dimensional image by a transmission-type direction-selective screen. (b) Imaging property of a convex lens.

As seen in Fig.5.23a, most of the transmission-type direction-selective screens like the double-lenticular screen also give an inversion effect. A convex lens has an imaging property but lacks such an inversion property (see Fig.5.23b).

5.5 OPTIMUM DESIGN OF A PROJECTION SYSTEM

As described at the beginning of Chapter 4, the optimum design problem of projection-type three-dimensional displays is entirely similar to that of lens-sheet three-dimensional pictures. In the former, we need only consider the LS in Fig.4.11b not as a lenticular-sheet picture but as a direction-selective screen, also having a discrete, mosaiclike structure.

As for the optimum design of the projector system, therefore, the statements concerning the primary picture taking for LS pictures remain valid. First, we should note that a scaling rule as described in Section 4.3.3.A holds between the primary picture taking and final projection parameters. Secondly, the optimum camera and projector separations are given by the formulas obtained in Section 4.3.3.B.

As pointed out in Section 4.3.6, the theoretical optimum camera separation (in most cases around 10 mm) is practically too small. In practice, technological limitations on preciseness in arranging cameras and projectors widen this optimum separation; slightly wider separations (20-60 mm) are more commonly employed in practical systems.

5.6 DIFFUSER-BACKED LENTICULAR SCREENS

5.6.1 OPTIMUM DESIGN

Obviously the projection-type display that has the greatest similarity with LS pictures is one using a diffuser-backed lenticular screen. Therefore, the optimum design theory for LS pictures described in Section 4.3.4 is also applicable to diffuser-backed lenticular screens. The only necessary modification is to replace the term *lenticular sheet* by *lenticular screen;* this difference requires no essential modification in the analysis.

5.6.2 MEASUREMENT OF DIRECTIVITY

In Section 4.3.5, measured characteristics of lenticular sheets were shown to demonstrate how actual directivity is degraded below the essential limitation, i.e., diffraction. The parameters of the four samples and their characteristics are tabulated in Table 4.4. In that table, r, p, and t denote the radius of curvature of the lens surface, lens pitch, and thickness, respectively.

In this section the details of the setup and methods of measurement will be described (8).

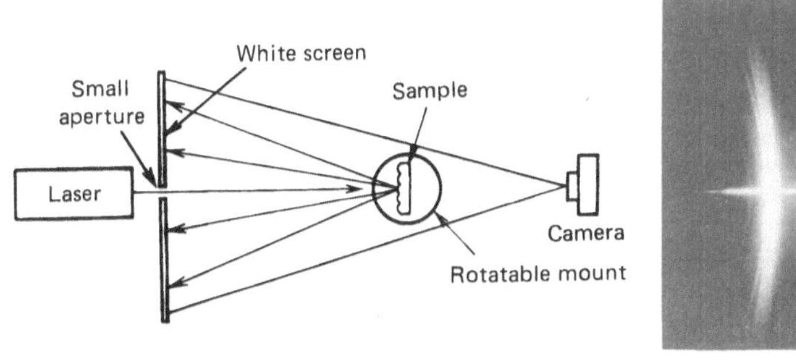

Fig.5.24

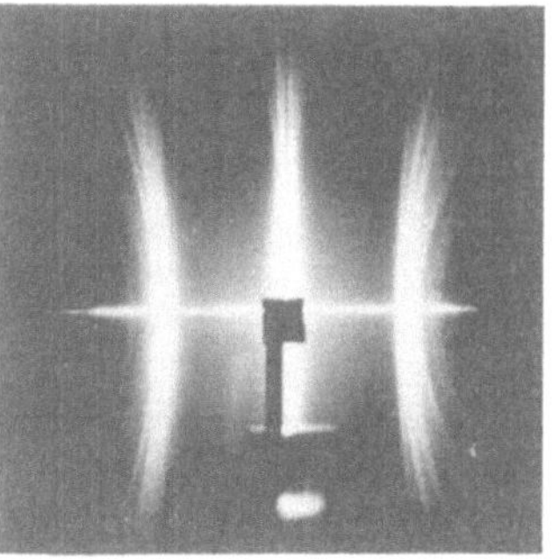

Fig.5.25

Fig.5.24 Optical setup for photographic recording of the reflection pattern.
Fig.5.25 An example of the reflection pattern of a diffuser-backed lenticular screen.

A. Preparation of the Sample

Four small samples (2 × 2 cm) of commercially available lenticular sheets were backed by white diffusing material (white vinyl paint).

B. Photographic Recording of the Overall Reflection Pattern

For the purpose of photographing, an optical setup as shown in Fig.5.24 is used. A He-Ne laser is used to provide a collimated light beam. The reflection pattern is projected upon a white screen and recorded by a camera behind the sample.

An example of the pattern obtained (sample 3 in Table 4.4) is shown in Fig.5.25. The bright, vertical stripe in the center shows the main lobe obtained with ordinary autocollimation. The slightly curved, vertical stripes on both sides show the side lobes (see Fig.5.4). The bright, horizontal line represents the surface reflection of the lenticular sheet. The sample and the rotating mount are seen in silhouette.

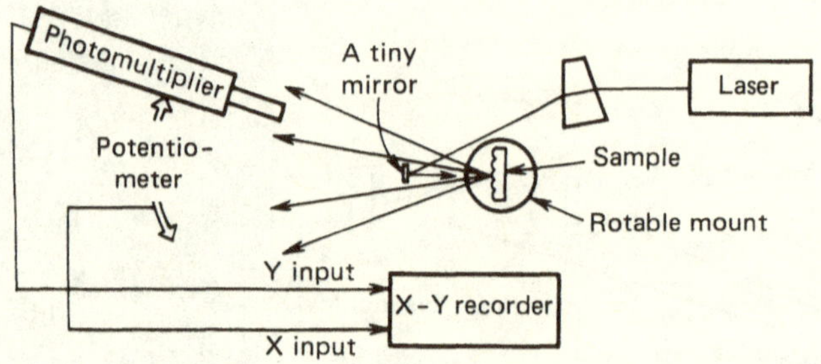

Fig.5.26 Setup for electronic measurement of the directivity.

C. *Electronic Measurement of the Directivity*

For quantitative measurement of directivity, a setup as shown in Fig.5.26 is used. The photomultiplier tube mounted on a rotation mechanism scans the reflection pattern horizontally, usually just above the bright horizontal line seen in Fig.5.25. The measured light intensity is given to the Y input of an X-Y recorder, whereas a voltage from the potentiometer ganged with the rotation mechanism provides the X input. The sample is also mounted on another rotation mechanism to vary the incident angle of the light beam.

An example of the recorded characteristics is shown in Fig.5.27. The symbols M and S denote the main and side lobes, respectively. Residual light of appreciable intensity is observed between M and S; this is called *background noise*. Probably the dominant source of background noise is the following optical process; some fraction of the light scattered upon the diffusing surface hits the surface of the lens from the rear with a high incident angle, and is subjected to total reflection. Some fraction of this reflected component will finally be emitted outward after experiencing multiple reflections between the diffuser and the lens surface.

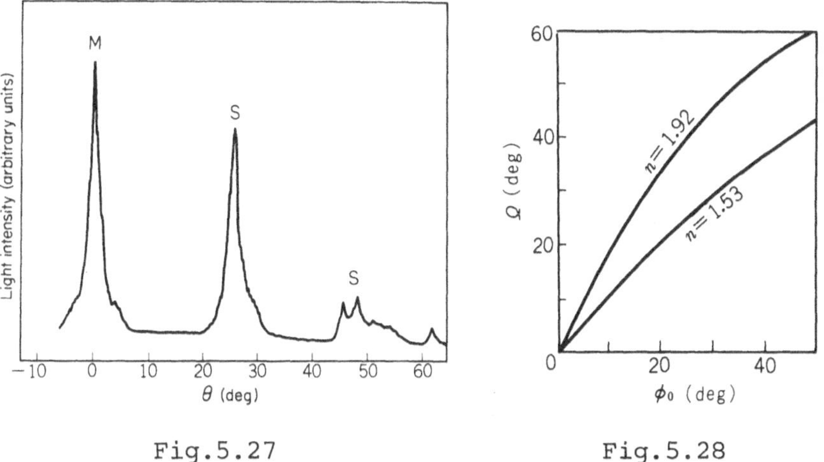

Fig.5.27 Fig.5.28

Fig.5.27 An example of the recorded reflection characteristics of a diffuser-backed lenticular screen.
Fig.5.28 Computed side-lobe angle.

D. *Side-Lobe Angle*

As described in Section 5.2.2, in many projection-type displays the side-lobe effect is used to widen the viewing zone. For such an application, the angle between the main and the side lobes, hereafter called the side-lobe angle Ω, is an important parameter for system design.

For normal incidence, the parameter Ω is given, after some simple geometrical considerations, as

$$\Omega = \sin^{-1}\left[n/\sqrt{1+\left(t/r\ \sin\phi_0\right)^2}\right], \quad (5.1)$$

where n, t, r, and ϕ_0 denote the refractive index of the lens material, lens thickness, radius of curvature of the lens, and the angle defined in Fig.4.19, respectively. Since the normalized lens thickness (t/r) is given as a function of ϕ_0 (see Fig.A.8 of the Appendix to Chapter 4), Ω is determined by ϕ_0 and n. The side-lobe angle thus obtained is shown in Fig.5.28 for n =

TABLE 5.2 Computed and Measured Side-Lobe Angles

Sample No.	Ω(computed)	Ω(measured)
1	25°	25.6°
2	14°	18.8°
3	35°	36.8°
4	49°	49.0°

1.53 and 1.92. The computed and measured values of Ω are compared in Table 5.2 for the four samples described in Section 4.3.5. The computation was performed using $n = 1.53$ and geometrical lens parameters r, t, and ϕ_0 measured under a microscope. Probably the discrepancy between the computed and measured Ω is mainly due to the error in measuring the geometrical lens parameters.

5.6.3 DESCRIPTION OF EXPERIMENTAL SYSTEMS

Some experimental projection-type three-dimensional display systems constructed in the author's laboratory will be described.

The first model (11) employed screens having dimensions 60 × 75 cm and optical characteristics shown as samples 1 and 3 in Table 4.4. The principal projection parameters were $f_1 = f_2 = 50$ mm, $a_1 = 70$ cm, $a_2 = 120$ cm, $d_1 = 3.0$ cm, $d_2 = 7.0$ cm, $D(= \alpha\beta/\gamma) = 1.46$, and N (number of projectors) = 12.

The second model employed a screen with dimensions 120 × 170 cm, p = 0.9 mm, and t = 3 mm. The half-intensity width of the directivity was about 3°. The projection parameters were $f_1 = f_2 = 50$ mm, $a_1 = 110$ cm, $a_2 = 250$ cm, $d_1 = 2.3$ cm, $d_2 = 7.0$ cm, and N = 10.

This second model was mainly used in a psychological test concerning the flipping effect. Three groups of the primary "slide" pictures giving different blurring-to-flipping ratios k_1 (refer to Section 4.3.3.B, item 2), $k_1 = 0.69$, 1.17, and 1.60, were prepared, by varying the F-number of the camera lens in the primary

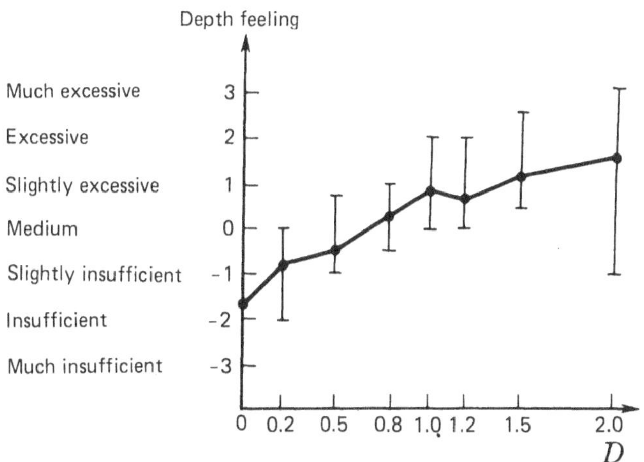

Fig.5.29 Relation between geometrical and subjective depths of a three-dimensional image.

picture-taking process. It was found that subjectively no appreciable difference could be perceived as to flipping.

Another psychological experiment was performed using an experimental projection system other than the above two models but similar to the second one. The experiment considered the measurement of the relation between geometrical depth and the subjective depth feeling of reconstructed three-dimensional images.

The result is shown in Fig.5.29, where the abscissa D is a parameter giving the geometrical depth-to-lateral ratio, defined by using parameters introduced in Section 4.3.3.A, as

$$D \triangleq \frac{(b_2/w_2)}{(b_1/w_1)} = \frac{\beta}{(b_1/a_1)(K-1)+K} . \qquad (5.2)$$

The latter relation is derived directly from Eq.(4.6). The parameter D might reasonably be called the *relative depth* or, as some specialists say, the *stereofactor*.

The ordinate in Fig.5.29 gives a subjective expression of depth feeling. The vertical bars in the figure show the distribution of opinions of the 20 people under test, and the dots show the average. What interests us most strongly is that the condition D = 1 giving the original depth-to-lateral ratio results in a slightly exaggerated depth feeling. This is a psychological effect found in artificial three-dimensional imaging.

5.7 TRIPLE-MIRROR SCREEN AND CURVED TRIPLE-MIRROR SCREEN

In this section the triple-mirror (TM) and curved triple-mirror (CTM) screens will be described in detail. For the ordinary triple mirror used as an optical target for surveying purposes, Yoder (12) and Fujii (13) have presented analyses and a design theory. Okoshi, Yano, and Fukumori presented analyses of the CTM screen (6). In this section the theories on the TM and CTM will first be presented on the bases of Refs.(13) and (6). Some experimental data will also be shown.

5.7.1 ANALYSIS OF TRIPLE MIRRORS

A. Basic Equations

We consider a triple mirror set upon the origin of an x,y,z coordinate system as shown in Fig.5.30. Suppose that a ray of light is incident upon the TM and is reflected three times in the following order:

S_x: x plane (defined as x = 0)

S_y: y plane (defined as y = 0)

S_z: z plane (defined as z = 0).

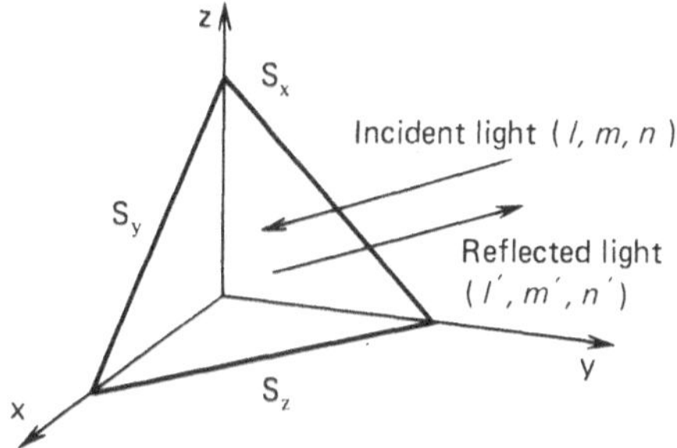

Fig.5.30 Retrodirective reflection on a triple mirror.

We define direction vectors I and I' representing the incident and reflected ray directions, respectively, both having unit magnitude, as

$$I = \begin{bmatrix} \ell \\ m \\ n \end{bmatrix} \tag{5.3}$$

$$I' = \begin{bmatrix} \ell' \\ m' \\ n' \end{bmatrix}, \tag{5.4}$$

where ℓ, m, n and ℓ', m', n' denote the direction cosines. Since the incident ray is directed toward the origin (see Fig.5.30), ℓ, m, and n are negative.

When the incident ray is reflected upon the x plane, the x component of the direction vector changes its sign. Upon the y plane and z plane, similar inversions of sign take place. Therefore, using reflection matrices defined as

$$R_x = \begin{bmatrix} -1 & 0 & 0 \\ 0 & 1 & 0 \\ 0 & 0 & 1 \end{bmatrix}, \quad R_y = \begin{bmatrix} 1 & 0 & 0 \\ 0 & -1 & 0 \\ 0 & 0 & 0 \end{bmatrix}, \quad R_z = \begin{bmatrix} 1 & 0 & 0 \\ 0 & 1 & 0 \\ 0 & 0 & -1 \end{bmatrix}, \tag{5.5}$$

we may write

$$I' = R_x R_y R_z I = -I. \tag{5.6}$$

The above equation shows that a triple mirror has the "retrodirective" property.

Equation (5.5) shows that for any permutation of R_x, R_y, and R_z, their product remains unchanged; in other words, R_x, R_y, and R_z are exchangeable. This means that $I' = -I$ holds for any order of reflections upon the three surfaces.

B. Effective Cross-Sectional Area (ECSA) of a Triple Mirror

Each surface of an actual triple mirror has a finite, triangular shape. As a result, an incident ray cannot always be reflected three times by the three surfaces; for specific points of incidence, only two reflections take place. Thus some fraction of the mirror area does not contribute to the effective cross-sectional area (ECSA) of the triple-mirror screen. Furthermore, for slant incidences, only one reflection takes place over some fraction of the mirror surface. In the following the ECSA of a triple mirror will be computed for various angles of incidence.

C. Transformation of Coordinates

For simplicity we assume that the edge lengths of the triple mirror are unity, and that these edges lie in the x,y,z coordinates as shown in Fig.5.31.

The x,y,z coordinate system, however, is not convenient for computing the reflection properties of a triple mirror. We assemble a large number of tiny triple mirrors to form a TM screen in a manner such that three apexes of each triple mirror lie on the screen surface. Therefore, for computing reflection properties such as the ECSA or the reflection pattern,

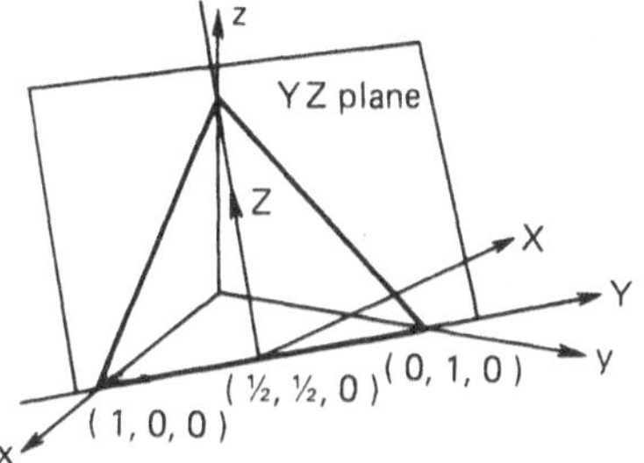

Fig.5.31 Coordinates used in the analysis.

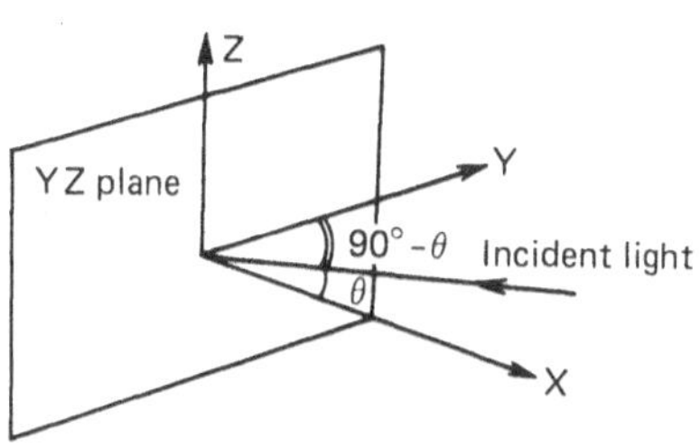

Fig.5.32 Direction of incidence in X,Y,Z coordinates.

it is more convenient to use a new coordinate system X,Y,Z as shown in Fig.5.31. The YZ plane shown in Fig.5.31 is expressed as

$$x + y + z = 1. \tag{5.7}$$

In an actual projection system, this YZ plane coincides with the plane of the screen.

We assume that the observer's eyes and the projectors are both located in the XY plane, and the incident ray (from a projector) has an incident angle of θ as shown in Fig.5.32. The direction vector in the X,Y,Z coordinates representing such an incident ray is expressed as

$$\begin{bmatrix} L \\ M \\ N \end{bmatrix} = \begin{bmatrix} -\cos\theta \\ -\sin\theta \\ 0 \end{bmatrix}. \tag{5.8}$$

On the other hand, the transformation between the x,y,z and X,Y,Z coordinates is given, from geometrical considerations, as

$$\begin{bmatrix} x-1/2 \\ y-1/2 \\ z \end{bmatrix} = \begin{bmatrix} 1/\sqrt{3} & -1/\sqrt{2} & -1/\sqrt{6} \\ 1/\sqrt{3} & 1/\sqrt{2} & -1/\sqrt{6} \\ 1/\sqrt{3} & 0 & 2/\sqrt{6} \end{bmatrix} \begin{bmatrix} X \\ Y \\ Z \end{bmatrix}. \tag{5.9}$$

The inverse transformation is

$$\begin{bmatrix} X \\ Y \\ Z \end{bmatrix} = \begin{bmatrix} 1/\sqrt{3} & 1/\sqrt{3} & 1/\sqrt{3} \\ -1/\sqrt{2} & 1/\sqrt{2} & 0 \\ -1/\sqrt{6} & -1/\sqrt{6} & 2/\sqrt{6} \end{bmatrix} \begin{bmatrix} x-1/2 \\ y-1/2 \\ z \end{bmatrix}. \tag{5.10}$$

Therefore, the incident ray with a slant angle θ as described above is expressed in x,y,z coordinates as

$$\begin{bmatrix} \ell \\ m \\ n \end{bmatrix} = \begin{bmatrix} -\cos\theta/\sqrt{3}+\sin\theta/\sqrt{2} \\ -\cos\theta/\sqrt{3}-\sin\theta/\sqrt{2} \\ -\cos\theta/\sqrt{3} \end{bmatrix}. \tag{5.11}$$

D. Computation of the Triple-Reflection Area

We compute the triple-reflection area (effective area) upon the YZ plane. We first define point O' as the point where a line having direction (l,m,n) and passing through the origin of the x,y,z coordinates intersects the XY plane. The coordinates of the point O' are expressed in the x,y,z system as

$$\left(\frac{\ell}{\ell+m+n}, \frac{m}{\ell+m+n}, \frac{n}{\ell+m+n}\right), \tag{5.12}$$

which, in the X,Y,Z system, are transformed into

$$\text{coordinates of P}' = \left(0, \frac{m-\ell}{\sqrt{2}(\ell+m+n)}, \sqrt{\frac{3}{2}}\,\frac{n}{\ell+m+n}\right). \tag{5.13}$$

On the other hand, since

$$m-\ell = -\sqrt{2}\,\sin\theta \tag{5.14}$$

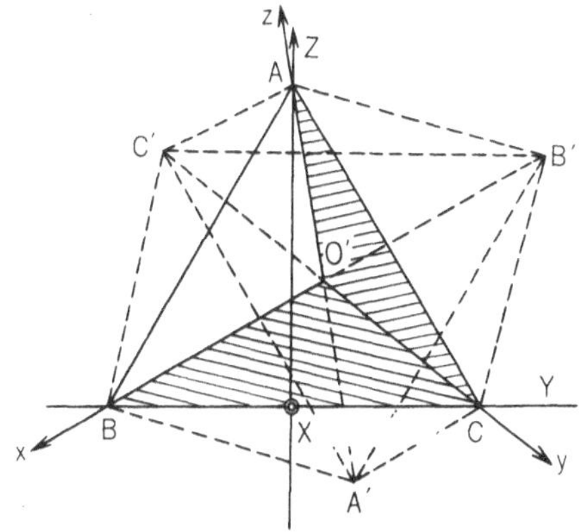

Fig.5.33 Projection of the triple mirror ABC upon the YZ plane.

$$\ell+m+n = -\sqrt{3}\cos\theta \tag{5.15}$$

$$n = -\cos\theta/\sqrt{3} \tag{5.16}$$

holds, Eq.(5.13) is simplified to

$$\text{coordinates of } P' = (0,\ \tan\theta/\sqrt{3},\ 1/\sqrt{6}). \tag{5.17}$$

Figure 5.33 shows the projection of the triple mirror ABC and the point O' upon the YZ plane. To compute the triple-reflection area, we further define points A', B', and C' which are the mirror images of A with respect to the z plane (z = 0), of B with respect to the x plane, and of C with respect to the y plane, respectively. In Fig.5.33, points A', B', C' are shown being projected upon the YZ plane. (Note that on the YZ plane A', B', and C' are given as symmetrical points of A, B, and C, respectively, with respect to O'.) The triple-reflection area is given as the common area of triangles ABC and A'B'C' as illustrated in Fig.5.34.

There are two ways to define the effective cross-sectional area (ECSA). In the first definition we

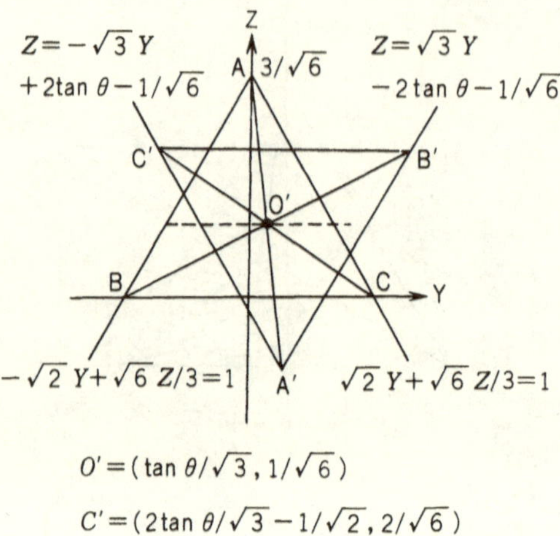

Fig.5.34 Computation of the effective cross-sectional area (ECSA) of a triple mirror.

define it as the ratio of the triple-reflection area (S_1) to the area of ABC(S_0), that is,

$$K_1 = S_1/S_0. \tag{5.18}$$

The above quantity might be called the relative ECSA. We should note that $S_0 = \sqrt{3}/2$ in the above case. In the second definition, we take into account the fact that when a ray is incident with a slant angle θ, the area of the TM seen from that direction is reduced by a factor of cosθ. Thus we define

$$K_2 = S_1 \cos\theta/S_0. \tag{5.19}$$

As a figure of merit of the triple-mirror optical target for distance measurements, such as the one left upon the moon surface by the astronauts, the second definition is more suitable because the irradiating laser light bundle is usually much wider than the target. For a TM screen used in a three-dimensional display system, the first definition is preferable because for skew incidences both the areas S_1 and S_0 are reduced, whereas the number of TMs irradiated by a light bundle increases. In the following discussion we principally use K_1 as the figure of merit.

To compute area S_1 as a function of the angle θ, we start with cases when point C' is located to the left of AB in Fig.5.34. In such cases the triple-reflection region will be a hexagon, and the conditions for θ and the area S_1 are expressed as

$$\theta \leqq \tan^{-1}(1/\sqrt{6}) \doteqdot 22°. \tag{5.20}$$

$$\begin{aligned} S_1 &= 2/\sqrt{6}(1/3\sqrt{2}+1/\sqrt{2}-2\tan\theta/\sqrt{3}) \\ &\quad -\sqrt{3}/2(1/\sqrt{2}-1/3\sqrt{2}-2\tan\theta/\sqrt{3})^2 \\ &= (1-2\tan^2\theta)/\sqrt{3}. \end{aligned} \tag{5.21}$$

Hence we obtain

$$K_1 = S_1/(\sqrt{3}/2) = 2(1-2\tan^2\theta)/3 \tag{5.22}$$

$$K_2 = 2(1-2\tan^2\theta)\cos\theta/3. \tag{5.23}$$

When point C' is located right of AB, that is, when $\theta > \tan^{-1}(1/\sqrt{6})$, the triple-reflection region becomes a parallelogram. When θ is further increased and finally point C' moves onto AC ($\theta = \tan^{-1}\sqrt{2/3}$), S_1 is reduced to zero. Thus, for

$$\tan^{-1}(1/\sqrt{6}) \leqq \theta \leqq \tan^{-1}(\sqrt{2/3}), \tag{5.24}$$

we obtain

$$S_1 = 4/3\sqrt{3} - 2\sqrt{2}\tan\theta/3. \tag{5.25}$$

Hence,

$$K_1 = 8/9 - 4\sqrt{2/3}\tan\theta/3, \tag{5.26}$$

$$K_2 = 8\cos\theta/9 - 4\sqrt{2/3}\sin\theta/3. \tag{5.27}$$

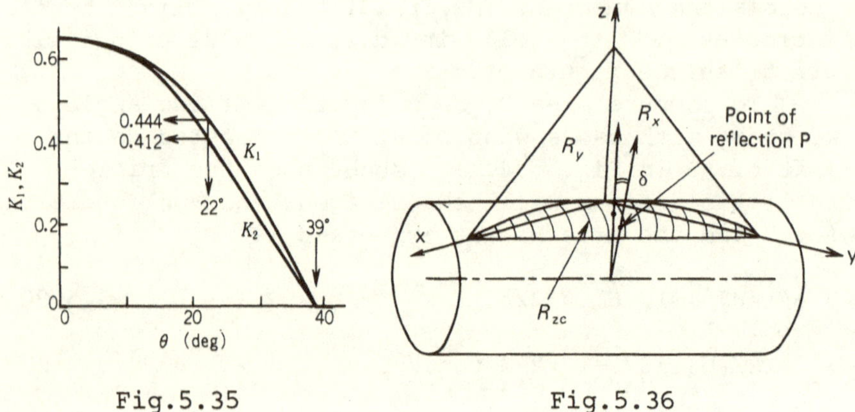

Fig.5.35 Fig.5.36

Fig.5.35 Effective cross-sectional area (ECSA) of a triple mirror.
Fig.5.36 A curved triple mirror with a cylindrical surface.

By using Eqs.(5.22), (5.23), (5.26), and (5.27), the two kinds of ECSA, K_1 and K_2, are shown in Fig.5.35 as functions of θ. These results agree with more general computations presented by Fujii, *et al.* (13), in which the incident angle of elevation (vertical deviation) ϕ is also considered. When we put ϕ = 0 in Fujii's results, we get the above formulas. The maximum K_1 and K_2 are both 2/3 (= 66.6%) as seen in Fig. 5.35.

5.7.2 ANALYSIS OF CURVED TRIPLE MIRROR

Next, we consider a curved triple mirror (hereafter abbreviated as CTM), in which one reflecting surface of a TM is curved slightly in a cylindrical form as shown in Fig.5.36. The axis of the cylinder lies parallel to the z plane, and intersects the y and x planes at 45°. The three reflecting surfaces and their reflection matrices will hereafter be denoted by S_x, S_y, S_{zc}, R_x, R_y, and R_{zc} (the subscript c implies "curved"), respectively.

A. Basic Equations

When a ray of light is incident upon a mirror surface whose normal unit vector is given as N = (A, B,C), the direction column vectors of the incident and reflected rays I (ℓ,m,n) and I' (ℓ',m',n') are related by the law of reflection (12)

$$I' = I - 2(NI)N. \tag{5.28}$$

The above equation is equivalent to

$$\begin{bmatrix} \ell' \\ m' \\ n' \end{bmatrix} = \begin{bmatrix} 1-2A^2 & -2AB & -2AC \\ -2BA & 1-2B^2 & -2BC \\ -2CA & -2CB & 1-2C^2 \end{bmatrix} \begin{bmatrix} \ell \\ m \\ n \end{bmatrix}. \tag{5.29}$$

By applying the above equation to each of three reflecting surfaces of a CTM successively, we can get its overall reflection characteristics. If δ denotes the angle between the normal of the cylindrical surface and the z axis at a point P upon S_{zc} as shown in Fig. 5.36, the normal vectors N_x, N_y, N_{zc} of the three reflecting surfaces may be written as

$$\begin{aligned} N_x &= (1,\ 0,\ 0), \\ N_y &= (0,\ 1,\ 0), \\ N_{zc} &= [(1/\sqrt{2})\sin\delta,\ (1/\sqrt{2})\sin\delta,\ \cos\delta]. \end{aligned} \tag{5.30}$$

The reflection matrices R_x and R_y shown in Eq. (5.5) remain unchanged, whereas matrix R_{zc} describing the reflection upon the curved z surface is given, from Eqs. (5.29) and (5.30), as

$$R_{zc} = \begin{bmatrix} \cos^2\delta & -\sin^2\delta & -\sqrt{2}\sin\delta\cos\delta \\ -\sin^2\delta & \cos^2\delta & -\sqrt{2}\sin\delta\cos\delta \\ -\sqrt{2}\sin\delta\cos\delta & -\sqrt{2}\sin\delta\cos\delta & 1-2\cos^2\delta \end{bmatrix}. \tag{5.31}$$

The triple reflection upon a CTM is expressed as the product of the three matrices R_x, R_y, and R_{zc}. All six permutations have to be computed to get the complete reflection pattern.

B. Classification of Reflection Components

A little algebra reveals that

$$R_{zc}R_x \neq R_xR_{zc}, \qquad R_{zc}R_y \neq R_yR_{zc}, \tag{5.32}$$

$$R_xR_y = R_yR_x. \tag{5.33}$$

Therefore, computations are required for the following four permutations:

Case 1: $S_x \rightarrow S_y \rightarrow S_{zc} \quad (= S_y \rightarrow S_x \rightarrow S_{zc})$

Case 2: $S_{zc} \rightarrow S_x \rightarrow S_y \quad (= S_{zc} \rightarrow S_y \rightarrow S_x)$

Case 3: $S_x \rightarrow S_{zc} \rightarrow S_y$

Case 4: $S_y \rightarrow S_{zc} \rightarrow S_x$.

C. Reflection-Ray Vectors

As in the discussion on TMs in the preceding section, we consider an incident ray of light lying in the XY plane and making an angle θ with respect to the X axis. To avoid confusion, only the results of computation of the reflection-ray vector in X,Y,Z coordinates (L',M',N') will be shown. For the details of the derivation, refer to the appendix of this chapter.

Case 1. For an incident ray given as Eq.(5.8), the reflection-ray vector is given as

$$\begin{bmatrix} L' \\ M' \\ N' \end{bmatrix} = \begin{bmatrix} \cos 2\delta \ \cos\theta \\ \sin\theta \\ \sin 2\delta \ \cos\theta \end{bmatrix}. \tag{5.34}$$

The autocollimating property in the horizontal direction is represented by the relation M' = -M.

Case 2.

$$\begin{bmatrix} L' \\ M' \\ N' \end{bmatrix} = \begin{bmatrix} \cos 2\delta \cos\theta \\ \sin\theta \\ -\sin 2\delta \cos\theta \end{bmatrix}. \qquad (5.35)$$

The only difference between Eqs.(5.34) and (5.35) is found in the sign of N'. Horizontal autocollimation, M' = -M, holds also in Eq.(5.35).

Case 3.

$$\begin{bmatrix} L' \\ M' \\ N' \end{bmatrix} = \begin{bmatrix} (2/3+\cos 2\delta/3)\cos\theta - \sin 2\delta \sin\theta/\sqrt{3} \\ \cos 2\delta \sin\theta + \sin 2\delta \cos\theta/\sqrt{3} \\ -\sqrt{2/3}\sin 2\delta \sin\theta + \sqrt{2/3}(\cos 2\delta - 1)\cos\theta \end{bmatrix}. \qquad (5.36)$$

Case 4.

$$\begin{bmatrix} L' \\ M' \\ N' \end{bmatrix} = \begin{bmatrix} (2/3+\cos 2\delta/3)\cos\theta + \sin 2\delta \sin\theta/\sqrt{3} \\ \cos 2\delta \sin\theta - \sin 2\delta \cos\theta/\sqrt{3} \\ \sqrt{2/3}\sin 2\delta \sin\theta + \sqrt{2/3}(\cos 2\delta - 1)\cos\theta \end{bmatrix}. \qquad (5.37)$$

D. *Summary of Computations*

Cases 1 and 2. Equations (5.34) and (5.35) show that M' = -M holds for these two cases. This is nothing but the desired autocollimating property with respect to the Y direction, that is, in the horizontal direction.

In the vertical direction, a diffusing property is found. The angle Θ made by the incident and reflected rays is given as

$$-\cos\Theta = (L,M,N)\begin{pmatrix} L' \\ M' \\ N' \end{pmatrix} = -\cos 2\delta \cos^2\theta - \sin^2\theta$$

$$= (1-\cos 2\delta)\cos^2\theta - 1 \qquad (5.38)$$

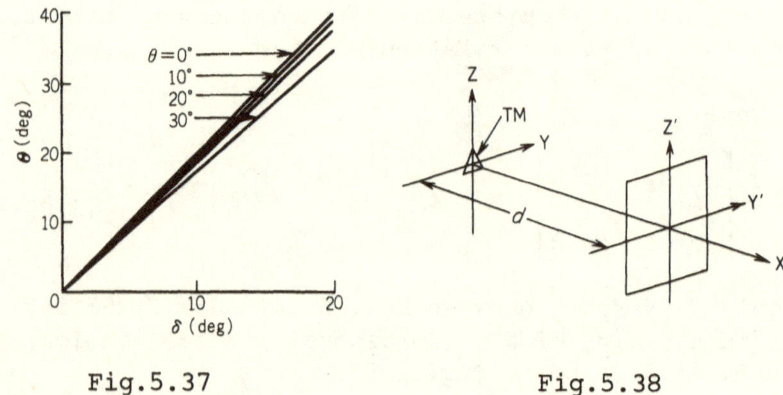

Fig.5.37 Fig.5.38

Fig.5.37 Relation between δ, θ, and Θ.
Fig.5.38 The Y',Z' coordinates (S plane) where the reflection pattern is observed.

Hence,

$$\cos\Theta = 1 - 2\sin^2\delta\, \cos^2\theta . \qquad (5.39)$$

The above relation is shown in Fig.5.37.

Equation (5.39) or Fig.5.37 shows that the relations

$$\Theta \doteqdot 2\delta, \quad \text{hence } \Theta_{max} \doteqdot 2\delta_{max} \qquad (5.40)$$

hold regardless of the angle θ. This means that if we are to design a CTM screen having, for example, a vertical diffusion of ±15°, we should give a curvature to the zc surface such that the angle δ ranges between -7.5 and +7.5°. The above statement may be understood intuitively from Fig.5.36.

Cases 3 and 4. In these cases M' ≠ -M; hence autocollimation is absent. This means that these permutations of reflections give unwanted, spurious reflection components.

E. Reflection Pattern of the Unwanted Components

Suppose that a plane S is located at X = d at a right angle to the X axis, and let (Y',Z') represent the coordinates (Y,Z) on plane S as shown in Fig.5.38. We first consider the reflection pattern of the unwanted, spurious component observed on this plane for normal incidence ($\theta = 0$). The distance d is assumed to be so long that from that plane the triple mirror is seen as a pinpoint.

For Case 3, the path of the reflected ray is given, from Eq.(5.36), as

$$\begin{aligned} X &= t(2/3+\cos 2\delta/3) \\ Y' &= t\ \sin 2\delta/\sqrt{3} \\ Z' &= \sqrt{2}t(\cos 2\delta-1)/\sqrt{3}, \end{aligned} \tag{5.41}$$

where t is a parameter. Normalizing Y' and Z' with respect to X (= d) leads to

$$\eta \triangleq \frac{Y'}{d} = \frac{1}{\sqrt{3}} \times \frac{\sin 2\delta}{2/3+\cos 2\delta/3}, \tag{5.42}$$

$$\zeta \triangleq \frac{Z'}{d} = \frac{\sqrt{2}}{\sqrt{3}} \times \frac{\cos 2\delta-1}{2/3+\cos 2\delta/3}. \tag{5.43}$$

When δ is small, we may approximate $(2/3)+(1/3)\cos 2\delta \doteqdot 1$. In such cases

$$\eta = \sin 2\delta/\sqrt{3}, \tag{5.44}$$

$$\zeta = \sqrt{2}(\cos 2\delta-1)/\sqrt{3}. \tag{5.45}$$

Eliminating 2δ from the above two equations, we obtain

$$\frac{(\zeta+\sqrt{2/3})^2}{(\sqrt{2/3})^2} + \frac{\eta^2}{(1/\sqrt{3})^2} = 1, \tag{5.46}$$

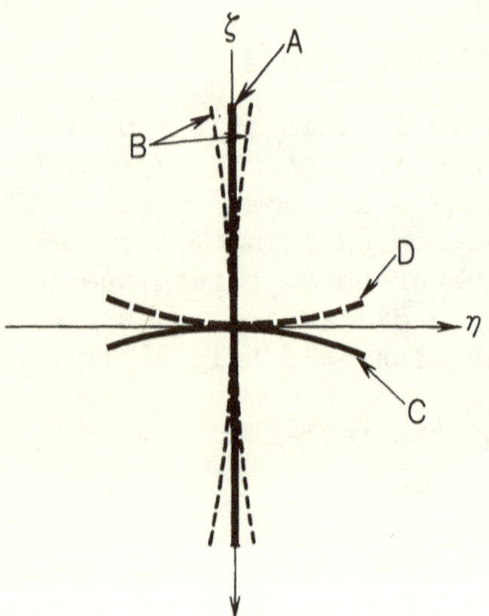

Fig.5.39 Reflection pattern of a curved triple-mirror screen [after Okoshi et al. (6)].

which represents an ellipse in the lower $\eta\zeta$ plane like pattern C in Fig.5.39.

The pattern actually observed is not a complete ellipse but merely a portion of it. Its width (in the η direction) can be determined, from Eq.(5.44), as

$$\eta_{max} = (1/\sqrt{3})\sin 2\delta_{max} \doteqdot (1/\sqrt{3})\Theta_{max}. \tag{5.47}$$

For Case 4, we obtain results identical to those for Case 3.

When many tiny CTMs are arranged to form a screen in a manner as shown in Fig.5.13, half of the CTMs will have their zc surfaces upside down. For these CTMs the spurious reflection components also result in an upside-down pattern as D shown in Fig.5.39. In reality, an algebraic computation similar to the above shows that for those reflection components

$$\frac{(\zeta-\sqrt{2/3})^2}{(\sqrt{2/3})^2} + \frac{\eta^2}{(1/\sqrt{3})^2} = 1 \tag{5.48}$$

holds, which represents an ellipse like D in Fig.5.39.

F. *Overall Reflection Pattern*

As will be described in Section 5.7.4, an overall reflection pattern like A, C, and D shown in Fig.5.39 is experimentally obtained for normal incidence. The vertical line A corresponds to the horizontally collimated components wanted.

When the ray is incident on the XY plane with an angle θ, the pattern is distorted. Component A is split into two hyperbolas like B shown in Fig.5.39. The unwanted components C and D are also distorted, but they remain ellipses passing through the origin of the $\eta\zeta$ coordinates; only their centers move. All of these properties are derived theoretically in Ref. (4), but these derivations are omitted here.

G. *Effective Cross-Sectional Area (ECSA) of a Curved Triple Mirror*

We consider the question of the ratio of the reflected light energies as distributed in patterns A (or B), C, and D in Fig.5.39. It is very difficult to compute this ratio rigorously; only an approximate analysis for the limit $\delta_{max} \to 0$ and for normal incidence ($\theta = 0$) will be presented.

When a unit CTM having very little curvature ($\delta_{max} \to 0$) is seen from the normal direction, six effective reflecting areas are observed as in an ordinary TM. They are indicated as 1-6 in Fig.5.40. We note that those light bundles that are first incident upon surfaces 1, 2, 4, and 5 are reflected in the following orders:

1: $S_x \to S_y \to S_{zc}$

2: $S_y \to S_x \to S_{zc}$

4: $S_{zc} \to S_y \to S_x$

5: $S_{zc} \to S_x \to S_y$,

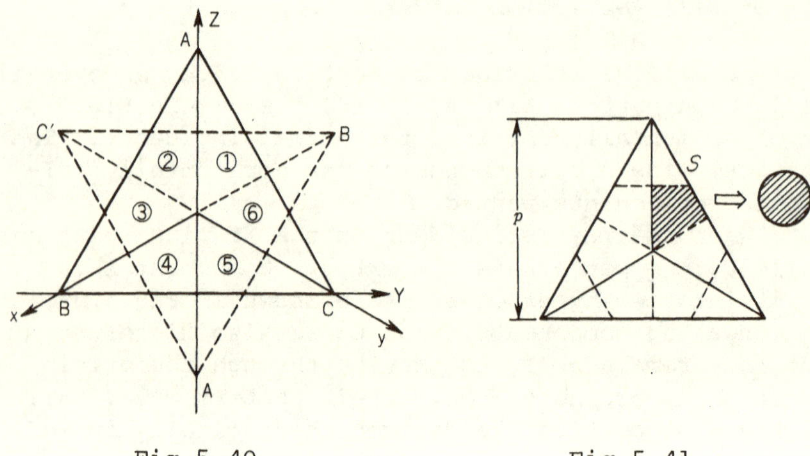

Fig.5.40 Fig.5.41

Fig.5.40 Computation of the effective cross-sectional area (ECSA) of a curved triple mirror.
Fig.5.41 Computation of the diffraction effect of a curved triple mirror.

and these components result in the collimated pattern A (or B). Those light bundles first incident upon 3 and 6 are reflected like

$$3:\quad S_y \rightarrow S_{zc} \rightarrow S_x$$

$$6:\quad S_x \rightarrow S_{zc} \rightarrow S_y,$$

and they constitute the unwanted patterns (C or D). From the above facts that in a CTM four out of six reflection permutations result in the wanted component and that the maximum ECSA K_1 of a TM is 2/3, the maximum ECSA of a CTM is given as

$$K_1 = 2/3 \times 4/6 = 4/9 \quad (\doteqdot 44\%). \tag{5.49}$$

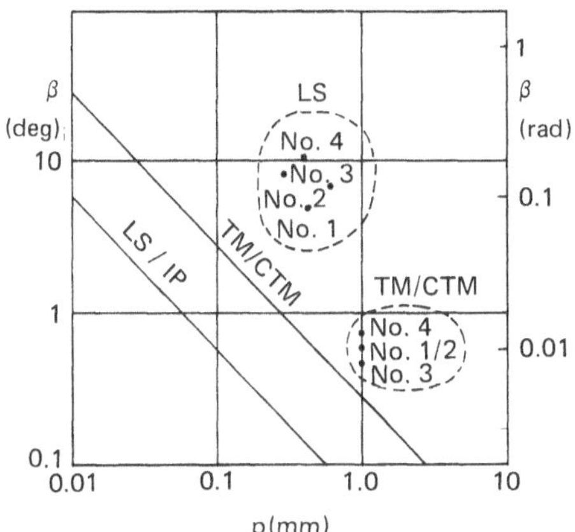

Fig.5.42 Diffraction limitation of the directivities of LS, IP, TM, and CTM. The measured directivities are also shown for comparison.

5.7.3 OPTIMUM PITCH AND RESOLUTION LIMITATION

The optimum design theory of LS pictures described in Section 4.3.4 is also applicable to projection-type displays using a TM or CTM screen, after a little modification of Eq.(4.21).

In a TM or CTM screen, rigorous computation of the directivity limitation due to diffraction is not easy since the diffraction in the unit triple mirror is rather complicated. An extremely simplified analysis will be as follows.

We first consider a CTM screen. When a slightly curved unit CTM is seen from the normal direction, six effective areas are observed as described above and illustrated in Figs.5.40 and 5.41. We assume that the diffraction given by a unit CTM is equal to the diffraction which would take place when the incident light passes through a circular aperture having the same area as S shown in Fig.5.41. The diameter of such an equivalent aperture is given as

TABLE 5.3 Optimum Pitch and Maximum Resolution of CTM Screen[a]

a_2 [m]	p_{opt} [mm]	$N_{D\ max}$
0.25	0.19_5	13.0
0.5	0.27_5	9.2
1.0	0.39	6.5
2.0	0.55	4.6

[a]Only the diffraction and sampling effect are considered.

$$D = \frac{2}{3^{5/4}\,\pi^{1/2}}\,p = 0.287p, \tag{5.50}$$

where p denotes the height of the CTM (see Fig.5.41). This height p also gives the "screen pitch" which determines the degree of picture deterioration due to sampling effects.

We note that the first minimum of the Fraunhofer diffraction pattern of a circular aperture appears at an angle $1.22\lambda/D$, whereas it appears at λ/p for a linear slit. The diffraction in a circular aperture integral photograph [Eq.(4.36) of Chapter 4] has been derived from Eq.(4.21) upon the above basis. If we use the same approximation, we again obtain from Eq.(4.21)

$$\beta_D\ (\mathrm{CTM}) = \frac{1.22}{\sqrt{2}} \times \frac{2.0\lambda}{D} = 1.70\lambda/D = 5.9\lambda/p. \tag{5.51}$$

In the first equation of Eq.(5.51), the factor $1/\sqrt{2}$ is present because Eq.(4.21) represents twofold diffractions in the recording and reconstruction processes. [Refer to the derivation of Eq.(4.21).]

The above diffraction limitation is shown in Fig. 5.42 (which is the same as Fig.4.26 in Chapter 4), together with that of the lenticular sheet [Eq.(4.21)],

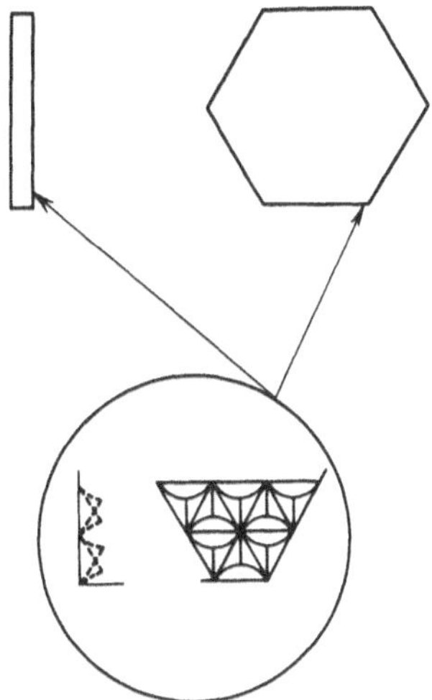

Fig.5.43 A unit curved triple-mirror screen.

both for $\lambda = 500$ nm. The values of p_{opt} and $N_{D\ max}$ for the diffraction-limited case can be computed by putting $k = 5.9$ into Eqs.(4.32) and (4.33). The results are tabulated in Table 5.3.

In the case of a CTM screen, the above statement and analysis remain valid in most practical cases. Theoretically, however, the situation is entirely different for the TM case (14). If the surface roughness and overall mirror distortion were both much less than the wavelength, all six reflected components would have a common phase regardless of the incident angle. This means that the diameter of the equivalent aperture should be multiplied by $\sqrt{6}$. Accordingly, the beam spread due to diffraction would be sharpened by $1/\sqrt{6}$ in the ideal case.

5.7.4 *MEASUREMENT OF DIRECTIVITY*

Some measurements on TMs and CTMs will be described, and the results will be compared with the theoretical limitation.

TABLE 5.4 The Samples of TM and CTM and Their Optical Characteristics

Samples	Pitch (mm)	Radius of curvature (mm)	Measured reflectance (%)	α (degree)	β (degree)	Comments
Triple mirror 1	1	∞	6.4	1.22	0.57	Spurious reflection upon the edge of each TM.
Triple mirror 2	1	∞	32	1.30	0.55	
Curved triple mirror 3	1	0.5	8.5	1.91	0.48	Large diffusing angle ($\sim 60°$) according to the small radius of curvature.
Curved triple mirror 4	1	2.2	32	1.15	0.69	

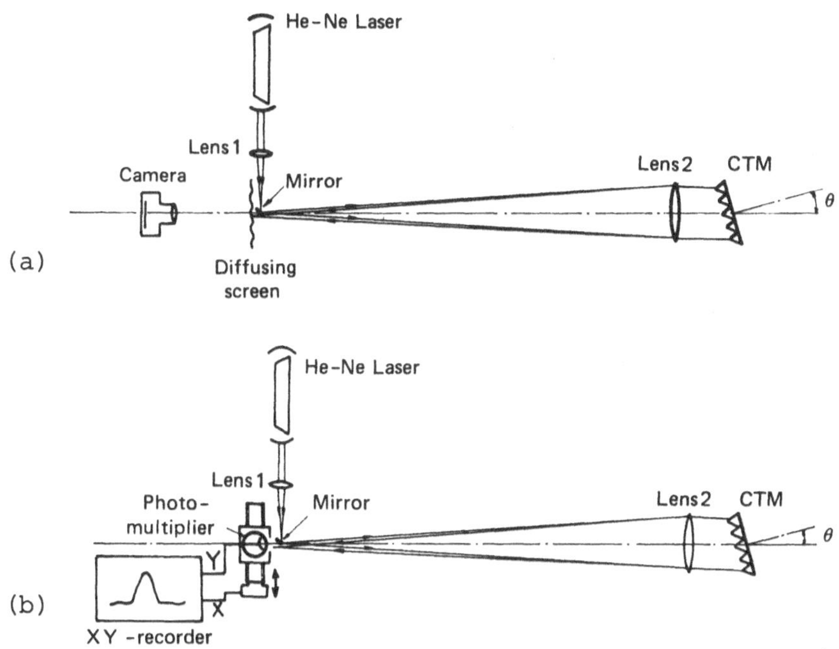

Fig.5.44(a) Optical setup for the photographic recording of the reflection pattern. (b) Setup for the electronic recording of the reflection pattern.

A. Samples and Optical Setup for Measurement

Four samples of TM and CTM screens as tabulated in Table 5.4 were prepared for measurement. All were of hexagonal shape as shown in Fig.5.43 and made of hard plastics. These are called unit screens. All the samples consist of 2400 unit triple mirrors whose pitch is 1 mm.

The measurement of the collimating characteristics of TM and CTM screens can be performed in a similar manner as with a lenticular screen; however, since TM and CTM screens usually have better directivity and larger pitch than does a typical lenticular screen, slightly modified setups are used both in photographic

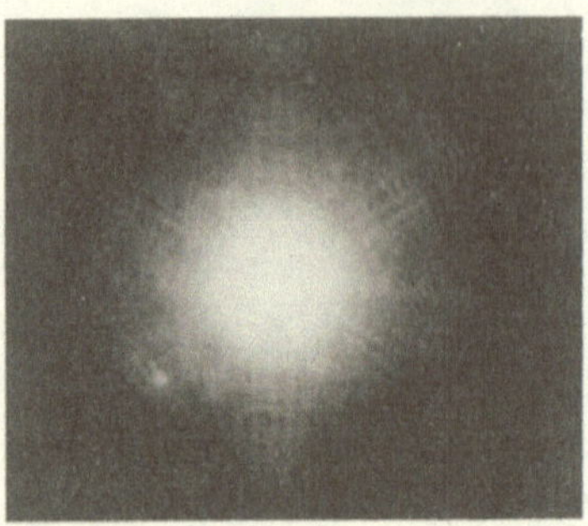

Fig.5.45 Reflection pattern of a TM screen for normal incidence [from Okoshi et al. (6)].

and electronic recording of the collimating characteristics.

The setup used in photographic recording of the overall reflection pattern is shown in Fig.5.44a. A He-Ne laser is used to prepare a collimated light beam. The laser beam is widened to a parallel beam having 3-cm diameter by lenses 1 and 2 to cover a unit screen, making it possible to obtain average characteristics of many unit triple mirrors. At the focus of both lenses 1 and 2 a tiny mirror (2 × 2 mm) is placed to deflect the input light beam by 90°. The reflected light again passes through lens 2, is projected upon the translucent screen, and photographed by a camera. The reflection pattern thus obtained is equivalent to the "far-field" pattern.

The setup used in the electronic recording of the collimating characteristics is shown in Fig.5.44b. In this arrangement, the far-field pattern is also recorded.

B. Results

Figures 5.45 and 5.46 show the reflection pattern of a unit TM (sample 2) for normal incidence and the overall reflection patterns of a unit CTM (sample 4) for normal and skew incidences. These patterns agree well with the theory described earlier.

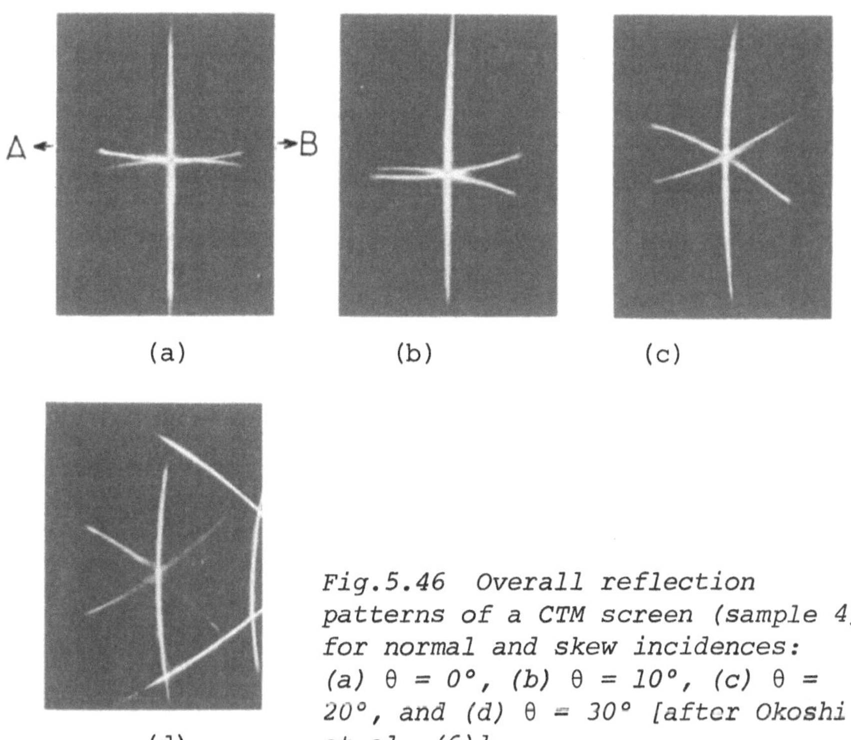

Fig.5.46 Overall reflection patterns of a CTM screen (sample 4) for normal and skew incidences: (a) θ = 0°, (b) θ = 10°, (c) θ = 20°, and (d) θ = 30° [after Okoshi et al. (6)].

Figure 5.47 shows examples of the intensity distribution of the reflection pattern as shown in Fig. 5.46, measured across the horizontally collimated, vertically diffusing component, just above the center of the pattern along the line AB shown in Fig.5.46a. The intensity distribution is sometimes triangular but sometimes flat-topped. Therefore, to characterize the intensity distribution, two quantities w_1 and w_2 as illustrated in Fig.5.47 are read upon the data and corresponding angle spreads are computed. The results are tabulated in Table 5.4 where $\alpha = w_1/d$ and $\beta = w_2/d$, the denominator d being the distance between the screen and the photomultiplier.

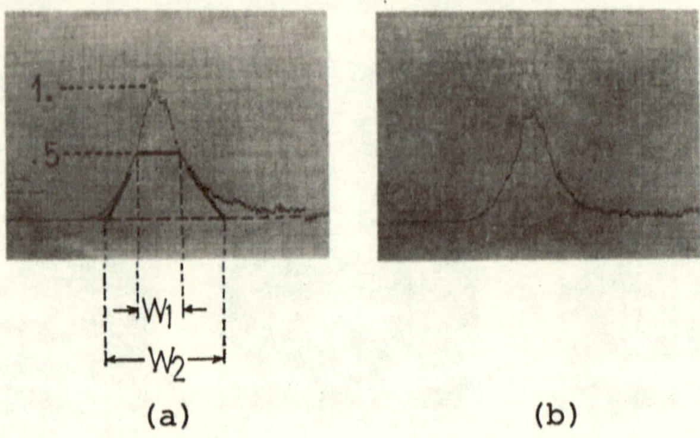

Fig.5.47 Energy distribution of the reflection pattern in cases (a) and (b) shown in Fig.5.46 (wanted component) [after Okoshi et al. (6)].

The intensity of the collimated component was also measured as a function of the incident angle. An example of the results is plotted in Fig.5.48 where the ordinate shows the relative intensity. The theoretical effective cross-sectional area (ECSA) K_1 is also shown for comparison.

C. Comparison between the Measured and Theoretical Directivities

Both the measured and theoretical directivities of TM and CTM screens are shown in Fig.5.42 for comparison with those of lenticular screens. It is seen that the discrepancy between the measured and theoretical directivities is much less than in lenticular screens.

5.7.5 DESCRIPTION OF AN EXPERIMENTAL SYSTEM

Many unit CTMs as described above were assembled to form a large CTM screen with dimensions 70 × 100 cm (6). This screen showed satisfactory characteristics when used in a three-dimensional imaging system, except for the discontinuity between unit CTMs. Development

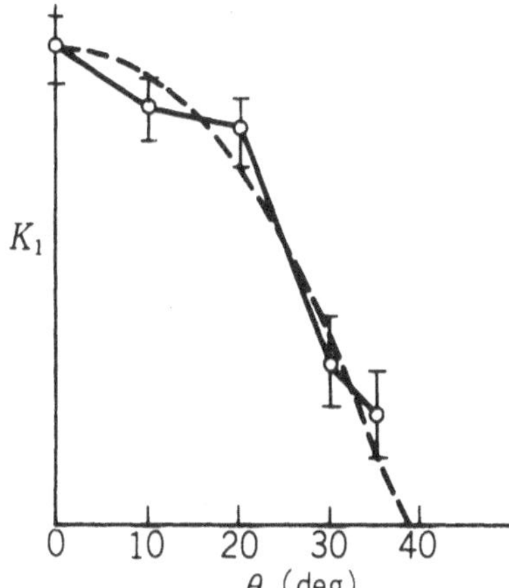

Fig.5.48 Comparison of measured (-o-) and theoretical (---) effective cross-sectional area.

of continuous screen-fabrication technology is needed. The advantages of the CTM screen can be summarized as follows:

(1) high reflection efficiency, practically above 30%;
(2) good directivity, practically less than 1°;
(3) controllable vertical diffusing angle;
(4) fewer spurious reflections;
(5) wide viewing angle (±39°) for horizontal rotation.

5.8 SUMMARY

In this chapter, the principles of projection-type displays and various screens have been described, and the two most practical and promising screens have been discussed in detail.

Projection-type three-dimensional displays are a promising field for future three-dimensional imaging engineering. It has not, however, been thoroughly exploited. The interesting tasks in the near future will be (1) the development of new screens, especially those made using holographic techniques; (2) improvement of conventional screens; and (3) combination with holographic projection systems. Items (1) and (3) will be described in Chapters 8 and 7, respectively.

APPENDIX. DERIVATION OF REFLECTION MATRICES AND REFLECTION-RAY VECTORS OF A CTM SCREEN

For computing the reflection pattern, X,Y,Z coordinates are more convenient than x,y,z coordinates since the incident ray lies in the XY plane. However, the original reflection matrices are given in terms of the x,y,z coordinates [see Eqs.(5.5) and (5.31)]. In the following we compute the reflection matrix R' in the X,Y,Z coordinates defined as

$$\begin{bmatrix} L' \\ M' \\ N' \end{bmatrix} = \begin{bmatrix} R' \end{bmatrix} \begin{bmatrix} L \\ M \\ N \end{bmatrix}. \tag{A.1}$$

Equations (5.9) and (5.10) relate the x,y,z and X,Y,Z coordinates. If we define a matrix T and its inverse T^{-1}:

$$\begin{bmatrix} x - 1/2 \\ y - 1/2 \\ z \end{bmatrix} = \begin{bmatrix} T^{-1} \end{bmatrix} \begin{bmatrix} X \\ Y \\ Z \end{bmatrix}, \tag{A.2}$$

$$\begin{bmatrix} X \\ Y \\ Z \end{bmatrix} = \begin{bmatrix} T \end{bmatrix} \begin{bmatrix} x - 1/2 \\ y - 1/2 \\ z \end{bmatrix}, \tag{A.3}$$

then, from Eqs.(5.9) and (5.10)

$$T = \begin{bmatrix} 1/\sqrt{3} & 1/\sqrt{3} & 1/\sqrt{3} \\ -1/\sqrt{2} & 1/\sqrt{2} & 0 \\ -1/\sqrt{6} & -1/\sqrt{6} & 2/\sqrt{6} \end{bmatrix}, \tag{A.4}$$

$$T^{-1} = \begin{bmatrix} 1/\sqrt{3} & -1/\sqrt{2} & -1/\sqrt{6} \\ 1/\sqrt{3} & 1/\sqrt{2} & -1/\sqrt{6} \\ 1/\sqrt{3} & 0 & 2/\sqrt{6} \end{bmatrix}. \quad (A.5)$$

Since

$$\begin{bmatrix} L' \\ M' \\ N' \end{bmatrix} = \begin{bmatrix} T \end{bmatrix} \begin{bmatrix} \ell' \\ m' \\ n' \end{bmatrix} \quad (A.6)$$

$$\begin{bmatrix} \ell \\ m \\ n \end{bmatrix} = \begin{bmatrix} T^{-1} \end{bmatrix} \begin{bmatrix} L \\ M \\ N \end{bmatrix} \quad (A.7)$$

hold, we obtain the reflection matrices for the four cases in the text as follows:

Case 1.

$$R' = TR_{zc}R_yR_xT^{-1} = \begin{bmatrix} -\cos 2\delta & 0 & -\sin 2\delta \\ 0 & -1 & 0 \\ +\sin 2\delta & 0 & +\cos 2\delta \end{bmatrix} \quad (A.8)$$

Equations (A.8) and (5.8) give Eq.(5.34) in the text.

Case 2.

$$R' = TR_yR_xR_{zc}T^{-1} = \begin{bmatrix} -\cos 2\delta & 0 & +\sin 2\delta \\ 0 & -1 & 0 \\ -\sin 2\delta & 0 & -\cos 2\delta \end{bmatrix} \quad (A.9)$$

Equations (A.9) and (5.8) give Eq.(5.35) in the text.

Case 3.

$$R' = TR_yR_{zc}R_xT^{-1}$$

(equation continues)

$$= \begin{bmatrix} -\frac{2}{3}-\frac{1}{3}\cos 2\delta & -\frac{1}{\sqrt{3}}\sin 2\delta & \frac{\sqrt{2}}{3}-\frac{\sqrt{2}}{3}\cos 2\delta \\ \frac{1}{\sqrt{3}}\sin 2\delta & -\cos 2\delta & \frac{2}{\sqrt{6}}\sin 2\delta \\ \frac{\sqrt{2}}{3}-\frac{\sqrt{3}}{3}\cos 2\delta & -\frac{2}{\sqrt{6}}\sin 2\delta & -\frac{1}{3}-\frac{2}{3}\cos 2\delta \end{bmatrix}. \quad \text{(A.10)}$$

Equations (A.10) and (5.8) give Eq.(5.36) in the text.

Case 4.

$$R' = TR_xR_{zc}R_yT^{-1}$$

$$= \begin{bmatrix} -\frac{2}{3}-\frac{1}{3}\cos 2\delta & \frac{1}{\sqrt{3}}\sin 2\delta & \frac{\sqrt{2}}{3}-\frac{\sqrt{2}}{3}\cos 2\delta \\ -\frac{1}{\sqrt{3}}\sin 2\delta & -\cos 2\delta & -\frac{2}{\sqrt{6}}\sin 2\delta \\ -\frac{\sqrt{2}}{3}-\frac{\sqrt{2}}{3}\cos 2\delta & \frac{2}{\sqrt{6}}\sin 2\delta & -\frac{1}{3}-\frac{2}{3}\cos 2\delta \end{bmatrix}. \quad \text{(A.11)}$$

Equations (A.11) and (5.8) give Eq.(5.37) in the text.

References

(1) Okoshi, T., and Yano, A., Reduced-Information Projection-Type Holography Using a Horizontally Direction-Selective Stereoscreen, *Opt. Commun. 3*, No. 2, 85-88 (April 1971).

(2) Okoshi, T., Projection-Type Holography, *Progr. Optics* (to appear).

(3) Ives, H. E., Parallax Panoramagrams for Viewing by Reflected Light, *J. Opt. Soc. Amer. 20,* 585-592 (Oct. 1930).

(4) Yano, A., Sayanagi, K., and Okoshi, T., Three-Dimensional Displays Using Various Screens, *Proc. Soc. Photoopt. Instrum. Eng.,* Japan Seminar (June 26-27, 1972).

(5) Valyus, N. A., "Stereoscopy." Focal Press, New York, 1966.

(6) Okoshi, T., Yano, A., and Fukumori, Y., Curved Triple Mirror Screen for Projection-Type Three-Dimensional Display, *Appl. Opt. 10,* No. 3, 482-489 (March 1971).

(7) Burckhardt, C. B., Collier, R. J., and Doherty, E. T., Formation and Inversion of Pseudoscopic Images, *Appl. Opt. 7,* No. 3, 627-631 (March 1968).

(8) Okoshi, T., Optimum Design and Depth Resolution of Lens-Sheet and Projection-Type Three-Dimensional Displays, *Appl. Opt. 10,* No. 10, 2284-2291 (Oct. 1971).

(9) Saint Genie, De Lassus, U.S. Patent No. 2,139,855 (1934).

(10) Ives, H. E., Reflecting Screens for Relief Picture Projection, *J. Opt. Soc. Amer. 21,* 109-118 (Feb. 1931).

(11) Okoshi, T., and Yoshida, M., Design and Construction of a Projection-Type Three-Dimensional Display, National Convention of the Institute of Television Engineers of Japan, Paper No. 10-14, Oct. 17, 1969.

(12) Yoder, P. R., Jr., Study of Light Deviation Errors in Triple Mirrors and Tetrahedral Prisms, *J. Opt. Soc. Amer. 48,* No. 7, 496-499 (July 1958).

(13) Fujii, Y., and Nakajima, K., Analysis of Optical Corner Reflectors, *Seisan Kenkyu* (monthly journal of the Institute of Industrial Science, University of Tokyo; in Japanese) *21,* No. 11, 611-616 (Nov. 1969).

(14) Chang, R. F., Currie, D. G., and Alley, C. O., Far-Field Diffraction Pattern for Corner Reflectors with Complex Reflection Coefficient, *J. Opt. Soc. Amer. 61,* No. 4, 431-438 (April 1971).

Holography

In this chapter, an outline of the holographic technique is presented. Emphasis is placed upon its features as a three-dimensional imaging technique. Since many good textbooks are now available on holography, detailed discussions on its technological aspects are omitted. Instead, efficiency, resolution, speckle noise, white-light recording, white-light reconstruction, and color holography will be described in more detail.

6.1 INTRODUCTION

6.1.1 FEATURES OF HOLOGRAPHY AS A THREE-DIMENSIONAL IMAGING TECHNIQUE

This subject has been discussed a little in Section 2.6.4. A more complete listing of the features will be as follows.

(1) Holography gives satisfactory three-dimensional sensation not only through binocular parallax and convergence but also through accommodation.

(2) In holographic recording of an optical image, the information of a spot on the image is "distributed" or "scattered" over a wide area of the hologram. In other words, a hologram is equivalent to a kind of integral transform of the image. Therefore, a scar or dust spot on the hologram never damages a specific point upon the image, but only degrades the overall image quality a little. This means that holography is a relatively invulnerable recording scheme.

(3) As a direct consequence of the above feature, holography enables us to store a huge amount of information in a limited area.

(4) The proportionality between the object-wave intensity and reconstructed-image intensity (overall linearity of the process) can be good provided that the reference-wave level is set high, even though the linearity of ordinary recording materials is rather poor. Moreover, that range of linearity (dynamic range) is wide if the noise level is properly suppressed.

6.1.2 DRAWBACKS OF HOLOGRAPHY AS A THREE-DIMENSIONAL IMAGING TECHNIQUE

This subject has also been discussed in Section 2.6.4. A more complete listing will be as follows.

A. Drawbacks Due to the Use of Coherent (or Monochromatic) Light in Recording and Reconstruction

(A-1) A dark room is needed for recording a hologram.

(A-2) The object should be still.

(A-3) Usually a human is not the object of holography for two reasons. First, he can hardly be still for the necessary exposure time except when a special pulse laser is used. Second, laser-light illumination is often dangerous for human eyes.

(A-4) From ordinary holograms only monochromatic images can be reconstructed.

(A-5) At present, recording of large objects and reconstruction of a large image are not practical because very high power lasers are not available. CW lasers with good coherence can deliver at most 1 or 2 W; even such medium-power lasers are usually very expensive.

(A-6) Speckle noise is present (see Section 6.9).

(A-7) Modulation noise is present (see Section 6.7.1).

(A-8) In the reconstruction process, efficiency is not usually very high. In most practical cases it is below 20% for a phase hologram and below 3% for an amplitude hologram.

B. Drawbacks Due to Causes Other Than the Use of Coherent Light

(B-1) At present, high-resolution recording material (photographic plate or film) is very expensive; a large-size hologram is not practical for most purposes because of its cost. In ordinary holography, the hologram size gives the viewing zone. When a wide viewing zone is required, therefore, the cost of the recording material becomes prohibitive.

(B-2) The amount of information recorded in ordinary holograms is usually tremendous. In most cases it is much larger than that really necessary to record the three-dimensional image; in other words, it is very redundant. Hence, holography is not an economical way to record or transmit three-dimensional image information.

Since 1964, much work has been performed to overcome the above difficulties. These efforts will be mentioned briefly in the following, together with the difficulties to be overcome and section numbers where they are described in detail.

(1) *White-light recording* aims at overcoming (A-1), (A-2), and (A-3). See Section 6.10.

(2) *White-light reconstruction* aims at overcoming (A-4), (A-5), and (A-6). See Section 6.11.

(3) *Color holography* aims at overcoming (A-4) [plus sometimes (A-5) and (A-6)]. See Section 6.12.

(4) *Use of a pulse laser* aims at overcoming (A-2) and (A-3). See Refs.(39) and (40).

(5) *Reduced-information holography* aims at overcoming (B-2). See Section 7.8.

(6) *Projection-type holography* aims at overcoming (B-1) and (B-2). See Section 7.8.6. Also see Ref. (2) of Chapter 5.

6.1.3 PURPOSE OF THIS CHAPTER

The discussions in this chapter will cover the principle, classification, imaging theory, noise, light sources, and recording materials of holography. Emphasis will be placed upon various techniques and proposals for overcoming the above-mentioned drawbacks of holography as a three-dimensional imaging technique. Analysis and reduction techniques for the tremendous amount of information recorded in a hologram will be described separately in Chapter 7. Some special topics on holographic techniques, such as computer-generated holograms, holographic stereoscreens, etc., will be discussed in Chapter 8.

6.2 PRINCIPLE OF HOLOGRAPHY

The basic principle of holography has been described briefly in Chapter 2 in connection with the history of holography; in this section, the subject will be discussed again in a more quantitative manner.

Various imaging theories of holography have been presented in the past decade. They may be classified into three major groups:

(1) theories based upon the wave equation,
(2) theories based upon the wavefront concept,
(3) theories based upon geometrical optics.

In this section, a treatment based upon the wavefront concept will first be presented to show in a quantitative, but simple manner how a three-dimensional image can be reconstructed from a hologram.

An object O is illuminated by coherent light (laser light) in a setup as shown in Fig.6.1a. An observer just in front of the object is looking at the object. What he perceives is a large number of spherical wavefronts generated upon the surface of the object, or in other words, scattered components of the illuminating wave.

Since the illuminating light is coherent, the scattered light is also coherent when the object is still. Therefore, the instantaneous electric field strength of the scattered light at a point (x,y,z) in space is expressed as

$$E_O(x,y,z,t) = \mathrm{Re}\{\sqrt{2}O(x,y,z)\exp(j\omega t)\} \tag{6.1}$$

where the subscript O denotes "object" and ω is the angular frequency of the light. As in the conventional manner in field theory, the *phasor* O(x,y,z) representing the object wave will mainly be considered in the following discussions.

Suppose that a photographic plate is placed between the object O and the observer at z = 0 as shown in Fig.6.1b. The photographic plate is exposed to the object wave plus the secondary, uniform illuminating wave originally obtained from the same light source. We call that uniform illumination the reference light or reference wave, and let it be expressed as

$$E_R(x,y,z,t) = \mathrm{Re}\{\sqrt{2}R(x,y,z)\exp(j\omega t)\}. \tag{6.2}$$

The light intensity upon the plate is then expressed as

$$I(x,y) = \overline{|E_O(x,y,0,t)+E_R(x,y,0,t)|^2}$$

(equation continues)

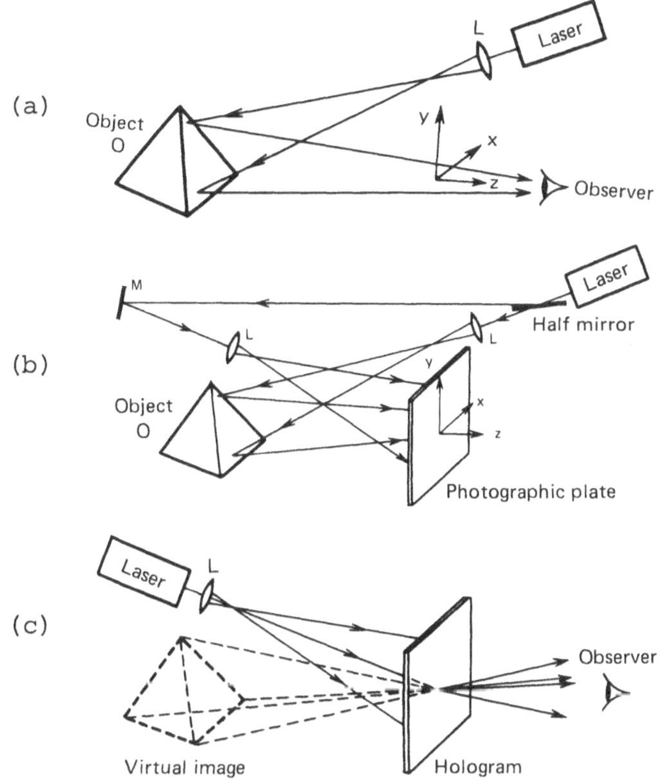

Fig.6.1 Principle of holography. (a) Observation of an object illuminated by a coherent light. (b) Recording of a hologram. (c) Observation of the image reconstructed from a hologram.

$$= |O(x,y)+R(x,y)|^2$$

$$= |O(x,y)|^2+|R(x,y)|^2+R^*O+RO^* \qquad (6.3)$$

In the above equations, a bar denotes time average and an asterisk denotes complex conjugation. If we write for simplicity, separating amplitude and phase

$$O(x,y) = O_0(x,y)\exp\{j\phi_0(x,y)\}, \qquad (6.4)$$

$$R(x,y) = R_0(x,y)\exp\{j\phi_R(x,y)\}, \qquad (6.5)$$

then we obtain

$$I(x,y) = \{O_0(x,y)\}^2 + \{R_0(x,y)\}^2 + 2R_0(x,y)O_0(x,y)\cos(\phi_0-\phi_R). \quad (6.6)$$

In any photographic recording the exposure time is very much longer than the period of the light wave. This means that a quantity corresponding to I(x,y) is recorded upon the plate. In Eq.(6.6) we see that the first and second terms correspond to the object-wave and reference-wave intensity and vary rather gradually with respect to position. However, the third term is proportional to the sinusoidal function of the phase difference between the two waves; it is readily understood that this term represents the interference fringe formed by the two waves. In reality this term plays the most important role in the image-reconstruction process, as will be seen later.

When the intensity of light incident upon a photographic plate increases, the transmittance of the plate obtained after exposure and development decreases, except for very special cases when so-called reversal development is performed. Let us express, for simplicity, the above relation as

$$T(x,y) = T_0 + KI(x,y), \quad (6.7)$$

where T is the transmittance, T_0 is a constant, and K is a negative quantity representing the sensitivity of the plate. (The real characteristics of a photographic plate are described in Section 6.15.2.)

Suppose that the plate is exposed to the object and reference waves having intensity given as in Eq. (6.3), developed, and then illuminated to reconstruct the image as shown in Fig.6.1c by a light expressed as

$$R'(x,y) = R_0'(x,y)\exp\{j\phi_{R'}(x,y)\}. \quad (6.8)$$

In such a case the wavefront that appears at the right-hand side of Fig.6.1c is given, from Eqs.(6.3), (6.7), and (6.8), as

$$R'(x,y)T(x,y) = T_0R' + KR'|O|^2 + KR'|R|^2$$
$$+ KR'R^*O + KR'RO^*$$
$$= W_1 + W_2 + W_3 + W_4 + W_5. \qquad (6.9)$$

It is seen in Eq.(6.9) that the terms W_1, W_2, and W_3 represent those illuminating light components that are subject to no critical change of the wavefront but pass through the hologram with mere attenuations; these are called directly transmitting waves. The most important term is the fourth term W_4. If we assume for simplicity that R' = R, then we obtain

$$W_4 = K|R|^2O \propto O(x,y), \qquad (6.10)$$

that is, W_4 is proportional to the original wavefront of the object wave. Therefore, if we observe the plate from the right in Fig.6.1c, we shall perceive a reconstructed three-dimensional image at exactly the original position of the object, as shown in the figure. This is nothing but the direct image observed in ordinary holography. The fifth term W_5 represents the conjugate image described in Section 2.6.3.

6.3 VARIOUS HOLOGRAPHY TECHNIQUES

The process described above is merely an example of various holographic processes. In reality, there are many versions according to the kind of recording material used and shape of the wavefronts of the object and reference waves. These are listed in condensed form in Table 6.1. The items in the table will be described in detail.

TABLE 6.1 Various Holography Techniques

	Classification
With respect to thickness of recording material	(1) Two-dimensional hologram (plane hologram) (2) Three-dimensional hologram (volume hologram)
Type of fringes recorded	(1) Amplitude hologram (absorption hologram) (2) Phase hologram (3) Blazed hologram (4) Complex-amplitude hologram
Shape of object wavefront	(1) Fresnel hologram (2) Fraunhofer hologram (3) Lensless Fourier-transform hologram (4) Image hologram

6.3.1 *Classification with Respect to the Thickness of Recording Material*

A. *Two-Dimensional Hologram*

The fringes of a hologram are recorded as a two-dimensional pattern upon the emulsion of the plate, as described in the preceding section. When the emulsion is thin, the hologram must be two-dimensional. Two-dimensional holograms are further classified as two-dimensional amplitude holograms and two-dimensional phase holograms (see the next subsection).

B. *Three-Dimensional Hologram*

A photographic plate having a thick emulsion layer is used, and the interference fringes are recorded in

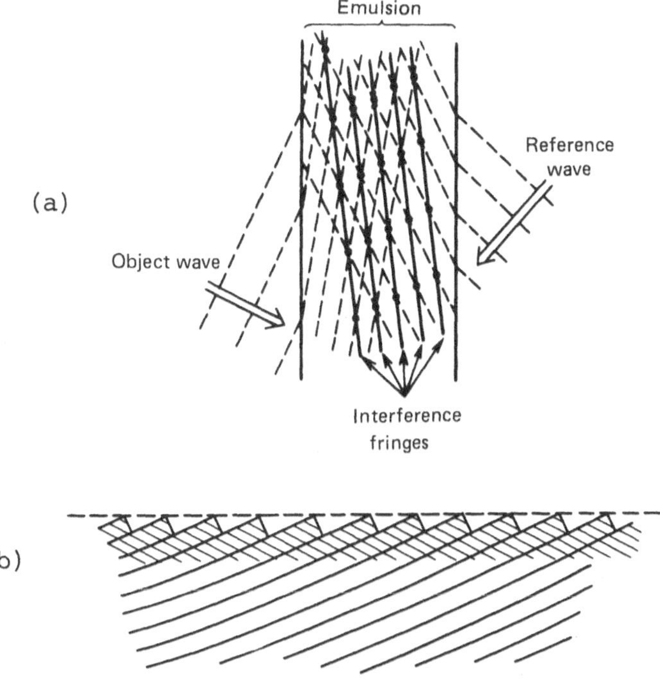

Fig.6.2(a) Lippmann hologram (volume hologram).
(b) Blazed hologram.

the three-dimensional medium. In most cases the object and reference waves are incident upon the hologram plate from opposite sides, as shown in Fig.6.2a.

Since the interference fringes formed in the emulsion are similar to those in Lippmann's color photography, this type of hologram is sometimes called a *Lippmann hologram*. When the image is reconstructed from a volume hologram, the so-called Bragg diffraction effect is used, as will be described in Section 6.6. Consequently, the volume hologram is sometimes also called a *Lippmann-Bragg* hologram, or even a *Bragg-type* hologram.

Bragg diffraction has a selectivity in wavelength as described in Section 6.6. This enables the use of the volume hologram in white-light reconstruction of three-dimensional color images, which will be discussed in Section 6.13.

6.3.2 CLASSIFICATION WITH RESPECT TO THE TYPE OF RECORDED FRINGES

A. Amplitude Hologram

The interference pattern between the object and reference waves in an amplitude hologram is recorded as the amplitude transmittance or the amplitude reflection factor of the recording medium. In most cases amplitude transmittance is used; use of the reflection factor is exceptional.

In reconstruction, the wavefront of the illuminating beam is "spatially amplitude-modulated" with respect to position upon the plate, and a component of the resulting diffracted waves reconstructs the image. The maximum theoretical efficiency (the reconstructing light energy divided by the illuminating light energy) is 6.25% (see Section 6.7).

B. Phase Hologram

The interference fringes in a phase hologram are recorded as the variation of the thickness or refractive index of the recording medium. In the latter case the hologram may either be two-dimensional or of volume type. In reconstruction, the wavefront of the illuminating beam is "spatially phase-modulated" with respect to position upon the plate, and the resulting diffraction gives the image. The greatest advantage over the amplitude hologram is that energy dissipation of light in the hologram is essentially absent. The maximum theoretical efficiency is 34% in a two-dimensional phase hologram and 100% in a three-dimensional phase hologram.

When an ordinary silver-halide photographic plate is used, the phase hologram is obtained by "bleaching" the developed silver particles in the gelatin (1).

C. Blazed Hologram

A blazed hologram may be called a reflection-type version of the phase hologram. The surface of the photosensitive material is shaped in a saw-tooth waveform as shown in Fig.6.2b using, for example, the photoresist technique. The shaped surface is then metallized and used as a reflection-type diffraction grating. When the surface is properly shaped, the diffraction efficiency approaches 100%. An efficiency as high as 73% has actually been obtained (2).

D. Complex-Amplitude Hologram

In this case both the amplitude and phase modulations contribute to diffraction for image formation. In actual amplitude holograms using silver halide recording material, the obtained grating is of a more-or-less complex nature because both the refractive index and the thickness of the emulsion are also modulated in the development process.

6.3.3 CLASSIFICATION WITH RESPECT TO THE SHAPE OF THE OBJECT WAVEFRONT

A. Fresnel Hologram

A hologram obtained by using an ordinary optical setup as shown in Fig.6.3a is called a Fresnel hologram. This term is derived from Fresnel diffraction, which denotes a diffraction pattern obtained at a finite distance from the aperture. (Refer to the third paragraph of Appendix I of this chapter.) If the object

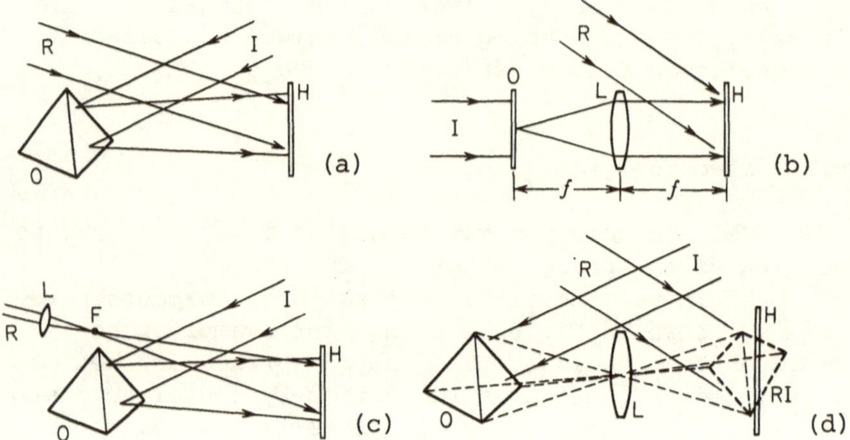

Fig.6.3(a) Fresnel hologram. (b) Fraunhofer hologram. (c) Lensless Fourier-transform hologram. (d) Image hologram. (R: reference wave; I: illuminating wave; O: object; H: hologram; L: lens; f: focal length; RI: real image.)

in Fig.6.3a is a two-dimensional transparency, the object wave arriving at the hologram plane will be the Fresnel diffraction pattern of the object.

B. Fraunhofer Hologram

When the distance a between the object and the hologram plate is sufficiently long as compared with the object size w so that $a \gg w^2/\lambda$ (λ is wavelength) holds, the obtained hologram is called a Fraunhofer hologram. This term is derived from Fraunhofer diffraction, denoting a diffraction pattern obtained at an infinite distance from an aperture (also called far-field diffraction pattern). If the object is a two-dimensional transparency, and the distance between the object and the hologram plate is effectively infinite in Fig.6.3a, the object wave arriving at the hologram plate gives the Fraunhofer diffraction pattern of the object.

When $w = 10$ cm and $\lambda = 500$ nm, the above condition gives $a \gg 40{,}000$ m. This example suggests that the above condition is far from practical. In practice, therefore, the Fraunhofer diffraction pattern (far-field pattern) is obtained equivalently by using a convex lens as shown in Fig.6.3b.

In the arrangement shown in Fig.6.3b, the separation between the object O and lens L, and that between the lens L and hologram plate H are both equal to the focal length of the lens f. In such an optical setup, the object O is seen at infinite distance from the hologram plate H, and H is also seen at infinite distance from O. Hence, the equivalent optical distance between H and O is infinite, and the Fraunhofer condition can be satisfied without difficulty.

A Fraunhofer hologram is sometimes called a Fourier-transform hologram because when the object is two-dimensional, the amplitude of the object wave upon the hologram plate is proportional to the Fourier-transform of the original amplitude distribution upon the object. This relation will be proved in Appendix I of this chapter.

C. *Lensless Fourier-Transform Hologram*

When a spherically diverging reference wave is used as shown in Fig.6.3c, and its point source F is located at the same distance from the hologram plate H as the median plane of the object O, the hologram produced is called a lensless Fourier-transform hologram. In such a case, although the object wave gives a Fresnel diffraction pattern, the interference-fringe pattern produced on the hologram plate is approximately identical to what is produced by a Fraunhofer diffraction pattern and a parallel reference wave. Therefore, a Fraunhofer hologram (Fourier-transform hologram) can be recorded at a finite distance and without use of a lens.

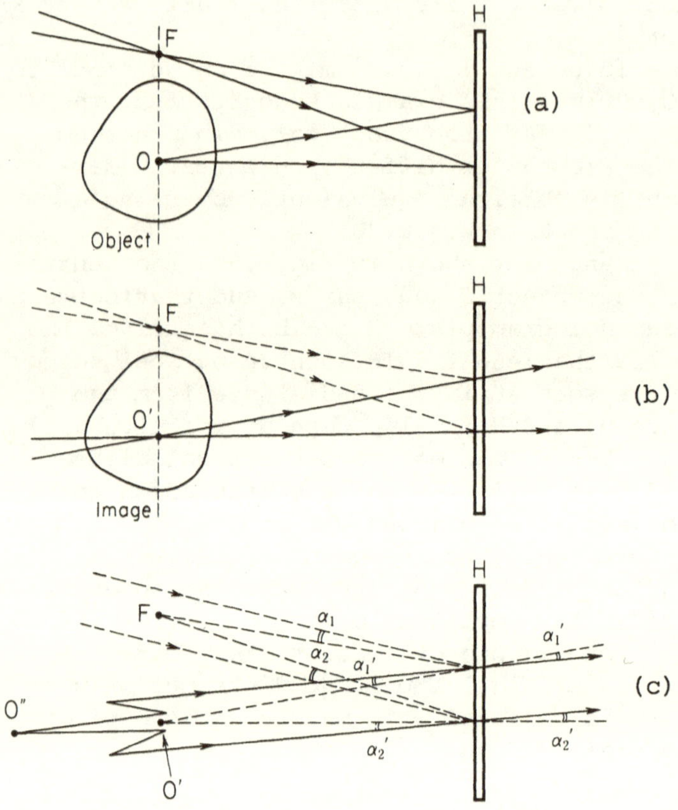

Fig.6.4 Equivalence of a lensless Fourier-transform hologram and a Fraunhofer hologram recorded by using a parallel reference beam.

The simplest proof of the above statement will be as follows. We consider a representative point O on the object and assume that O and F are of equal distance from H, as illustrated in Fig.6.4a. In reconstruction, when an illuminating beam diverging from F is used, the reconstructed point O' is observed at exactly the same position as O (see Fig.6.4b). If we use, instead of the diverging beam, a parallel illuminating beam from the direction of F as shown in Fig.6.4c, since $\alpha_1' = \alpha_1$ and $\alpha_2' = \alpha_2$ holds in the "paraxial" approximation, a point image O'' will be reconstructed at infinite distance. This means that the hologram obtained in a setup like Fig.6.4a is equivalent to the Fraunhofer hologram recorded using a parallel reference beam.

[We may give a more direct proof of the above statement if we rely upon the equations to be derived later. Putting $\mu = m = 1$, $z_O = z_R$ and $z_C = \infty$ into Eqs.(6.35a) and (6.36a), we immediately obtain $z_1 = \infty$ and $z_2 = \infty$, which are the relations to be proved. For the symbols used, refer to Sections 6.4.1 and 6.4.2.]

D. Image Hologram

When a convex lens is inserted between the object O and the hologram plate H as shown in Fig.6.3d so that the real image of O is produced in the vicinity of H, the hologram will record that real image instead of the object itself.

In reconstruction from such a hologram the image is observed in the vicinity of the plate; this fact decreases the color dispersion in the reconstructed image and enables us to use white light. On the other hand, when the object is two-dimensional, the image hologram is similar to an ordinary photograph and some of the features of holography are entirely lost.

The image hologram will be discussed in more detail in Section 6.12.

6.3.4 APPLICATIONS OF HOLOGRAPHY

Another classification of holography techniques can be made according to application. Table 6.2 is a rather simplified, short-form table showing various applications of holography. Detailed discussions on individual applications will be omitted due to space restrictions. However, it should be emphasized that

TABLE 6.2 Applications of Holography

Hologram interferometry	(1) Real-time interferometry (2) Double-exposure interferometry (3) Interferometry of vibrating surfaces (4) Contour generation
Microscopy and related imaging techniques	(1) Microscopy (2) Observation of aerosol particles
Digital information storage	(1) Holographic memories using thin media (2) Holographic memories using thick media (3) Associated-readout memory
Image processing	(1) Image improvement (2) Noise removal (3) Differentiation (4) Integration (5) Integral transforms
Image storage and display	(1) Two-dimensional imaging (2) Three-dimensional imaging
Nonoptical holography techniques	(1) Acoustical holography (2) Microwave holography (3) Computer-generated holograms

there are many interesting and promising applications of holography in addition to imaging which is the main subject of this book.

6.4 THEORY OF IMAGING IN A HOLOGRAPHIC SYSTEM

In Section 6.2, the basic principle of imaging in holography was described in the simplest way by

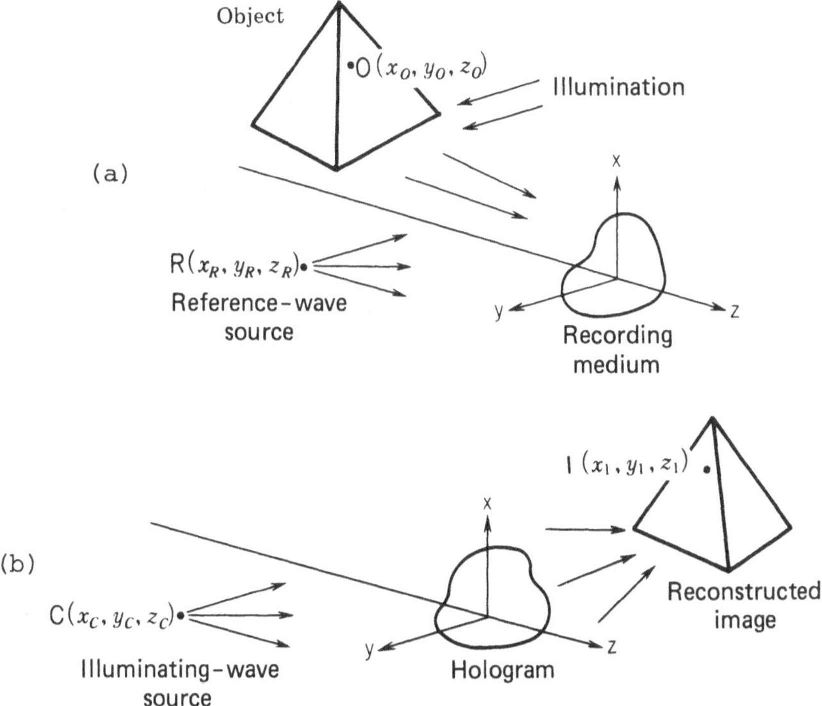

Fig.6.5 Optical arrangements for recording of and reconstruction from a hologram. (a) Recording. (b) Reconstruction.

assuming that the reference and the illuminating waves have wavefronts identical to each other. In practice, however, the above condition is not always satisfied. The wavelengths of the recording and illuminating lights may not be equal. As a result, the image is not always reconstructed at exactly the original position, but generally both its size and position are different from those of the original.

In this section, the imaging properties described above (except for aberration) are discussed for a two-dimensional amplitude hologram according to Meier's treatment based upon the wavefront concept (3)-(5).

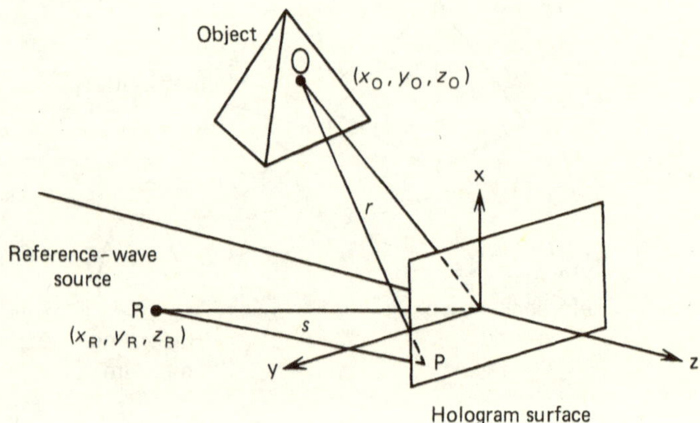

Fig.6.6 Symbols used in the analysis.

6.4.1 BASIC EQUATIONS

We consider a recording system as shown in Fig. 6.5a, in which an object represented by a point $O(x_O, y_O, z_O)$ is illuminated and its hologram recorded using a reference wave having its point source at $R(x_R, y_R, z_R)$. In reconstruction, the source of the illuminating wave is located at $C(x_C, y_C, z_C)$, and the resulting image is produced at $I(x_1, y_1, z_1)$ as illustrated in Fig.6.5b.

We first consider the recording process. Suppose that the hologram plate is located at $z = 0$, and let the coordinates of a point P upon the hologram be $(x,y,0)$. We use symbols $\overline{OP} = r$ and $\overline{RP} = s$ as shown in Fig.6.6. Then we obtain

$$r = \sqrt{(x-x_O)^2 + (y-y_O)^2 + z_O^2} \tag{6.11}$$

$$s = \sqrt{(x-x_R)^2 + (y-y_R)^2 + z_R^2}. \tag{6.12}$$

If the instantaneous amplitudes of the object and reference waves at P are denoted by $O(x,y,t)$ and $R(x,y,t)$, respectively, we may write

$$O(x,y,t) = (O_0/r)\exp(j\phi_O + j\omega t) \tag{6.13}$$

$$R(x,y,t) = (R_0/s)\exp(j\phi_R + j\omega t), \tag{6.14}$$

$$\phi_O = -(2\pi/\lambda)r + \psi_O \tag{6.15}$$

$$\phi_R = -(2\pi/\lambda)s + \psi_R, \tag{6.16}$$

where ω and λ are the angular frequency and wavelength of the recording light, respectively, and O_0, R_0, ψ_O, ψ_R are constants. From Eqs.(6.13) and (6.14), the resultant amplitude at P is given as

$$H(x,y,t) = O + R$$

$$= \left[(O_0/r)\exp(j\phi_O) + (R_0/s)\exp(j\phi_R)\right]\exp(j\omega t). \tag{6.17}$$

As described in Section 6.2, what is recorded upon the hologram plate is the total intensity of light:

$$I(x,y,t) = \overline{|H(x,y,t)|^2}\,. \tag{6.18}$$

In the discussion of Section 6.2, the amplitude transmittance obtained after the development was assumed as in Eq.(6.7); however, in the present discussion we will use a further simplified relation:

$$T(x,y) = KI(x,y) \tag{6.19}$$

where $K > 0$. We can prove without difficulty that the above simplification can be made without great loss of generality. From Eqs.(6.17)-(6.19), we obtain

$$T(x,y) = K\Big(\left[O_0{}^2/r^2\right] + \left[R_0{}^2/s^2\right] + \left[O_0R_0/rs\right] \left[\exp\{j(\phi_O-\phi_R)\} + \exp\{-j(\phi_O-\phi_R)\}\right]\Big). \tag{6.20}$$

Next we consider the reconstruction process. The hologram produced by the above method is illuminated by light having a wavelength of $\mu\lambda$ and diverging from point $C(x_C, y_C, z_C)$, as shown in Fig.6.5b. Then, at the left-hand side of the hologram the amplitude of the illuminating light is given as

$$C(x,y,z=-0,t) = (C_0/\ell)\exp(j\phi_C)+j(\omega/\mu)t, \tag{6.21}$$

$$\phi_C = -(2\pi/\mu\lambda)\ell+\psi_C, \tag{6.22}$$

$$\ell = \sqrt{(x-x_C)^2+(y-y_C)^2+z_C^2}. \tag{6.23}$$

Therefore, at the right-hand side of the hologram, the amplitude is

$$\begin{aligned} C(x,y,z=+0,t) &= T(x,y)C(x,y,z=-0,t) \\ &= K\left(\frac{O_0^2}{r}+\frac{R_0^2}{s}\right)\frac{C_0}{\ell}\exp\left[j\phi_C+j(\omega/\mu)t\right] \\ &+\frac{KO_0R_0C_0}{rs\ell}\exp\left[j(\phi_O-\phi_R+\phi_C)+j(\omega/\mu)t\right] \\ &+\frac{KO_0R_0C_0}{rs\ell}\exp\left[j(-\phi_O+\phi_R+\phi_C)+j(\omega/\mu)t\right] \end{aligned} \tag{6.24}$$

6.4.2 DIRECT AND CONJUGATE IMAGES

Equation (6.24) is much too complicated to allow an immediate physical interpretation. Therefore, we first investigate the physical implications of each term for the simplest case. If we set $\mu = 1$ and $C = R$ and further assume that both C and R are located very far so that $s \doteqdot$ const and $\ell \doteqdot$ const holds, then

$$\text{the second term of Eq.(6.24)} \propto (O_0/r)\exp(j\phi_O+j\omega t), \tag{6.25}$$

the third term of Eq.(6.24) $\propto \left(O_0/r\right)\exp\left(-j\phi_0+j\omega t\right).$ (6.26)

Equation (6.25) represents a wavefront entirely identical to that of Eq.(6.13). This means that a virtual image entirely identical to the object (except for the brightness level) is reconstructed. This is called a *direct image*.

The wavefront represented by Eq.(6.26) has an inverse phase distribution with respect to position as compared with the object wave; this corresponds to the conjugate image mentioned in Sections 2.6.3 and 6.2.

The first term in Eq.(6.24) represents the directly transmitting wave as is immediately understood by comparing that term with W_2 and W_3 in Eq.(6.9).

6.4.3 *AMPLITUDE HOLOGRAM AS AN IMAGING SYSTEM*

Since we are now familiar with the physical implications of Eq.(6.24), let us next investigate in more quantitative detail what is meant by that equation. In the following discussions, we assume for simplicity the so-called paraxial approximation; that is, points O, R, and C are all located relatively close to the axis so that the following approximations hold:

$$r = z_O\left\{1+\frac{(x-x_O)^2+(y-y_O)^2}{z_O^2}\right\}^{1/2}$$

$$\simeq \frac{x^2+y^2-2x_Ox-2y_Oy}{2z_O}+z_O+\frac{x_O^2+y_O^2}{2z_O} \tag{6.27}$$

$$s \simeq \frac{x^2+y^2-2x_Rx-2y_Ry}{2z_R}+z_R+\frac{x_R^2+y_R^2}{2z_R}. \tag{6.28}$$

Using the above approximations, we may write

$$\phi_O \simeq -\frac{2\pi}{\lambda}\left\{\frac{1}{2z_O}(x^2+y^2)-\frac{x_O}{z_O}x-\frac{y_O}{z_O}y\right\}+\text{const} \tag{6.29}$$

$$\phi_O-\phi_R \simeq -\frac{2\pi}{\lambda}\left\{\frac{1}{2}\left(\frac{1}{z_O}-\frac{1}{z_R}\right)(x^2+y^2) -\left(\frac{x_O}{z_O}-\frac{x_R}{z_R}\right)x-\left(\frac{y_O}{z_O}-\frac{y_R}{z_R}\right)y\right\}+\text{const.} \tag{6.30}$$

For generality in the following discussions, we now assume that the hologram is enlarged with a ratio m before it is used in reconstruction of the image. Since the transmittance of the new enlarged hologram T'(x,y) is given as T(x/m, y/m), the phase distributions of the direct and conjugate images, ϕ_1 and ϕ_2, can be written as

$$\begin{aligned}\Phi_1 &= \phi_O\left(\frac{x}{m},\frac{y}{m}\right)-\phi_R\left(\frac{x}{m},\frac{y}{m}\right)+\phi_C(x,y)\\ &\simeq -\frac{2\pi}{\mu\lambda}\left\{\frac{1}{2}\left(\frac{\mu}{m^2z_O}-\frac{\mu}{m^2z_R}+\frac{1}{z_C}\right)(x^2+y^2)\right.\\ &\quad\left.-\left(\frac{\mu x_O}{mz_O}-\frac{\mu x_R}{mz_R}+\frac{x_C}{z_C}\right)x-\left(\frac{\mu y_O}{mz_O}-\frac{\mu y_R}{mz_R}+\frac{y_C}{z_C}\right)y\right\}+\text{const}\end{aligned} \tag{6.31}$$

$$\begin{aligned}\Phi_2 &\simeq -\frac{2\pi}{\mu\lambda}\left\{\frac{1}{2}\left(-\frac{\mu}{m^2z_O}+\frac{\mu}{m^2z_R}+\frac{1}{z_C}\right)(x^2+y^2)\right.\\ &\quad\left.-\left(-\frac{\mu x_O}{mz_O}+\frac{\mu x_R}{mz_R}+\frac{x_C}{z_C}\right)x-\left(-\frac{\mu y_O}{mz_O}+\frac{\mu y_R}{mz_R}+\frac{y_C}{z_C}\right)y\right\}+\text{const.}\end{aligned} \tag{6.32}$$

Note that both Eqs.(6.31) and (6.32) have the form $A(x^2+y^2)+Bx+Cy$. As seen in Eq.(6.29), such a form represents nothing but the phase distribution of a spherical wavefront. Therefore, the sources or focuses of the spherical waves represented by Eqs. (6.31) and (6.32) give the positions of the direct and conjugate images, respectively. If the wavefront represents a diverging spherical wave, its source will be found on the same side as C and it gives a virtual image. If the wavefront represents a converging spherical wave, its focus will be found on the side opposite C and it gives a real image.

If we write the positions of the above-mentioned direct and virtual images as (x_1,y_1,z_1) and (x_2,y_2,z_2), respectively, we obtain by comparing Eqs.(6.31), (6.32) and (6.29),

$$z_1 = \frac{1}{\mu/m^2 z_O - \mu/m^2 z_R + 1/z_C}, \tag{6.33a}$$

$$x_1 = \frac{\mu x_O/m z_O - \mu x_R/m z_R + x_C/z_C}{\mu/m^2 z_O - \mu/m^2 z_R + 1/z_C}, \tag{6.33b}$$

$$y_1 = \frac{\mu y_O/m z_O - \mu y_R/m z_R + y_C/z_C}{\mu/m^2 z_O - \mu/m^2 z_R + 1/z_C}, \tag{6.33c}$$

$$z_2 = \frac{1}{-\mu/m^2 z_O + \mu/m^2 z_R + 1/z_C}, \tag{6.34a}$$

$$x_2 = \frac{-\mu x_O/m z_O + \mu x_R/m z_R + x_C/z_C}{-\mu/m^2 z_O + \mu/m^2 z_R + 1/z_C}, \tag{6.34b}$$

$$y_2 = \frac{-\mu y_O/m z_O + \mu y_R/m z_R + y_C/z_C}{-\mu/m^2 z_O + \mu/m^2 z_R + 1/z_C}. \tag{6.34c}$$

The above equations do not allow intuitive understanding of their physical implications; to facilitate this, we rewrite these equations as

$$\frac{m^2}{\mu z_1} - \frac{1}{z_O} = \frac{m^2}{\mu z_C} - \frac{1}{z_R}, \tag{6.35a}$$

$$\frac{x_1}{z_1} = \frac{\mu}{m}\frac{x_O}{z_O} + \left(\frac{x_C}{z_C} - \frac{\mu}{m}\frac{x_R}{z_R}\right), \tag{6.35b}$$

$$\frac{y_1}{z_1} = \frac{\mu}{m}\frac{y_O}{z_O} + \left(\frac{y_C}{z_C} - \frac{\mu}{m}\frac{y_R}{z_R}\right), \tag{6.35c}$$

$$\frac{m^2}{\mu z_2} + \frac{1}{z_O} = \frac{m^2}{\mu z_C} + \frac{1}{z_R}, \tag{6.36a}$$

$$\frac{x_2}{z_2} = -\frac{\mu}{m}\frac{x_O}{z_O} + \left(\frac{x_C}{z_C} + \frac{\mu}{m}\frac{x_R}{z_R}\right), \tag{6.36b}$$

$$\frac{y_2}{z_2} = -\frac{\mu}{m}\frac{y_O}{z_O} + \left(\frac{y_C}{z_C} + \frac{\mu}{m}\frac{y_R}{z_R}\right). \tag{6.36c}$$

We first consider Eqs.(6.35a) and (6.36a). If we assume temporarily that $\mu = m = 1$, then both of these equations are identical to the so-called lens formula. Therefore, we may say that a hologram is a peculiar lens which has two focal lengths: $f_1 = (1/z_C - 1/z_R)^{-1}$ [from Eq.(6.35a)] and $f_2 = (1/z_C + 1/z_R)$ [from Eq.(6.36a)]. When $m^2/\mu \neq 1$, the above statement is still valid if appropriate scalings are made for z_1 and z_2.

Next, the following statements may be derived from Eqs.(6.35b), (6.35c), (6.36b), and (6.36c).

(1) Generally the lateral and longitudinal magnifications are not equal to each other except when $\mu/m = 1$. The lateral magnification M_{lat}, longitudinal magnification M_{long}, and angular magnification M_{ang} are derived from the above four equations as follows:

For the direct image,

$$M_{lat} = \frac{dx_1}{dx_O} = \frac{dy_1}{dy_O} = \frac{\mu}{m}\frac{z_1}{z_O} = \frac{m}{1+(m^2/\mu)(z_O/z_C)-z_O/z_R}, \tag{6.37}$$

$$M_{long} = \frac{dz_1}{dz_O} = \frac{1}{\mu} M_{lat}^2, \tag{6.38}$$

$$M_{ang} = \frac{d(x_1/z_1)}{d(x_O/z_O)} = \frac{\mu}{m}. \tag{6.39}$$

For the conjugate image,

$$M_{lat} = -\frac{\mu}{m}\frac{z_2}{z_O} = \frac{m}{1-(m^2/\mu)(z_O/z_C)-z_O/z_R}, \tag{6.40}$$

$$M_{long} = \frac{1}{\mu} M_{lat}^2, \tag{6.41}$$

$$M_{ang} = -\frac{\mu}{m}. \tag{6.42}$$

(2) The terms in parentheses in Eqs.(6.35b), (6.35c), (6.36b), and (6.36c) represent constant displacement of the image in the x or y directions. This means that generally a holographic imaging system is equivalent to a "decentered" lens system. The condition for cancelling the displacement is given as follows:

For the direct image,

$$\frac{x_C}{z_C} = \frac{\mu}{m}\frac{x_R}{z_R}, \quad \frac{y_C}{z_C} = \frac{\mu}{m}\frac{y_R}{z_R}. \tag{6.43}$$

For the virtual image,

$$\frac{x_C}{z_C} = -\frac{\mu}{m}\frac{x_R}{z_R}, \quad \frac{y_C}{z_C} = -\frac{\mu}{m}\frac{y_R}{z_R}. \tag{6.44}$$

In most holographic processes $\mu = m = 1$ holds. Even in such simple cases the magnifications are not equal to unity but are dependent upon relations between z_O, z_C, and z_R. The holographic microscope proposed by Gabor in 1948 (see Section 2.6.1) took advantage of the angular magnification of 10^5 obtained for $m = 1$ and $\mu = 10^5$.

6.4.4 *PHASE HOLOGRAM AS AN IMAGING SYSTEM*

For phase holograms, we may derive imaging properties similar to those obtained above for amplitude holograms.

We now assume that a phase hologram gives to the illuminating beam in the reconstruction process a phase variation proportional to Eq.(6.18). Then the transfer function of the hologram may be expressed as

$$T(x,y) = \exp\left[j\phi_H(x,y)\right], \tag{6.45}$$

$$\phi_H(x,y) = \frac{\alpha}{2} I(x,y)$$

$$= \frac{\alpha}{2}\left(\frac{O_0^2}{r^2} + \frac{R_0^2}{s^2}\right) + \frac{\alpha O_0 R_0}{2rs} \left\{\exp\left[j(\phi_O - \phi_R)\right] + \exp\left[-j(\phi_O - \phi_R)\right]\right\}, \tag{6.46}$$

where α is a real constant. On the other hand, we note a mathematical relationship:

$$\exp\left\{j\frac{\beta}{2}[\exp(j\theta)+\exp(-j\theta)]\right\} = \exp\{j\beta\cos\theta\}$$
$$= \cos(\beta\cos\theta)+j\sin(\beta\cos\theta)$$
$$= J_0(\beta)+2\sum_{n=1}^{\infty}(-1)^n J_{2n}(\beta)\cos(2n\theta)$$
$$+2j\sum_{n=1}^{\infty}(-1)^{n+1}J_{2n-1}(\beta)\cos\{(2n-1)\theta\} \tag{6.47}$$

where $J_m(\beta)$ denotes the mth order Bessel function of the first kind. Hence, if we write $\alpha O_0 R_0/rs = \beta$, we obtain from Eqs. (6.45) and (6.46),

$$T(x,y) = \exp\left\{j\alpha\left(\frac{O_0{}^2}{r^2}+\frac{R_0{}^2}{s^2}\right)\right\}\left[J_0\left(\frac{\alpha O_0 R_0}{rs}\right)+\sum_{n=1}^{\infty}(-1)^n J_{2n}\right.$$
$$\times\left(\frac{\alpha O_0 R_0}{rs}\right)\left\{\exp\left[j2n(\phi_O-\phi_R)\right]+\exp\left[-j2n(\phi_O-\phi_R)\right]\right\}$$
$$+j\sum_{n=1}^{\infty}(-1)^{n+1}J_{2n-1}\left(\frac{\alpha O_0 R_0}{rs}\right)\left\{\exp\left[j(2n-1)(\phi_O-\phi_R)\right]\right.$$
$$\left.\left.+\exp\left[-j(2n-1)(\phi_O-\phi_R)\right]\right\}\right] \tag{6.48}$$

Careful comparison of the above computation with that performed for deriving Eqs. (6.25) and (6.26) from Eq. (6.20) for the amplitude hologram will suggest that the term $\exp\{j(\phi_O-\phi_R)\}$ in Eq. (6.48) contributes to the reconstruction of the direct image, whereas the term

$\exp\{-j(\phi-\phi_0)\}$ corresponds to the conjugate image. However, Eq.(6.48) has a large number of higher order terms, which in the reconstruction process generate many "ghost" images in addition to the genuine image desired.

The positions of the reconstructed images including the ghosts are computed as follows. If the terms in Eq.(6.48) including $\exp\{\pm jk(\phi_O-\phi_R)\}$ are denoted by T_k, they may be written as

$$T_k(x,y) = \begin{cases} (-1)^{k/2} J_k\left(\dfrac{\alpha O_0 R_0}{rs}\right)\exp\left\{j\alpha\left(\dfrac{O_0{}^2}{r^2}+\dfrac{R_0{}^2}{s^2}\right)\right\} \\ \times\{\exp[jk(\phi_O-\phi_R)]+\text{c.c.}\} & (6.49) \\ (-1)^{(k-1)/2} J_k\left(\dfrac{\alpha O_0 R_0}{rs}\right)\exp\left\{j\alpha\left(\dfrac{O_0{}^2}{r^2}+\dfrac{R_0{}^2}{s^2}\right)+j\dfrac{\pi}{2}\right\} \\ \times\{\exp[jk(\phi_O-\phi_R)]+\text{c.c.}\} & (6.50) \end{cases}$$

where c.c. denotes the complex conjugate, and Eqs. (6.49) and (6.50) should be applied for even and odd integers k, respectively. Computations similar to those performed for the amplitude hologram show that the positions of the kth order direct images reconstructed from a phase hologram are given as

$$z_{1k} = \frac{1}{k\mu/m^2z_O-k\mu/m^2z_R+1/z_C}, \tag{6.51a}$$

$$x_{1k} = \frac{k\mu x_O/mz_O-k\mu x_R/mz_R+x_C/z_C}{k\mu/m^2z_O-k\mu/m^2z_R+1/z_C}, \tag{6.51b}$$

$$y_{1k} = \frac{k\mu y_O/mz_O-k\mu y_R/mz_R+y_C/z_C}{k\mu/m^2z_O-k\mu/m^2z_R+1/z_C}. \tag{6.51c}$$

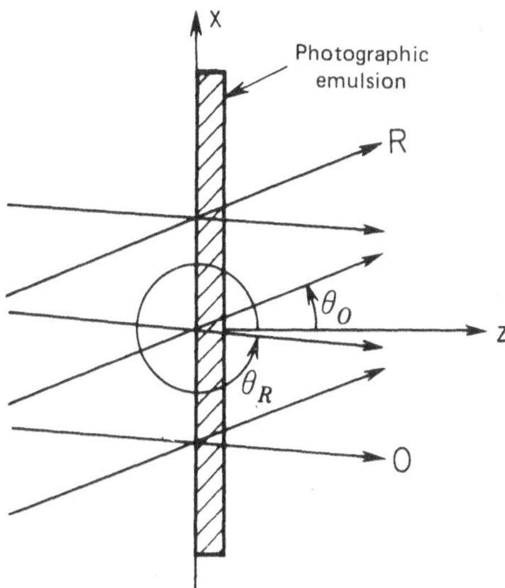

Fig.6.7 Symbols used in the analysis of the fringe formation.

The positions of the kth order conjugate images are obtained if z_O and z_R in the above three equations are replaced by $-z_O$ and $-z_R$, respectively.

6.5 INTERFERENCE FRINGES FORMED IN A HOLOGRAM

In this section, we investigate in further detail the configurations of the interference fringes formed *upon* a two-dimensional hologram or *in* a volume hologram.

6.5.1 BASIC EQUATIONS

For simplicity, we consider a case as shown in Fig.6.7 in which a plane object wave O and a plane reference wave R are incident upon a photographic plate with angles θ_O and θ_R, respectively. Both of the rays are assumed to lie in the xz plane. The assumption of plane waves is reasonable as long as a small portion of a hologram is to be considered. In the case of a Lippmann hologram, the reference wave is incident upon the hologram from the side opposite to the object wave. For such a case, however, assuming only $90° < \theta_R < 270°$ in Fig.6.7 is enough to make the following analysis applicable.

Let the unit vectors in the z and x directions be denoted by i and j. Then the vector r representing a point P(z,x) and the wavenumber vectors of the object and reference waves k_O and k_R are expressed as

$$r = iz + jx\,, \tag{6.52}$$

$$k_O = k(i\cos\theta_O + j\sin\theta_O)\,, \tag{6.53}$$

$$k_R = k(i\cos\theta_R + j\sin\theta_R)\,, \tag{6.54}$$

where $k = 2\pi/\lambda'$ and λ' denotes the reduced wavelength in the photographic emulsion. From the above equations the amplitudes of the object and reference waves at P(x,y) are given as

$$\begin{cases} O(z,x,t) = O_0 \exp\{-j\phi_O(z,x)+j\omega t\}\,, & (6.55) \\ R(z,x,t) = R_0 \exp\{-j\phi_R(z,x)+j\omega t\}\,, & (6.56) \end{cases}$$

$$\begin{cases} \phi_O(z,x) = k_O\cdot r = k(z\cos\theta_O + x\sin\theta_O)\,, & (6.57) \\ \phi_R(z,x) = k_R\cdot r = k(z\cos\theta_R + x\sin\theta_R)\,. & (6.58) \end{cases}$$

If the phase difference between the two waves is denoted by δ,

$$\delta(z,x) = kz(\cos\theta_R - \cos\theta_O) + kx(\sin\theta_R - \sin\theta_O)\,. \tag{6.59}$$

Evidently, the photographic emulsion is subject to the strongest photochemical reaction at positions where

$$\delta(x,y) = 2m\pi \quad (m\text{, an integer})\,. \tag{6.60}$$

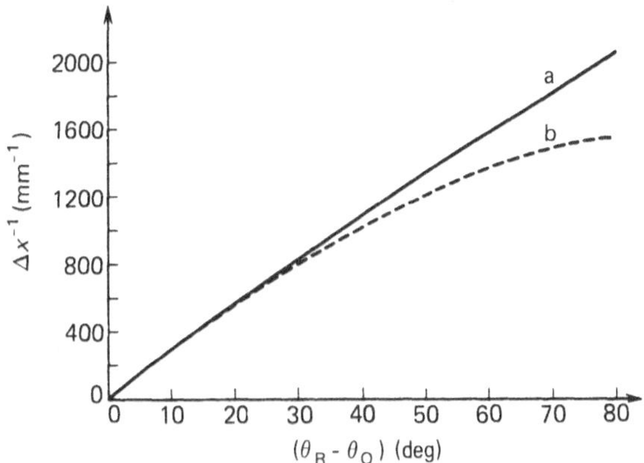

Fig.6.8 The fringe spacing given by Eq.(6.62) for λ = 632.8 nm. (a) When $\theta_R = -\theta_O(\Delta x^{-1} = 2\sin[(\theta_R-\theta_O)/2\lambda])$. (b) When $\theta_O = 0(\Delta x^{-1} = \sin[(\theta_R-\theta_O)/\lambda])$.

6.5.2 CASE OF A PLANE HOLOGRAM

If the emulsion thickness is much smaller than the wavelength (although this is not the case with ordinary photographic plates), the positions of the strongest photochemical reaction are obtained by equating Eqs.(6.59) and (6.60), and letting z = 0. The result is

$$x = 2m\pi/k(\sin\theta_R-\sin\theta_O). \tag{6.61}$$

Therefore, the spacing between closest fringes measured in the x direction is given as

$$\Delta x = \lambda/(\sin\theta_R-\sin\theta_O). \tag{6.62}$$

The above relation is shown for two typical cases in Fig.6.8 where the abscissa shows $\theta_R-\theta_O$. This figure shows that the fringe spacing Δx is principally determined by the angle made by the object and reference waves when $\theta < 45°$.

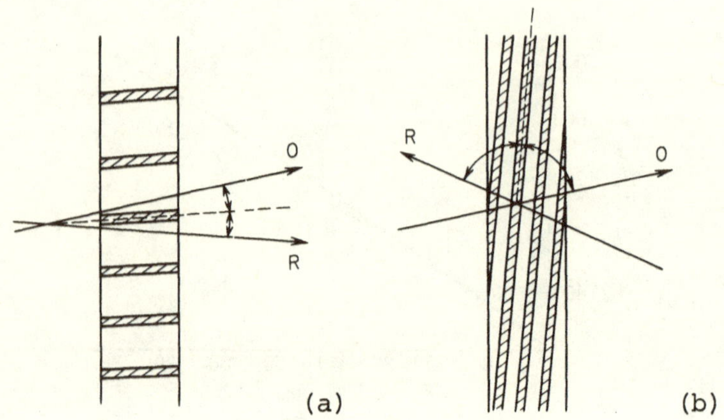

Fig.6.9 Direction of the interference fringes. (a) When the object and reference waves are incident from the same side. (b) When the object and reference waves are incident from the opposite sides.

6.5.3 CASE OF A VOLUME HOLOGRAM

High-resolution plates now being used in holography have thickness t ranging between 5 and 20 μm. On the other hand, Fig.6.8 shows that when λ = 632.8 nm, the fringe spacing $\Delta x \doteqdot 3$ μm for $\theta_R - \theta_O = 10°$ and $\Delta x \doteqdot 1.5$ μm for $\theta_R - \theta_O = 20°$. Therefore, the aforementioned assumption for the plane (two-dimensional) hologram, $\Delta x \gg t$, holds only when $\theta_R - \theta_O$ is exceptionally small. If we want to be entirely strict, we must treat a hologram as three-dimensional even when it is originally intended to be two-dimensional. However, to our surprise, the simplified, two-dimensional analysis usually gives satisfactory results when the object and reference waves are incident upon the plate from a common side. In such a case fringes are made almost perpendicular to the emulsion surface as shown in Fig.6.9a, and the obtained hologram is of transmission-type, as will be described in Section 6.6.2. (The thickness effect is still important in a transmission-type hologram when we consider multiple recording and wavelength selectivity.)

When the object and reference waves are incident from sides opposite each other, fringes are made almost in parallel with the emulsion surface as shown in Fig. 6.9b, and the obtained hologram is of reflection-type, as will be described in Section 6.6.2. In such a case three-dimensional analysis is needed. Equating Eqs. (6.59) and (6.60) immediately gives the fringe pattern as

$$x = \left[\frac{\cos\theta_O - \cos\theta_R}{\sin\theta_R - \sin\theta_O}\right] z + \frac{m\lambda}{\sin\theta_R - \sin\theta_O}, \tag{6.63}$$

where m is an integer. From this equation we may compute the angle α made by the fringes and the z axis as

$$\alpha = \tan^{-1}\left[\frac{\cos\theta_O - \cos\theta_R}{\sin\theta_R - \sin\theta_O}\right] = \frac{\theta_O + \theta_R}{2} \tag{6.64}$$

by using the trigonometric formulas

$$\cos x - \cos y = -2\sin\{(x+y)/2\}\sin\{(x-y)/2\},$$

$$\sin x - \sin y = +2\cos\{(x+y)/2\}\sin\{(x-y)/2\}.$$

Equation (6.64) shows that the volume interference fringes are always formed in a direction separated from the object and reference wave directions by equal angles.

When "almost paraxial" object and reference waves are used, $\alpha \simeq 0$ or 180° as seen in Fig.6.9a. On the contrary, in ordinary volume holograms $\theta_O \simeq 0$ whereas $\theta_R \simeq 180$; thus $\alpha \simeq 90°$ is obtained, and we see that fringes are formed almost in parallel with the emulsion surface.

The spacing between two neighboring fringes measured in the direction normal to the fringe planes is given as

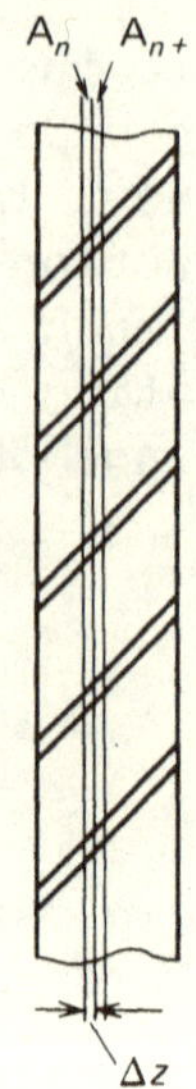

Fig.6.10 Basic concept of volume-hologram analysis.

$$d = \Delta x \cdot \cos\alpha = \frac{\lambda}{2\sin[(\theta_R - \theta_O)/2]} . \qquad (6.65)$$

In the case shown in Fig.6.9b, d is approximately equal to $\lambda/2$.

6.6 IMAGE RECONSTRUCTION FROM A VOLUME HOLOGRAM

So far we have discussed the recording of and reconstruction from various kinds of holograms; however, the reconstruction process from a volume hologram has not been considered.

To treat this subject in a strict manner, we first divide the volume hologram into a number of thin layers, as A_1, A_2, A_3,... shown in Fig.6.10, and then integrate the diffractions in all these layers. However, such a treatment is rather complex and does not allow direct insight into the problem. In this section, therefore, we derive the basic properties of Bragg diffraction in a more qualitative manner.

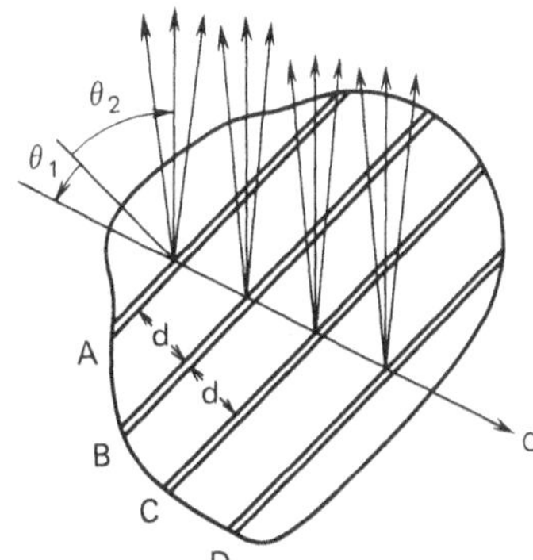

Fig.6.11 Conditions for the Bragg diffraction, I.

6.6.1 BASIC PROPERTIES OF BRAGG DIFFRACTION

We consider a small portion of a volume hologram and assume that equally spaced, flat layers A, B, C, D,..., where the property of the recording material (the refractive index, for example) has been changed, are present in it. The spacing of these layers is denoted by d as shown in Fig.6.11. For simplicity we further assume that these layers have infinitesimal thickness and act as plane "scatterers" against the incident light.

Bragg showed that when a ray of light is incident upon such a multilayer structure with an angle θ_1, it reflects the light in a direction-selective as well as wavelength-selective manner.

There are two conditions under which reflection is enhanced. In the first place, components scattered upon two different spots on the same layer must enhance each other (see Fig.6.12a). This condition is expressed in terms of the distance ℓ between the two spots as

$$\ell(\sin\theta_1-\sin\theta_2) = m\lambda \tag{6.66}$$

where m is an integer. Since the above equation must hold for an arbitrary ℓ,

$$\theta_1 = \theta_2 \tag{6.67}$$

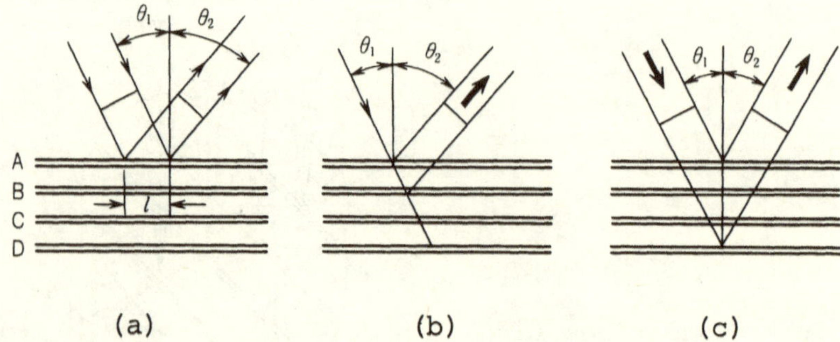

Fig.6.12 Conditions for the Bragg diffraction, II.

is required. This is nothing but the condition for mirror reflection. Secondly, components scattered upon two layers must enhance each other (see Fig.6.12b). This condition is given as

$$d\cos\theta_1 + d\cos\theta_2 = m\lambda. \tag{6.68}$$

Combining the first and second conditions, we obtain

$$d = m\lambda/2\cos\theta_1. \tag{6.69}$$

For the case m = 1,

$$d = \lambda/2\cos\theta_1. \tag{6.70}$$

Equation (6.70) shows that only a specific wavelength can satisfy the conditions of Bragg diffraction; the wavelength of efficient reflection is determined by the incident angle θ_1 and the spacing of layers d. Thus we may conclude, combining the above statement with Eq.(6.67), that Bragg diffraction is wavelength-selective mirror reflection, the wavelength being dependent upon the incident angle. This relation is illustrated in Fig.6.12c.

6.6.2 RELATIONS BETWEEN OBJECT WAVE, REFERENCE WAVE, AND ILLUMINATING WAVE

We consider a hologram as shown in Fig.6.9b which has been produced by using object and reference waves incident from opposite directions. Comparison of Figs. 6.9b and 6.12c leads to the following statements:

(1) When a volume hologram is illuminated by a wavefront C that is identical to that of the reference wave R, then a wavefront identical to that of the object wave O is obtained due to Bragg diffraction.

(2) When a volume hologram is illuminated by a wavefront C which is identical to that of the object wave O, then a wavefront identical to that of the reference wave R is obtained.

Actually an object wave consists of a large number of spherical wavefronts. Therefore, it is usually impossible to realize the condition described above in statement (2) over an appreciable area of a hologram. The condition of statement (1), on the other hand, is realizable because the reference wave usually has a spherical or planar wavefront. When a volume hologram is illuminated in this manner, an image of the original object can be reconstructed. The use of an illuminating wave identical to the reference wave is, in reality, a common practice in Lippmann-Bragg holography.

When the object and reference waves are incident upon the hologram plate from a common side as shown in Fig.6.9a, the hologram becomes transmission type. In such a case, the reconstruction process is essentially identical with that of the plane hologram even when the recording material has an appreciable thickness, because the diffractions in each layer in Fig. 6.10 are merely added in a common phase. (As described earlier, most ordinary holograms originally intended to be plane holograms have more or less such a three-dimensional structure because the thickness of the silver-halide recording material is usually 10 μm or even more.)

However, some of the properties of Bragg diffraction, especially the wavelength-selective property, are more or less preserved. The wavelength-selective property is readily noticeable when a hologram is recorded using, for example, a He-Ne laser, and the image is reconstructed using an argon laser. Reduction in efficiency and blurring in the image are observed. However, when the wavelengths are identical in the recording and reconstruction processes, the Bragg-diffraction property (in this case, the mirror reflection property) is not appreciable, and the theory of the plane hologram holds with good accuracy.

Therefore, the volume-type hologram illustrated in Fig.6.9a is usually neither treated as a three-dimensional volume hologram nor called a Lippmann-Bragg hologram. When such a hologram is intentionally produced to take advantage of the thickness of the recording material, for example for the purpose of multicolor reconstruction, it is sometimes called a *Bragg-effect hologram*. The terminology *Lippmann-Bragg hologram* or *Lippmann hologram* is usually used solely for the type shown in Fig.6.9b. The Lippmann-Bragg hologram is always a reflection-type hologram because the image is reconstructed on the side opposite to the illuminating light source.

6.6.3 APPLICATION OF THE WAVELENGTH SELECTIVITY

The wavelength-selective property is an important feature of the Lippmann-Bragg hologram when it is used in three-dimensional imaging. When a volume hologram is recorded using a single coherent light source, the image can be seen in the same color by illuminating it with merely a white light, although the picture quality will be to some degree degraded by background noise. Furthermore, if the volume hologram is originally recorded using two or three coherent light sources (green, blue, and red, for example) and the corresponding fringes are produced superposed upon each other, white light illumination will reconstruct the image in full color. Those techniques will be described in detail in Section 6.13.

6.7 EFFICIENCY OF HOLOGRAPHY

In the early stages of research and development on holography, the interest of the investigators was concentrated upon the imaging properties of holography, and the low efficiency of the reconstruction process was left unnoticed. After 1966, however, people began to endeavor to raise actual hologram efficiency toward its theoretical limit. Such efforts are important because a serious drawback of holography as a three-dimensional imaging technique is that the use of a laser is practically indispensable in the reconstruction of the image, whereas the efficiency of lasers is still very low in the present state of the art.

In most cases, the object wave recorded in a hologram consists of a very large number of spherical waves. Therefore, the efficiency obtainable in a practical case is critically dependent upon the shape, size, and surface optical properties of the object; it is almost impossible to predict the maximum obtainable efficiency. In this section, therefore, the discussions will be restricted to simple diffraction gratings of various types. The maximum theoretical efficiencies and some corresponding experimental data will be described.

6.7.1 DIFFRACTION EFFICIENCY OF A PLANE AMPLITUDE HOLOGRAM

Suppose that a plane amplitude (absorption) grating is placed at $z = 0$ as shown in Fig.6.13a, and its transmittance is given as

$$T(x,y) = [1-(m/2)]+(m/2)\cos(2\pi x/d) \tag{6.71}$$

as shown in Fig.6.13b. When $m = 1$, Eq.(6.71) represents

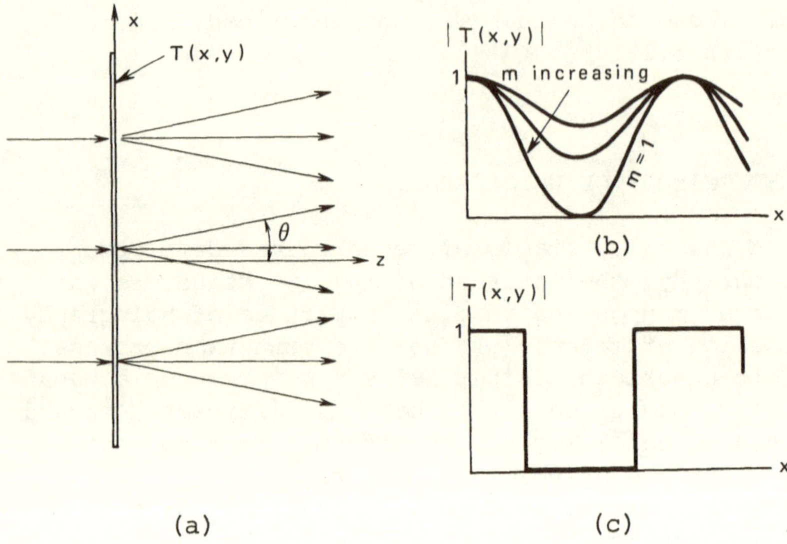

Fig.6.13(a) Diffraction upon a two-dimensional hologram. (b) An example of the amplitude-transmittance variation (sinusoidal variation). (c) An example of the amplitude-transmittance variation (rectangular variation).

a "fully modulated" grating, whereas $m \ll 1$ corresponds to "shallow" modulation. When a plane wave

$$C(x,y,z=-0,t) = C_0 \exp(j\omega t) \tag{6.72}$$

is incident upon the grating from the left, the amplitude of the secondary light source generated on the right-hand surface of the grating is expressed as

$$C(x,y,z=+0,t) = C_0[1-(m/2)] \exp(j\omega t) + (mC_0/2)\cos(2\pi x/d) \exp(j\omega t) . \tag{6.73}$$

Such a light source produces a number of plane waves heading for the right, the angle θ shown in Fig.6.13a ranging between -90° and +90°. Generally, if we write the amplitude of the wave component between θ and θ+dθ as p(θ)d, for the region z > 0,

$$C(x,y,z>0,t) = \int_{-\pi/2}^{+\pi/2} p(\theta)\exp[-j(kx\sin\theta+kz\cos\theta)+j\omega t]d\theta \tag{6.74}$$

holds. The distribution function $p(\theta)$ can be determined from the condition that Eq. (6.74) agrees with Eq. (6.73) at $z = 0$. In the present case, the function $p(\theta)$ need not be continuous; if we rewrite Eq. (6.74) into summation form and equate it with Eq. (6.73) at $z = 0$, we obtain

$$\sum_{i=1} p(\theta_i)\exp(-jkx\sin\theta_i) = C_0\left(1-\frac{m}{2}\right)+\frac{mC_0}{2}\cos\frac{2\pi x}{d}. \tag{6.75}$$

Careful inspection of the above equation shows that only three terms

$$\begin{aligned} p(\theta_0) &= C_0(1-m/2) && \text{at} \quad \theta = \theta_0 = 0 \\ p(\theta_{+1}) &= mC_0/4 && \text{at} \quad \theta = \theta_{+1} = \sin^{-1}(2\pi/kd) \\ p(\theta_{-1}) &= mC_0/4 && \text{at} \quad \theta = \theta_{-1} = -\sin^{-1}(2\pi/kd) \end{aligned} \tag{6.76}$$

are enough to get a "match" between both sides of Eq. (6.75), or in other words between Eqs. (6.73) and (6.74) at $z = 0$. The first component at $\theta = \theta_0$ corresponds to the directly transmitting wave, whereas the second ($\theta = \theta_{+1}$) and third ($\theta = \theta_{-1}$) terms represent the diffracted waves of orders +1 and -1, respectively. Equations (6.76) show that the efficiency of the first-order diffraction is given as

$$\eta = [p(\theta_+)/C_0]^2 = (m/4)^2. \tag{6.77}$$

Since $0 \leq T(x) \leq 1$ must hold everywhere, m must be ≤ 1 for sinusoidal modulation of the amplitude transmittance. Thus $\eta_{max} = (1/4)^2 = 6.25\%$ is obtained for full modulation ($m = 1$). Equations (6.76) show that in such a case, the incident light energy is distributed to the various components in the following ratio:

(1) 1/4 of the incident energy is directly transmitted,

(2) 1/16 is distributed to the diffracted wave of order +1,

(3) another 1/16 is distributed to the diffracted wave of order -1, and

(4) hence, the remaining 5/8 is dissipated in the gratings.

In practice, if we carry out so-called "hard development" of the photographic plate to obtain T(x) of the rectangular waveform as shown in Fig.6.13c, we may raise the efficiency up to $\eta_{max} = (1/\pi)^2 = 10\%$. (This computation is not difficult if one follows the above process.) In such a case, however, diffracted wave components of higher order terms are also generated, producing ghost images around the desired image.

6.7.2 DIFFRACTION EFFICIENCY OF A PLANE PHASE HOLOGRAM

We consider again the same arrangement as shown in Fig.6.13a, in which, however, the transmittance of the grating is given as

$$T(x,y) = \exp\{j[\phi_0+\phi_1 \cos(2x/d)]\}. \tag{6.78}$$

Without loss of generality, we may assume that $\phi_0 = 0$. If a plane wave as expressed by Eq.(6.72) is incident from the left, the wavefront after transmission through the grating is derived, using the relation of Eq.(6.47), to be

$$C(z = +0) = C_0 \cos(\phi_1 \cos 2\pi x/d)+jC_0 \sin(\phi_1 \cos 2\pi x/d) \tag{6.79}$$

$$= C_0 J_0(\phi_1)+2C_0 \sum_{n=1}^{\infty} (-1)^n J_{2n}(\phi_1)\cos(2n2\pi x/d)$$
$$+2jC_0 \sum_{n=1}^{\infty} (-1)^{n+1} J_{2n-1}(\phi_1)\cos[(2n-1)2\pi x/d] \tag{6.80}$$

The distribution function $p(\theta)$ satisfying Eq.(6.80) at $z = +0$ is again discontinuous. To obtain components of $p(\theta)$, we again write an equation similar to Eq.(6.75) and compare the two sides. Then we may obtain by inspection the following components:

$$
\begin{aligned}
p(\theta_0) &= C_0 J_0(\phi_1) && \text{at} \quad \theta = \theta_0 = 0 \\
p(\theta_{\pm 1}) &= jC_0 J_1(\phi_1) && \text{at} \quad \theta = \theta_{\pm 1} = \pm\sin^{-1}(2\pi/kd) \\
p(\theta_{\pm 2}) &= -C_0 J_2(\phi_1) && \text{at} \quad \theta = \theta_{\pm 2} = \pm\sin^{-1}(2\times 2\pi/kd) \\
&\vdots && \qquad \vdots
\end{aligned}
\tag{6.81}
$$

and so forth. The first component is the directly transmitting wave. The components at $\theta = \theta_{\pm m}$ are the diffracted waves of orders $\pm m$. The efficiency of the first-order diffracted wave is given as

$$\eta = |p(\theta_{\pm 1})/C_0|^2 = [J_1(\phi_1)]^2, \tag{6.82}$$

which gives

$$\eta_{max} = (0.582)^2 = 33.9\% \tag{6.83}$$

at $\phi_{1\ opt} = 1.84$ rad. Note that this maximum efficiency is about five times higher than that of the absorption grating (6.25%).

It has been shown that in an absorption grating the efficiency is improved by using the transmittance distribution of a rectangular waveform. This fact naturally urges us to consider whether the efficiency of a phase grating can also be improved by choosing the transmittance distribution more freely. This presumption is valid; however, the maximum theoretical efficiency thus obtained will be 100% for a phase grating; such a value could be obtained if we make an ideal phase grating having a cross section of saw-tooth waveform as shown in Fig.6.14a. Such an ideal grating is nothing but an ensemble of tiny, triangular prisms. In the case of a reflection-type phase grating, 100% efficiency is also achievable if we make its cross section like that in Fig.6.14b. As described in Section

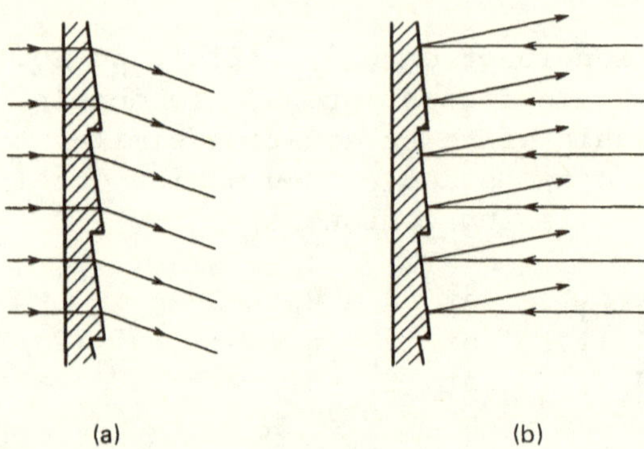

Fig.6.14 Two-dimensional phase hologram. (a) Transmission type. (The thickness variation shown in the figure can be replaced by a refractive-index variation.) (b) Reflection type (blazed hologram).

6.3.2, when a hologram is made in the form of such a reflection-type phase grating, it is called a blazed hologram.

6.7.3 DIFFRACTION EFFICIENCY OF A VOLUME HOLOGRAM

When a volume grating is made with the modulation of the refractive index, the theoretical maximum efficiency approaches unity, for both reflection and transmission modes. This conclusion is natural because no energy dissipation will take place in the hologram, and all the incident energy can be converted into the diffraction component wanted.

When the grating is of absorption type, in other words when the local transmittance is modulated to record the hologram information (6), the maximum theoretical efficiency is 3.7% (= 1/27) for the transmission mode and 7.2% [= $(2+\sqrt{3})^{-2}$] for the reflection mode. The details of the analyses are omitted here because these types of hologram have less practical importance. [Refer to the original literature (6) or textbooks on holography, for example, Ref.(7)].

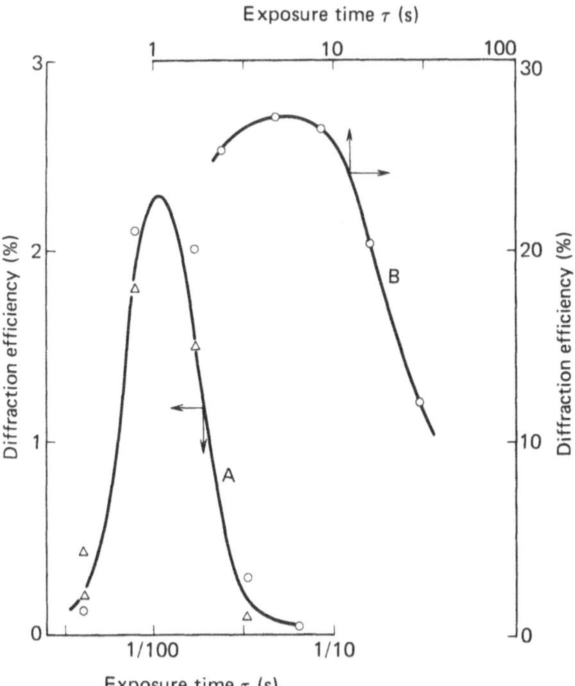

Fig.6.15 Examples of the measured diffraction efficiency of a hologram as functions of the exposure time. A: amplitude hologram; B: phase hologram.

6.7.4 MEASURED EFFICIENCIES OF VARIOUS GRATINGS

Some good experimental data that have appeared in the literature will be cited, together with some measured efficiency-versus-exposure characteristics.

A. Plane Amplitude Grating

Lee and Greer reported an efficiency of 4.5% with an ordinary silver-halide photographic plate (8). Amodei and Mezrich reported that they obtained 6.0% using a bismuth film (9). An example of the efficiency-versus-exposure characteristics of a silver-halide plate (Kodak 649F) obtained in the author's laboratory is shown in Fig.6.15 as curve A. The maximum efficiency of 2.7% was obtained in the same series of experiments (10).

B. *Plane Phase Grating*

Pennington and Harper reported 25.5% (11). The highest efficiency ever reported is probably that of Takeda *et al.* (12). They obtained the highest efficiency of 70% using a Kodak 649F plate as the recording material, chromium intensifier as the bleaching agent, and thioglycollic acid as the stabilizer. However, the very high efficiency obtained by Takeda *et al.* might be attributed to the three-dimensional aspects of the recording.

An example of the efficiency-versus-exposure characteristics obtained in the author's laboratory is shown in Fig.6.15 as curve B. In this case the highest efficiency is 27% (10).

C. *Blazed Grating*

Bryngdahl reported an efficiency of 56% (13).

D. *Volume-Type Gratings*

Collier *et al.* (7) presented a table concerning the maximum efficiencies obtained experimentally. According to that table, the best data reported before 1969 are:

Absorption-transmission type: 3.0% reported by George and Mathews (14).

Refraction-transmission type: 90% reported by Lin (15). (Later, also 90% by Colburn and Haines (17).)

Absorption-reflection type: 3.8% reported by Lin and LoBianco (16).

Refraction-reflection type: 80% reported by Lin (15).

6.8 RESOLUTION OF HOLOGRAPHY

So far we have assumed that holography has unlimited good resolution; however, in reality, in any holographic scheme the image of a point object has a finite spread. In other words any holographic scheme has a finite resolution.

There are several factors that limit resolution: hologram size, granularity (or turbidity) of the recording material used, speckle noise, and aberration. Among these the resolution limitation due to finite hologram size is the most essential. In this section this subject will be considered.

We consider a plane amplitude hologram as described in Section 6.7.1, having, however, finite size D along the x direction. In such a case, Eq.(6.71) must be replaced by

$$T(x,y) = T(x) = \begin{cases} [1-(m/2)]+(m/2)\cos 2\pi x/d, & \\ & (-D/2 < x < +D/2) \\ 0 & (x < -D/2,\ D/2 < x). \end{cases} \quad (6.84)$$

If we assume that D >> d holds, the diffracted wave components are given, in place of Eq.(6.76), as

$$p(\theta) \doteqdot C_0\left(1-\frac{m}{2}\right)\frac{\sin(\pi D\theta/\lambda)}{(\pi D\theta/\lambda)} \qquad \text{around} \quad \theta = 0$$

$$p(\theta) \doteqdot \frac{mC_0}{4}\,\frac{\sin\{\pi D(\theta-\theta_{+1})/\lambda\}}{\{\pi D(\theta-\theta_{+1})/\lambda\}} \qquad \text{around} \quad \theta = \theta_{+1} = \sin^{-1}(2\pi/kd)$$

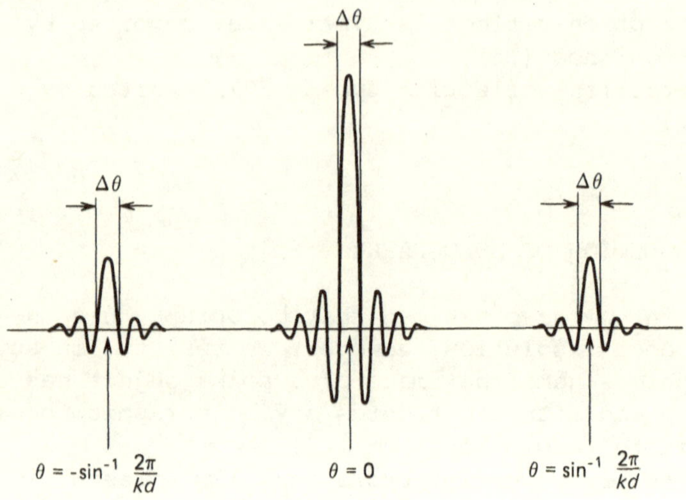

Fig.6.16 The amplitude distribution of the diffracted wave from a finite-size hologram.

$$p(\theta) \doteqdot \frac{mC_0}{4} \frac{\sin\{\pi D(\theta-\theta_{-1})/\lambda\}}{\{\pi D(\theta-\theta_{-1})/\lambda\}} \quad \text{around} \quad \theta = \theta_{-1} = -\sin^{-1}(2\pi/kd). \tag{6.85}$$

The derivation of the above equations is shown in Appendix II to this chapter.

The spectra given by Eq.(6.85) are illustrated in Fig.6.16, which shows that each of three components given by Eq.(6.76) now has a spread of the sampling-function form (= sin x/x). If the *spread angle* of the sampling function is defined as the separation of the two closest zeros, it is given as

$$\Delta\theta = 2\lambda/D. \tag{6.86}$$

In the above discussion a plane amplitude hologram has been considered; however, the above result [Eq. (6.86)] remains valid for other types of holograms, i.e., plane phase holograms and Lippmann (or Bragg-type) holograms provided that the hologram thickness is much smaller than the lateral size D. In all these cases, the diffracted light beam always has an angular spread comparable to the ratio (wavelength)/(hologram size); this spread gives an essential limitation to the resolution of holography.

6.9 NOISE IN HOLOGRAPHY

6.9.1 MEANING OF NOISE

The term *noise* used in the title of this section might require some comments. In the early history of electrical communication, the signals transmitted via communication channels were always voice signals. Therefore, in those days most amplifiers were audio amplifiers; fluctuation of the output voltage was hazardous because it was superposed upon the voice signal as an audible "noise." Since then, it became a common rule to call fluctuations of voltage and current in general *noise,* regardless of the frequency range of the fluctuation. Nowadays the terms *video noise* and *microwave noise* are rather commonly used; those terms denote voltage or current fluctuations in several megahertz (the so-called video frequency) and in microwave frequencies, respectively, and have nothing to do with audible noise. This custom is common among most of the modern languages; the words expressing audible fluctuations [noise (English), Rauschen (German), bruit (French), and Zatsuon (Japanese)] are also used to denote fluctuations having any frequency.

The word *noise* used in connection with holography is the result of a further extension of the above terminology. It means a random, spatial fluctuation of the brightness upon the image obtained from a hologram. Therefore, the quantity corresponding to the frequency $[s^{-1}]$ of the noise in time-sequential signals is the "spatial frequency" having dimensions of $[m^{-1}]$. Television engineers were familiar with the concept of such "spatial noise" long before the noise study of holography started. We should note that in television, the fluctuation of the image brightness is given as a function of both space and time.

6.9.2 VARIOUS NOISES IN HOLOGRAPHY

The most important noise-generation process in holography is speckle noise, which stems from the use of coherent light in the recording and reconstruction processes. In addition to speckle noise, there are other sources of noise: modulation noise due to non-linear characteristics of the recording material, disturbances given by ghost (spurious) image, presence of white light in the recording process, and the granularity of the recording material. In this section only speckle noise will be discussed.

6.9.3 SPECKLE NOISE

Speckle noise is a kind of spatial noise often observed in coherent optical systems. It is defined in general as spatial noise generated by mutual interference between light beams arriving at the surface of an object or at an image space from various, random directions. This phenomenon degrades the quality of the image reconstructed from a hologram when coherent light is used in the reconstruction process, as is usually the case.

It is difficult to discuss in general in a quantitative manner how much speckle noise is observed in the final image reconstructed from a hologram because the magnitude of the speckle noise is a function of many factors such as:

(1) how the hologram is illuminated;

(2) whether the image is real or virtual;

(3) whether the image is observed directly by human eyes or projected onto a screen (as in projection-type holography);

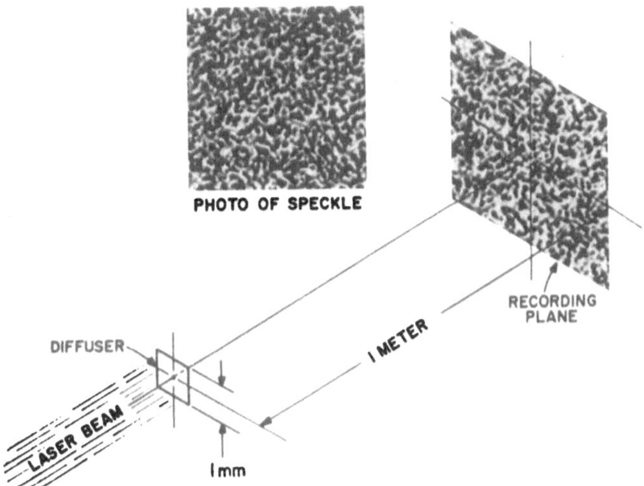

Fig.6.17 A speckle-noise model [after Gerritsen et al. (18)].

(4) if the image is observed directly, what is the diameter of the pupil;

(5) if thc image is recorded by a camera, what is the diameter (or the f-number) of the lens; and

(6) what is the viewing or recording distance.

In this section, therefore, the experiment by Gerritsen, *et al.* (18) on a simplified speckle-noise model will be described first, which may give good insight into the nature of speckle noise. Next, a theoretical analysis of such a simplified noise model (19) will be presented. Finally, we will consider how this theory can be applied to various stages of the holographic process.

The noise model dealt with by Gerritsen *et al.* is shown in Fig.6.17. This figure shows that when a coherent laser light is incident upon a tiny translucent diffuser (a ground glass plate, for example) and the diffused light is projected onto a white screen, a spotted pattern (speckle noise) is observed on the screen.

Reinheimer and Copple measured the spatial frequency spectra of such patterns (20). Figures 6.18a and 6.18b

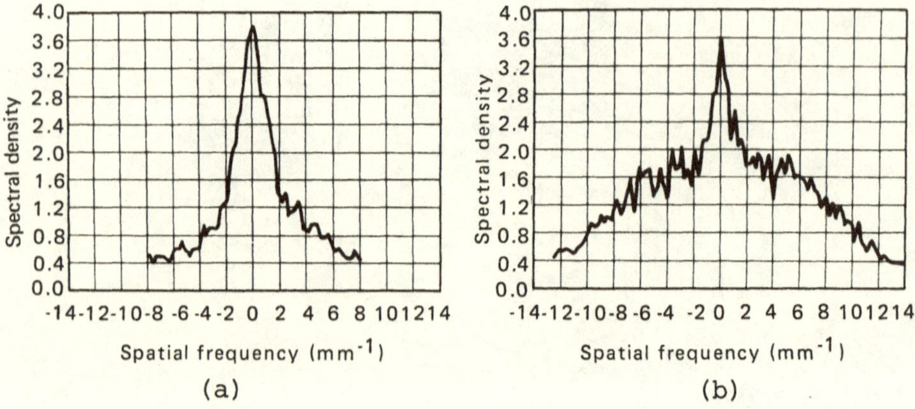

Fig.6.18 Spatial frequency spectrum of speckle patterns [after Reinheimer and Copple (20)]. (a) Case of incoherent illumination. (b) Case of coherent illumination.

show the measured spectra for cases when the illuminating beam is incoherent and coherent, respectively. If spatial noise were absent, the spectrum would be a spike at the center. The spread of the spectrum toward the high-frequency regions corresponds to spatial noise generation. Comparing Figs.6.18a and 6.18b, we may conclude that the use of coherent light appreciably enhances the spectral components in the frequency range 2-10 mm^{-1}.

6.9.4 ANALYSIS OF SPECKLE NOISE

Following Gabor (19), the spatial frequency spectrum and magnitude of the speckle noise will be analyzed on a simplified model.

We consider a two-dimensional noise model as shown in Fig.6.19. A projection screen is placed at $z = 0$, where N plane waves are incident, approximately from the -z direction but from directions slightly different from each other. If the amplitude of the nth wave, the wave number of the nth wave in the x direction,

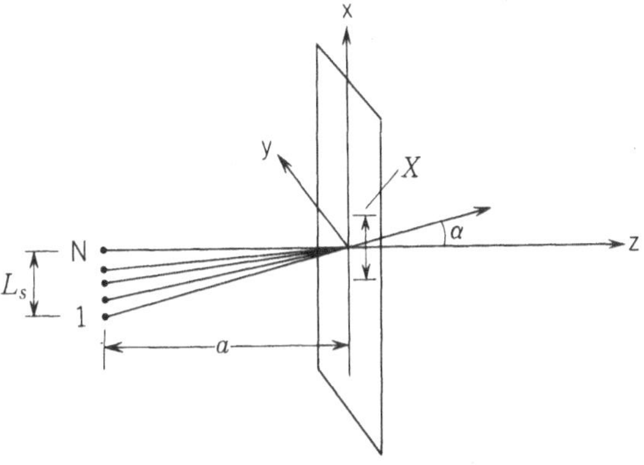

Fig.6.19 A model for speckle-noise analysis [after Gabor (19)].

and the phase of the nth wave at x = z = 0 are denoted by A_n, k_n, and ϕ_n, respectively, the amplitude distribution at z = 0 is expressed as

$$A(x) = \sum_{n=1}^{N} A_n \exp(-jk_n x + j\phi_n + j\omega t). \tag{6.87}$$

Hence, the corresponding light intensity is given as

$$I(x) = A(x)A^*(x) = \sum_n \sum_m A_n A_m^* \exp\{-j(k_n - k_m)x + j(\phi_n - \phi_m)\}. \tag{6.88}$$

We now compute the energy $E(X,x_0)$ included in a sampling width X centered at $x = x_0$, and further compute the fluctuation of $E(X,x_0)$ with respect to the sampled position x_0. We may write

$$E(X,x_0) = \int_{x_0-(1/2)X}^{x_0+(1/2)X} I(x)\,dx$$

(equation continues)

$$= X \sum_n \sum_m A_n A_m^* \operatorname{sin c}\left\{\frac{1}{2}(k_n-k_m)X\right\}$$

$$\times \exp\left\{-j(k_n-k_m)x_0+j(\phi_n-\phi_m)\right\} \qquad (6.89)$$

where $\operatorname{sin c}(Z) \triangleq \sin z/z$. The deviation of $E(X,x_0)$ from its average value is given as

$$F(X) = \frac{<[E(X,x_0)-<E>]^2>_{x_0}}{<E>^2} = \frac{<E^2>_{x_0}-[<E>_{x_0}]^2}{<E>^2} \qquad (6.90)$$

where the symbol < > denotes an ensemble average. We put Eq.(6.89) into Eq.(6.90) and further assume that in reality the incident light comes from directions distributed continuously in a certain range of angles, and hence, the wave number k is distributed uniformly between k = 0 and k = K. Then, after some computations whose details are described in Appendix III of this chapter, we obtain

$$F(X) = \left(\frac{1}{2}KX\right)^{-2}\left[(KX)\mathrm{Si}(KX)-\mathrm{Cin}(KX)-2\sin^2\left(\frac{1}{2}KX\right)\right], \qquad (6.91)$$

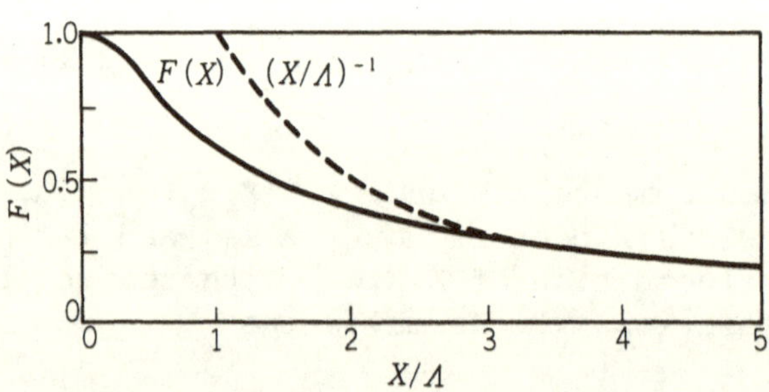

Fig.6.20 The fluctuation of the image brightness due to the speckle noise as a function of the sampled range X.

where functions Si and Cin are defined by Eq.(A-25) in the appendix. The above relation is shown in Fig.6.20 as a solid curve. In this figure, a normalizing parameter Λ is used, defined as

$$\Lambda = 2\pi/K, \tag{6.92}$$

and has the dimensions of length. The broken curve in the figure shows a simple function $(X/\Lambda)^{-1}$. Figure 6.20 shows that:

(1) the fluctuation of the sampled light is very large for $X \lesssim \Lambda$; and
(2) when $X \gtrsim 4$, $F(X) \simeq (X/\Lambda)^{-1}$ holds.

If we assume, as shown in Fig.6.19, that the component of light incident with angle α gives the maximum wave number in the x direction X, then

$$K = (2\pi/\lambda)\sin\alpha \fallingdotseq 2\pi\alpha/\lambda. \tag{6.93}$$

On the other hand,

$$\alpha \fallingdotseq L_S/a, \tag{6.94}$$

where L_S and a are the width and the distance to the source. Hence we obtain from Eqs.(6.92)-(6.94),

$$\Lambda = \lambda/\alpha = \lambda a/L_S. \tag{6.95}$$

In Section 6.8, it was shown that a light beam from a source of size L_S (note that in Section 6.8 it was denoted by D) has an angular resolution limit of $2\lambda/L_S$ due to diffraction. Hence at a distance a, the lateral resolution limit is given as $2\lambda a/L_S$, and we may conclude from Eq.(6.95) that Λ is equal to half the lateral resolution limit due to diffraction of light from a source of size L_S.

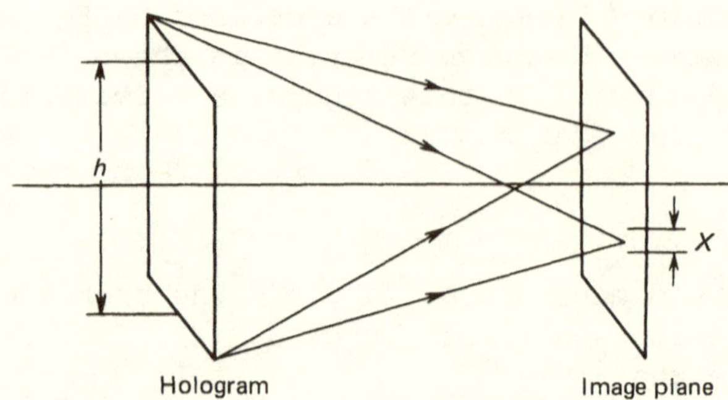

Fig.6.21 Reconstruction of a two-dimensional image from a hologram.

6.9.5 APPLICATION OF THE SPECKLE-NOISE THEORY TO HOLOGRAPHY

Let us apply the above theory to some holographic processes.

The model shown in Fig.6.21 permits the most direct application of the theory. In this model a two-dimensional real image reconstructed from a hologram is projected onto a screen, or onto an image sensor (as in holographic picture-memory devices), or onto a planar array of light sensors (as in holographic digital memory devices). It is assumed that the hologram does not record the entire object-wave bundle, but records only a portion of it. (When the entire object wave is recorded as in some holographic digital memories, a speckle-free image may be reconstructed.) In such a case, the reconstructing light intensity inevitably exhibits fluctuation upon the image, and the hologram shown in Fig. 6.21 may be compared to the diffuser in the above discussion. If we replace L_S by the hologram size D, the above theory also remains valid for such a holographic process.

As described above, $F(X) \simeq (x/\Lambda)^{-1}$ holds for sampling widths greater than 4Λ. This means, for example, if we want to get an S/N (signal-to-noise) ratio greater than 20 dB, we should either:

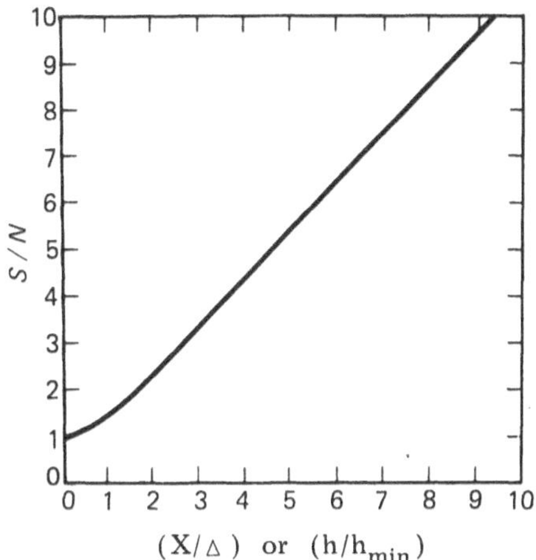

Fig.6.22 The S/N ratio due to the speckle noise as a function of the hologram size [after Gerritsen et al. (18)].

(1) receive (or observe) the projected image after blurring it within sampling area of $100\Lambda \times 100\Lambda$; or

(2) use a hologram which is 100×100 times larger than the minimum hologram size D_{min} giving the required resolution (18).

The above relation is shown in Fig.6.22 (18). The fact that the curve in this figure shows a difference from a simple proportionality relation $S/N = X/\Lambda$ in the vicinity of the origin corresponds to the fact that in Fig.6.20 $F(X)$ and $(X/\Lambda)^{-1}$ show a difference from each other. [The details of the derivation of the relation shown in Fig.6.22 are omitted. The reader should refer to Ref.(18) if necessary.]

Next we consider the cases shown in Fig.6.23, where a three-dimensional virtual image (Fig.6.23a) or a real image (Fig.6.23b) is directly observed by an eye. In these figures we readily see that the sampling range X is given as

$$X = Ab/(a+b) \tag{6.96}$$

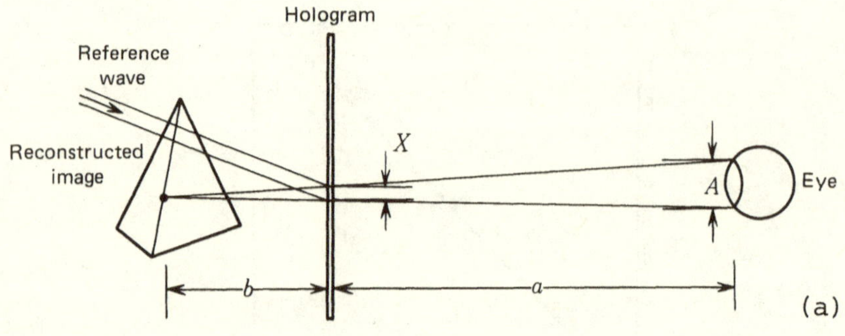

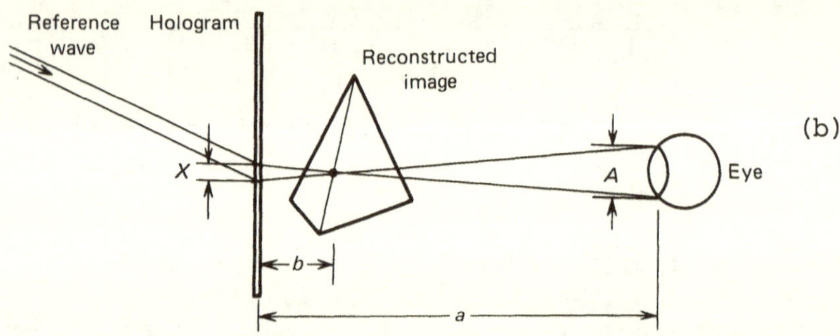

Fig.6.23 Direct observation of the image reconstructed from a hologram.

when a virtual image is observed, and

$$X = Ab/(a-b) \tag{6.97}$$

when a real image is observed, where A denotes pupil diameter. For typical parameters of the former case, a = b = 250 mm and A = 3 mm, we obtain X = 1.5 mm.

Note that Eqs.(6.96) and (6.97) give the quantity same as the minimum hologram size for giving the resolution limit of a human eye, as shown in the following. If we assume for simplicity that the pupil has a rectangular aperture, the resolution limit due to diffraction of the pupil is given as

$$\delta = 2(a+b)\lambda/A. \tag{6.98}$$

On the other hand, the minimum size of a hologram giving the same resolution is given, in the case of Fig.6.23a, as

$$D_{min} = 2b\lambda/\delta = Ab/(a+b) \qquad (6.99)$$

which is identical to Eq.(6.96).

Therefore, we may conclude that when an image reconstructed from a hologram is observed directly by a human eye, speckle noise having a roughness whose wavelength is comparable to the resolution limit of the pupil (or some components of lower spatial frequencies as suggested by the slowly decreasing skirt of Fig. 6.20) is observed inevitably. If we want to reduce speckle noise, one possible way is to reduce the resolution below the diffraction limitation. This statement may be understood from Fig.6.22 for the case of a two-dimensional image.

6.10 WHITE-LIGHT RECORDING OF A HOLOGRAM

In the introduction to this chapter, it was stated that holography has many practical drawbacks when used in three-dimensional imaging, and that most of these stem from the use of coherent light in the recording of (and in most cases, also in the reconstruction from) the hologram. It has also been stated that several special holography techniques such as white-light recording, white-light reconstruction, and color recording/reconstruction are being investigated to overcome these drawbacks. In the following four sections (Sections 6.10-6.13), these techniques are described.

As to white-light recording techniques, two methods have been proposed and experimented with and are described in the following.

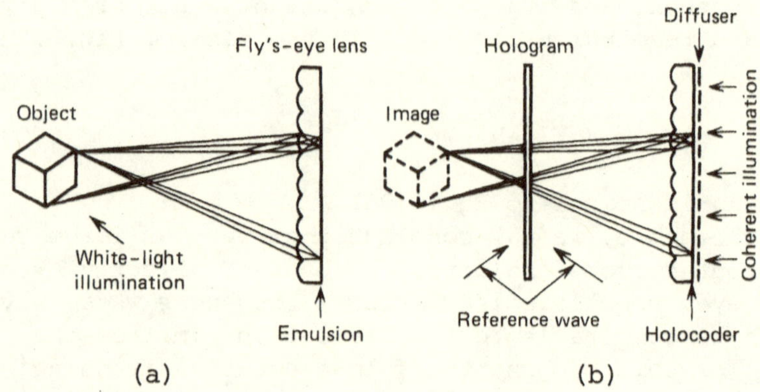

Fig.6.24 Holocoder-hologram scheme [after Pole (21)]. (a) Recording of the holocoder. (b) Recording of the hologram.

6.10.1 HOLOCODER-HOLOGRAM

The earliest proposal for white-light recording holography was Pole's "holocoder-hologram" method (21); this is a combination of integral photography (IP) and holography. It might be looked at either as a version of two-step IP in which the second step is replaced by holography, or as a version of holography in which the image is first recorded as an integral photograph. It should be noted, however, that Pole was ignorant of Lippmann's IP at the time he invented the holocoder-hologram method.

Pole's method is shown in Fig.6.24. The object is first recorded in an integral photograph using white-light illumination as shown in Fig.6.24a. The integral photograph is developed, and then illuminated by coherent light from the right as shown in Fig.6.24b, to reconstruct a three-dimensional real image in front of it. To make a hologram, we place a photographic plate just in front of the integral photograph and illuminate it with a reference beam obtained from the same source as the beam illuminating the integral photograph. The image thus recorded in the hologram is identical to the original one except that the resolution is somewhat degraded by the preliminary IP process.

Pole named the preliminary IP process the *holocoder* in the sense that the *hologram* information is recorded in it in somewhat *coded* form.

In Ref.(21), a photograph of an image reconstructed from such a hologram is shown; however, the image quality is not excellent. This is a rather natural consequence of the fact that an integral photograph is suited for close reconstruction of a spatial image but not suited for remote reconstruction as in the present case. One should use larger lenses in the primary stage to maximize the resolution. (See the discussions in Section 4.4.3.E.) The use of bigger lenses in the primary stage will also make the practical use of such an imaging process much easier, as described in the following subsection.

6.10.2 HOLOGRAPHIC STEREOGRAM

McCrickerd and George proposed and experimented with a system that is similar to the holocoder-hologram method in principle but uses larger lenses in the primary stage than in IP (22). They called this system a *holographic stereogram*. (It would better be called a *holographic panoramagram* since it allows observation from almost continuously different directions. The term *holographic stereogram*, however, is now widely used.)

The primary picture-taking and hologram-recording processes are illustrated in Fig.6.25. The primary pictures are $n \times m$ ordinary photographs. In their experiment, these photographs were not taken all at once, but taken successively by a moving camera. The hologram recording is also performed in a successive manner; $n \times m$ segmental holograms are made from the $n \times m$ primary pictures in a setup like that shown in Fig.6.25b. The time required for the primary picture taking and the hologram recording might be appreciable. However, the finally reconstructed image may in practice be much better than that of the holocoder-hologram.

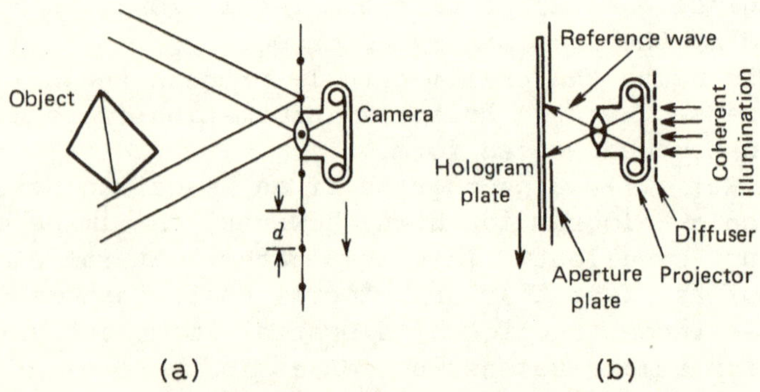

Fig.6.25 The holographic stereogram scheme [after McCrickerd and George (22)]. (a) Primary recording. (b) Secondary hologram recording.

After McCrickerd's proposal, Kasahara *et al.* (23) reported a simplified version of the holographic stereogram in which vertical parallax information was discarded; they employed, for example, n (horizontal picture number) = 20 and m (vertical picture number) = 1. Such a "unidirectional" holographic stereogram was later applied to three-dimensional x-ray imaging (by Kasahara *et al.* and Groh and Kock; see Section 8.6) and computer-generated three-dimensional imaging [by King *et al.*; see Section 8.2.4].

6.10.3 RECORDING OF A HOLOGRAPHIC STEREOGRAM USING A DIRECTION-SELECTIVE SCREEN

The biggest drawback of the holographic stereogram is the tedious step-by-step recording of n × m primary pictures and also of n × m holograms. To reduce this tediousness, Yano proposed and experimented with a one-shot method of recording a holographic stereogram from n × m primary pictures, using the direction-selective screen described in Chapter 5 (24).

An example of the optical setup for the hologram recording is shown in Fig.6.26a, in which a unidirectionally direction-selective screen consisting of a

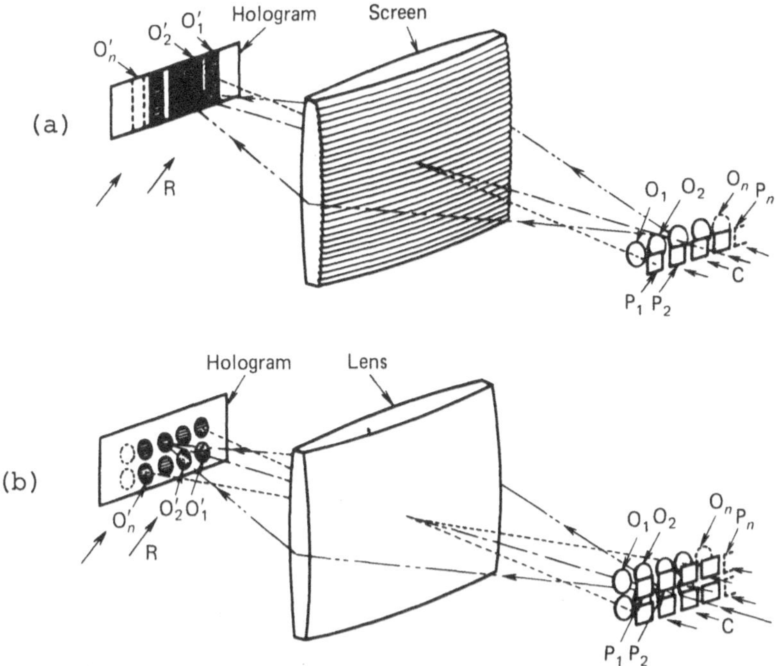

Fig.6.26 One-shot recording of a holographic stereogram, (a) using a unidirectionally direction-selective screen, and (b) using an ordinary direction-selective screen.

large convex lens and a lenticular screen is used (see Section 5.3.3). In practice, a large Fresnel lens may be used instead of the heavy convex lens shown in the figure. Either of such lens systems collimates light bundles from a point source toward a vertical stripe. As shown in Fig.6.26a, we place n primary pictures on the right of the lens system, illuminate them with coherent light C, place a hologram plate at the focusing position on the left, and expose it to the object wave from O_1, O_2,... as well as to the reference wave R. Thus a holographic stereogram provided with unidirectional (only horizontal) parallax information is recorded.

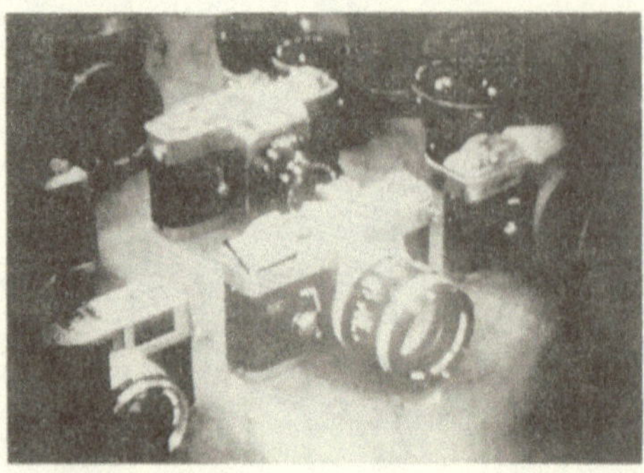

Fig.6.27 Photograph of an image reconstructed from a holographic stereogram recorded with the method of Fig.6.26a.

If we also wish to record vertical parallax information, we just remove the lenticular sheet and use a two-dimensional array of primary pictures as shown in Fig.6.26b. However, such a system might not be practical because the projection system would become expensive. When vertical parallax information is needed, we would better use a more-or-less successive method, in other words, use only a single projector and shift it step by step in the vertical and horizontal directions to record the holographic stereogram.

Figure 6.27 shows a photograph of the image reconstructed upon a unidirectionally direction-selective screen from a holographic stereogram recorded with the method of Fig.6.26a. The holographic stereogram had seven elements within a width of 90 mm. Light from an Ar-ion laser (514.5 nm) was used in the reconstruction of the image.

6.11 WHITE-LIGHT RECONSTRUCTION OF AN IMAGE FROM A HOLOGRAM

There are three major approaches toward white-light reconstruction of a holographic image. The best known

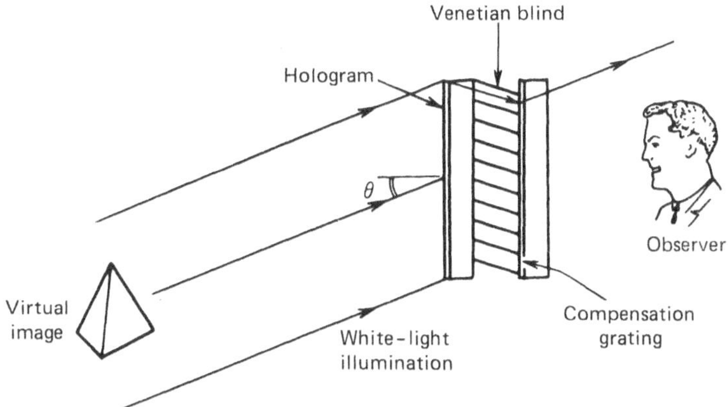

Fig.6.28 Venetian blind method [after Burckhardt (25)].

and most widely experimented with is Lippmann-type volume holography. This approach, however, gives a color image and would better be described in another section as a technique of color holography (see Section 6.13). The second approach is image holography; this has been described briefly in Section 6.3.3. However, image holography will be discussed in the next section separately because it has some other features in addition to white-light reconstruction. The third method is the use of a *compensation grating*. This method was first proposed by Burckhardt (25), and a little later by DeBitetto (26) independently. In this section, Burckhardt's proposal and experiment will be described.

The hologram is recorded in an entirely ordinary manner. The reference beam is preferably a parallel beam and will hereafter be assumed to be so. The incident angle of the reference beam is denoted by θ.

The optical setup for image reconstruction is shown in Fig.6.28. The hologram is illuminated by white light from the left with an angle θ. The illuminating beam is then diffracted upon the hologram to generate components of orders of -2, -1, 0, +1, +2, and so forth. The venetian-blind structure placed just behind the hologram intercepts all the components except for order +1 which is supposed to be the component wanted for reconstructing the virtual image. This component passes through the venetian blind and is incident upon the compensation grating.

This compensation grating has been produced in a holographic manner using two plane waves and has a pitch equal to the average pitch of the hologram on the left. (It should preferably be a phase grating to assure proper efficiency.) Therefore, the light from the virtual image is again diffracted by the compensation grating approximately toward the original direction.

The possibility of white-light reconstruction in such a system may be understood in a qualitative manner. We first consider the case in which monochromatic light having the same wavelength as the recording light is used in reconstruction. Since the pitch of the grating is equal to the average pitch of the hologram, the observer on the right will find the image in the direction of the illuminating light source. However, the image is not masked by strong background light because the venetian-blind structure intercepts the directly transmitted component.

Next we consider the case in which the reconstructing wavelength is different from the recording wavelength. In such a case, the direction of the light paths between the hologram and the grating depends upon the reconstructing wavelength because of the color dispersion of the hologram. This dispersion, however, will be compensated for by the compensation grating on the right; the light paths after passing through the grating will not be dependent, within a first-order approximation, upon the wavelength. Therefore, when we use white-light illumination consisting of a number of components having different wavelengths, we can observe a black-and-white image.

We may repeat the above discussion in a more quantitative manner as follows. We assume that the object and reference waves upon the hologram plate in the recording process are expressed as

$$O(x,y) = O_0(x,y)\exp(j\omega_O x) \tag{6.100}$$

$$R(x,y) = R_0 \exp(-j\omega_R x), \tag{6.101}$$

where the temporal term $\exp(j\omega t)$ is omitted. Note that $O_0(x,y)$ is not a real quantity but includes a phase perturbation representing the object information. When the interference fringes between $O(x,y)$ and $R(x,y)$ are recorded as a hologram, the term contributing to the reconstruction of the virtual image is

$$OR^* = O_0(x,y)R_0 \exp[j(\omega_O+\omega_R)x], \tag{6.102}$$

where the asterisk denotes complex conjugation. Next, we assume that in producing the compensation grating, we have used the interference of two waves expressed as $G_1 = G_0 \exp(j\omega_O x)$ and $G_2 = G_0 \exp(-j\omega_R x)$. Then the intensity upon the plate is given as

$$|G_1+G_2|^2 = G_0^2\{2+\exp[j(\omega_O+\omega_R)x]+\exp[-j(\omega_O+\omega_R)x]\}. \tag{6.103}$$

Suppose for simplicity that both the hologram and the compensation grating are processed so that their amplitude transmittances are proportional to Eqs.(6.102) and (6.103), respectively.

Next, we consider the reconstruction process. When we use an illuminating beam of an arbitrary wavelength λ_C, its phase distribution upon the plate will be

$$C(x,y) = C_0 \exp(-j\omega_C x), \tag{6.104}$$

where $\omega_C = \omega_R(\lambda_C/\lambda_R)^{-1}$ and λ_R is the recording wavelength. We note that when the wavelength is changed, the angle of diffraction is approximately multiplied by λ_C/λ_R. Therefore, the beam diffracted by the hologram is approximately given, from Eqs.(6.102) and (6.104), as

$$C' = O_0(x,y)R_0C_0 \exp[j(\lambda_C/\lambda_R)(\omega_O+\omega_R)x-j\omega_C x], \tag{6.105}$$

which shows a color dispersion. The virtual image obtained finally after the second diffraction upon the compensation grating is given as the product of Eq.(6.105) and the second term of Eq.(6.103), the latter being modified a little so that the exponent is again multiplied by λ_C/λ_R. The result thus obtained is

$$C'' = C'G_0^2 \exp[-j(\lambda_C/\lambda_R)(\omega_O+\omega_R)x]$$

$$= O_0(x,y)R_0C_0G_0^2 \exp(-j\omega_C x). \tag{6.106}$$

This final formula represents a dispersion-free image because the last exponential term gives a phase shift proportional to λ_C^{-1}, which results in an achromatic angle deviation of the image.

6.12 IMAGE HOLOGRAPHY

6.12.1 PRINCIPLE AND FEATURES

The principle and features of image holography have been described briefly in Section 6.3.3. If we use the terminology introduced in Section 4.4.3.E, image holography may be called holography of close-imaging type. A real image of the object made by a lens very close to the hologram plate is recorded in the hologram as shown in Fig.6.29.

The principal features of image holography are as follows:

(1) White-light reconstruction of the final image is possible if a certain amount of image deterioration is tolerated because the smearing of the reconstructed image due to color dispersion is relatively small.

(2) Smearing of the reconstructed image due to the use of a spatially incoherent, extended light source is relatively small.

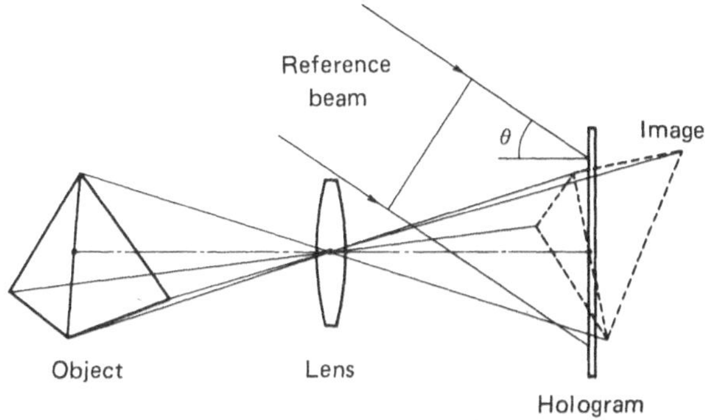

Fig.6.29 Optical setup for recording an image hologram.

These advantages stem from the fact that, as shown in the following subsections, both of the above smearings are proportional to the distance between the hologram plate and the reconstructed image, which is usually very short in image holography. The first and second of the above features were found by Stroke (27) and Rosen (28), respectively, who proposed a *hologram of a focused image*, as they named it, almost simultaneously in 1966.

In the following we discuss the above two features quantitatively. The image hologram is assumed to be recorded in a setup like that shown in Fig.6.29. Whether a real object or a spatial image is recorded on a hologram is not an essential problem for consideration of smearing. (However, this factor gives an essential difference when considering the viewing zone; see Section 6.12.5.) Therefore, in the following subsections, properties of a hologram recording a very close object will be considered.

6.12.2 BASIC EQUATIONS

The symbols used in the following discussions are shown in Fig.6.30. The beam from point O upon the fictitious object (actually not an object but a focused

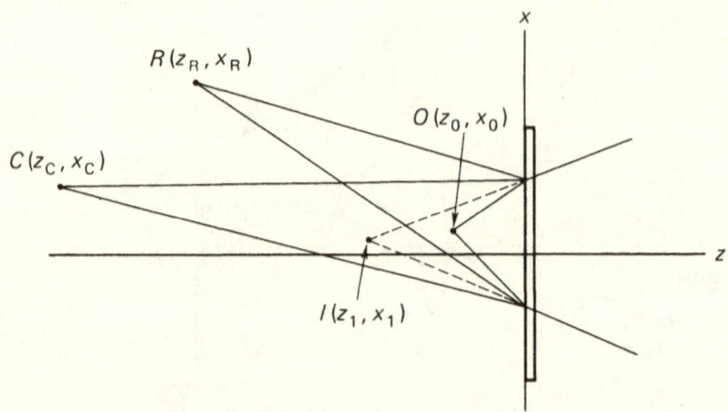

Fig.6.30 Symbols used in the discussions on the properties of image hologram.

image) is recorded using a reference beam from a point source R. In reconstruction, an illuminating beam from a point source C is used, which reconstructs the image at point 1. The coordinates of each point are denoted, as in Section 6.4, by $O(x_O, y_O, z_O)$, $R(x_R, y_R, z_R)$, $C(x_C, y_C, z_C)$, and $I(x_1, y_1, z_1)$.

We further assume for simplicity that $y_O = y_R = y_C = y_1 = 0$. Putting m = 1 into Eqs.(6.35a) and (6.35b) in Section 6.4.3, we obtain

$$\frac{1}{\mu z_1} - \frac{1}{z_O} = \frac{1}{\mu z_C} - \frac{1}{z_R} \qquad (6.107a)$$

$$\frac{x_1}{z_1} = \mu\frac{x_O}{z_O} + \left(\frac{x_C}{z_C} - \mu\frac{x_R}{z_R}\right), \qquad (6.107b)$$

where $\mu = \lambda_C/\lambda_R$ and λ_R and λ_C are the recording and reconstructing light wavelengths, respectively.

The above two equations are the basis for the following discussions. Objection might arise against the use of the equations derived in Section 6.4 where the paraxial approximation is made because the following discussion must deal with very small z_O and z_1, which seemingly do not allow the paraxial approach. In reality, however, the rays in image holography are not very far from paraxial since their directions are limited by the lens diameter. This problem will be discussed in Section 6.12.5 in connection with the viewing zone of the image hologram.

6.12.3 SMEARING DUE TO COLOR DISPERSION

We first compute the smearing of the reconstructed image when the wavelength of the reconstructing light has a certain spread. For simplicity, we assume identical plane waves are used as the reference and reconstructing beams, and set $z_C = z_R = \infty$, $x_C = x_R$, and $x_R/z_R = \theta$. Then, after some computations, we obtain from the above two equations

$$z_1 = (1/\mu) z_O \qquad (6.108a)$$

$$x_1 = x_O + [(1/\mu)-1] z_1 \theta. \qquad (6.108b)$$

We assume that λ_C has a certain spread around λ_R; in other words, μ is distributed between $1 \pm \Delta\mu$. Then, for both ends of the distribution, the positions of the reconstructed image are given as

$$x_1 - x_O = \pm z_1 \theta \Delta\mu. \qquad (6.109)$$

Hence smearing due to color dispersion is proportional to z_1, and we may conclude that image deterioration in achromatic reconstruction is less in image holography than in ordinary holography.

6.12.4 SMEARING DUE TO EXTENDED ILLUMINATING SOURCE

Next we consider the case in which the reconstructing light is monochromatic ($\mu = 1$) but comes from many mutually incoherent point sources distributed in the range $x_C = x_R \pm \Delta s$. Putting $\mu = 1$ into Eqs.(6.107a) and (6.107b), and further assuming that $z_C = z_R$ for simplicity, we obtain

$$z_1 = z_O \tag{6.110a}$$

$$x_1 = x_O + (z_1/z_R)(x_C - x_R). \tag{6.110b}$$

Equation (6.110b) tells that the x coordinate of both ends of the smearing due to the extension of the illuminating source is given as

$$x_1 - x_O = \pm(z_1/z_R)\Delta s = \pm z_1 \Delta\theta, \tag{6.111}$$

which is also proportional to z_1. Therefore, we may conclude that image holography gives less image deterioration due to source extension.

6.12.5 VIEWING ZONE

So far we have discussed only favorable features of image holography. It has, however, an essential limitation as a three-dimensional imaging technique; the viewing zone is limited by the lens diameter.

In the preceding subsections, it has been assumed that a real object placed very close to the hologram plate is recorded, instead of an image once focused by a lens. This replacement has been reasonable in the discussion of smearings. However, as to the viewing zone of the reconstructed image, the difference between recording an object and recording an image is essential.

In recording an image hologram, the actual object light comes from the focusing lens. Therefore, in reconstruction, if we observe the image from the right in Fig.6.30, we may observe the image only within the directions where the focusing lens was in the recording stage. To widen this viewing zone, we should use a lens having as large an F-number as available.

6.13 COLOR HOLOGRAPHY

6.13.1 PROPOSALS IN THE EARLY STAGES

The first proposal for multicolor reconstruction of a holographic image was made by Leith and Upatnieks in 1964 (29). Their basic idea was to record the holographic fringes for various colors (three primary colors, for example) upon a single photographic plate, each fringe pattern being superposed upon others. Later, Lohmann (30) proposed a method of color holography using six reference waves, corresponding to the three primary colors and two (vertical and horizontal) polarizations. His proposal, however, seems to have stemmed principally from a theoretical interest in recording "complete" image information, and is less significant practically.

The first experiment demonstrating the feasibility of the volume hologram for multicolor reconstruction was reported by Pennington and Lin in 1965 (31). As described in Section 6.6.3, the volume hologram has an advantage in that it permits use of white-light illumination for multicolor reconstruction. Since their first experiment, there have been two major streams in color holography:

(1) one stream consisting of methods using a plane hologram and several monochromatic illuminations for reconstruction,

(2) another consisting of those using a volume hologram and achromatic illumination for reconstruction.

Those various proposals and experiments will be described in Section 6.13.4.

6.13.2 CHROMATICITY AND CHROMATICITY DIAGRAM

The concept of "color" is difficult to define because it is a psychological as well as a physical concept.

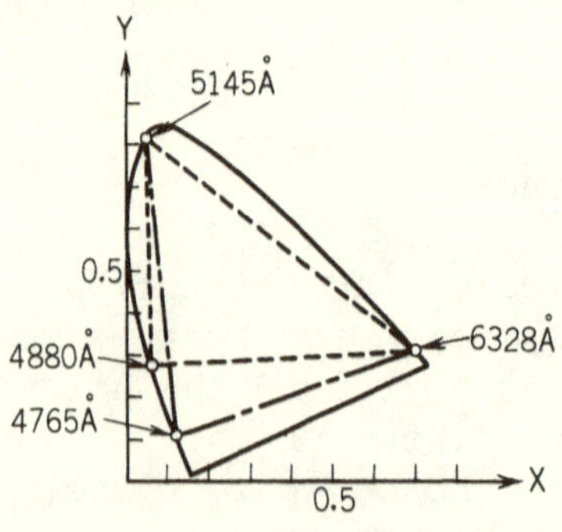

Fig.6.31 The CIE chromaticity diagram.

To simplify the following considerations, we first separate the concept of "brightness" from color. When we feel that two colors give sensations entirely identical to each other except for brightness, we say those two colors have the same chromaticity.

Any chromaticity can be synthesized by monochromatic light. If the wavelengths of the monochromatic light can be chosen arbitrarily, two monochromatic light sources are enough. However, since such an arbitrary choice is not possible in practice, usually three "primary" colors of fixed wavelengths are used to synthesize the desired chromaticity.

The above description may best be understood upon the so-called CIE (Comité International d'Eclairage) chromaticity diagram shown in Fig.6.31. Each chromaticity is assigned coordinates (x,y) in the diagram. Monochromatic visible light is represented by the horseshoelike curve, from 400 through 700 nm in the clockwise direction. When three primary colors, blue (x_B,y_B), green (x_G,y_G), and red (x_R,y_R) are available, the synthesized chromaticity (x_s,y_s) is represented by a weighted linear average in the diagram:

$$x_s = \alpha x_B + \beta x_G + \gamma x_R \tag{6.112}$$

$$y_s = \alpha y_B + \beta y_G + \gamma y_R, \tag{6.113}$$

where $\alpha + \beta + \gamma = 1$. Therefore, when the primary colors are monochromatic as in color holography, the region

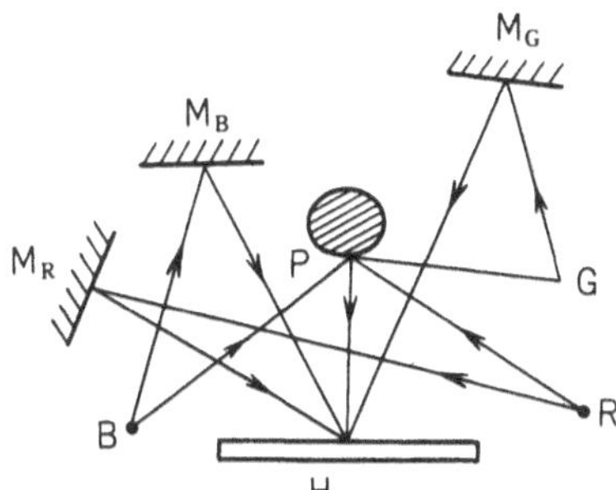

Fig.6.32 Color holography by spatial frequency multiplexing method.

of reproducible chromaticity is represented by a triangle having its vertexes on the horseshoe.

In color holography, the wavelengths most commonly used as the primary colors are:

blue: 488.0 nm (or 476.5 nm) from an argon-ion laser and 441.6 nm from a He-Cd laser,
green: 514.5 nm from an argon-ion laser, and
red: 632.8 nm from a He-Ne laser and 647.1 nm from a krypton-ion laser.

The reproducible regions for some combinations of these wavelengths are also shown in Fig.6.31. These regions do not cover the entire visible chromaticity, the region inside the horseshoe, but cover most of it. Sometimes only two primary colors (488.0 and 632.8 nm, for example) are used for experimental ease. In such a case the reproducible chromaticity is represented by a straight line connecting two points upon the horseshoe.

6.13.3 GHOST IMAGES

The generation of ghost images is the biggest problem in using a plane hologram in multicolor holography.

Let us consider the optical arrangement for hologram recording as shown in Fig.6.32, proposed first by Leith and Upatnieks. Points B, G, R in the figure represent light sources of the three primary colors.

As shown in the figure, three reference waves are assumed to be incident upon the hologram plate from sufficiently different directions. In such a case, the three fringe patterns recorded may have average spatial frequencies much different from one another, while these spatial frequencies are modulated by the object wave information in corresponding colors. Such a scheme is called *spatial frequency multiplexing*.

In reality, the reference-beam paths need not be in a common plane as they seem to be in Fig.6.32. They may be arranged "spatially" to reduce mutual interference. For example, the three interference-fringe patterns may be in directions separated by 120° from each other, while these fringes are modulated by the object-wave information in corresponding colors. Such a scheme is also called spatial frequency multiplexing.

In both cases it may happen that, for example, the fringes for the green light diffract the blue reconstruction light to generate a parasitic blue image. The generation of such a ghost image is sometimes called *color cross talk*.

The number of ghost images (of the desired order of diffraction) is given by $n(n-1)$ where n denotes the number of the recording (and reconstructing) colors. For example, when three colors are used, the three fringe patterns generate nine images, three of them being overlapped to form the wanted image, and the other six constituting ghost images. Two of the ghost images are blue, two are green, and two are red. All of them are enlarged (or reduced) approximately by the factor (reconstruction wavelength)/(recording wavelength), and are shifted in space from the original image position.

The problem of color cross talk makes the method of spatial frequency multiplexing less practical. Some means for overcoming this difficulty have been proposed, such as limiting the visual field so that only the wanted image becomes visible (32), or extinguishing the ghost images by making the reference-beam angle very wide (33). However, both of these methods are accompanied by other drawbacks. Therefore, investigations after 1965 were mainly concentrated on more

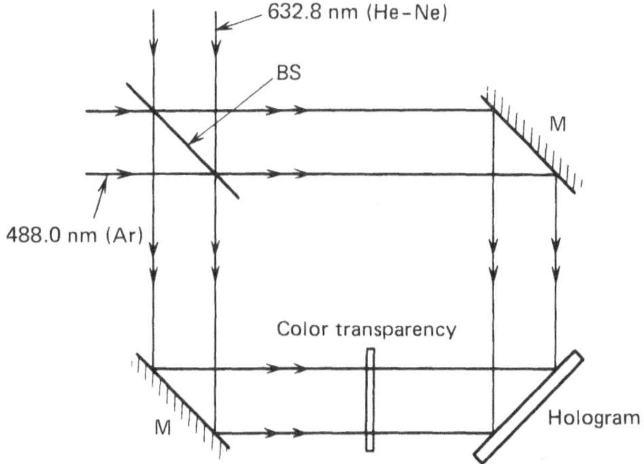

Fig.6.33 The optical setup used in the experiment of Pennington and Lin (31).

practical approaches: white-light reconstruction from volume holograms and multicolor reconstruction from plane holograms of "nonoverlapping spatial multiplexing."

6.13.4 VARIOUS EXPERIMENTS REPORTED IN 1965-1967

In this section, significant experiments reported in the literature in 1965-1967 will be described in chronological order. The author believes that the problems inherent in color holography may best be understood in this manner. Most of the basic ideas of color holography appeared during this short period.

A. Utilization of the Emulsion Thickness

As described before, the use of a volume hologram for color holography was first reported by Pennington and Lin (31). Their optical setup is shown in Fig.6.33. Monochromatic light sources with wavelengths of 488.0 and 632.8 nm were used both in recording and reconstruction.

To remove (or minimize the effect of) ghost images, they utilized the wavelength selectivity of Bragg diffraction by making a Bragg-effect hologram (see Section 6.6.2) in a thick photographic emulsion. The object and reference waves are incident, as shown in Fig.6.33, upon the hologram plate from the same side and with angles of approximately ±45°. Therefore, the fringes are primarily perpendicular to the surface of the plate; the hologram produced is of the volume-type but is not a Lippmann hologram.

The wavelength selectivity of Bragg diffraction, however, is still effective to a certain extent. In the present case, the separation of two wavelengths giving diffraction of half the maximum intensity is given as

$$\Delta\lambda = \frac{\log_e 2}{\pi} \frac{\lambda^2 \cos\theta_O}{\sin\{(\theta_O+\theta_R)/2\}t}, \tag{6.114}$$

where λ is the reconstruction wavelength, t is the thickness of the emulsion, θ_O and θ_R are the incident angles of the object and reference waves, respectively. Note that $\Delta\lambda$ is inversely proportional to the thickness t. If we put, as an example, λ = 500 nm, $\theta_O = \theta_R = 45°$, and t = 15 µm into Eq.(6.114), we obtain $\Delta\lambda$ = 10 nm. Pennington and Lin obtained nearly the same value experimentally.

B. *White-Light Reconstruction of a Multicolor Image*

White-light reconstruction of a color image from a Lippmann-type hologram was reported first by Lin *et al.* (34). The optical setup for recording the hologram is shown in Fig.6.34, where two light sources giving wavelengths of 488.0 and 632.8 nm are used. The object is a color transparency.

It was pointed out by Lin *et al.* that a serious problem in color holography using a Lippmann-type hologram was shrinkage of the emulsion layer of the hologram

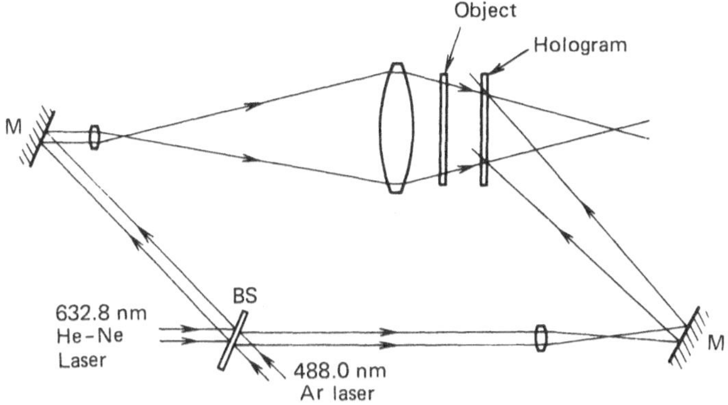

Fig.6.34 The optical setup used in the experiment of Lin et al. (34).

plate; this is caused by removal of the undeveloped silver-halide particles from the emulsion during the fixation process. In Bragg-effect holograms in which the fringes are formed almost perpendicular to the surface of the plate, the effect of that shrinkage is not serious because it takes place in parallel with the fringes. In a Lippmann-type volume hologram, however, the shrinkage of the emulsion brings about a reduction of the fringe spacings in the same proportion. Thus, in white-light reconstruction of an image, the image is reconstructed with colors different from the original ones. In Kodak 649F emulsion which is most commonly used in holography, this shrinkage is about 20% when ordinary developing and fixing processes are performed. Therefore, the red image originally recorded at 632.8 nm (He-Ne laser light) is observed as a green image at about 530 nm (38).

In the experiment reported by Lin *et al.*, the fixation process is omitted to minimize shrinkage of the emulsion. However, some problems such as chemical instability and scattering of light by the remaining silver-halide particles arose instead.

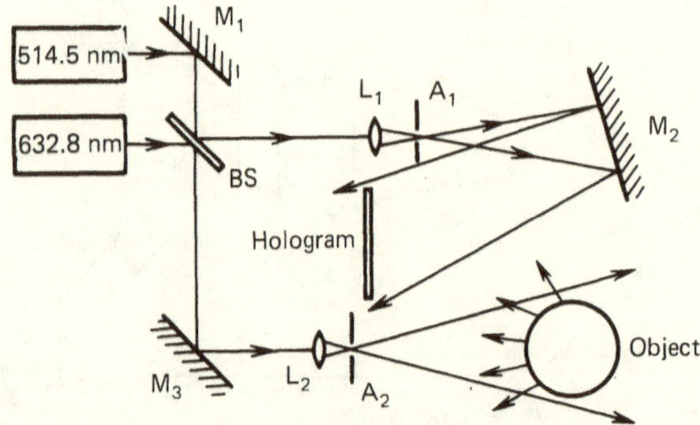

Fig.6.35 The optical setup used in the experiment of Friesem et al. (36).

C. First Tricolor Experiment

In both of the above two experiments, two-color images were reconstructed. The reconstruction of a tricolor image was first reported by Upatnieks, *et al.* (35), who used a Lippmann-type volume hologram with recording wavelengths of 488.0, 514.5, and 632.8 nm. Since their experiment, this combination has become the most common one in tricolor holography. Shrinkage of the emulsion was minimized by omitting the fixation process. Wavelength selectivity (half-intensity width) was about 10 nm.

D. Synthesized Multicolor Method

No photograph of a tricolor image was shown in Ref.(35), but one appears in their following report (36). In the latter report they proposed and showed experiments of another method, called the *synthesized multicolor* method, in which a monochromatic light was used in recording, whereas multicolor lights were used in reconstruction of the image. This method, however, is probably not practical. They also reported experiments of Bragg-effect two-color holography for a three-

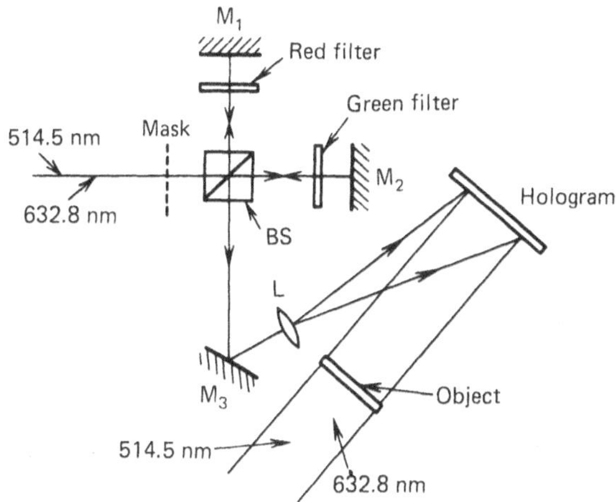

Fig.6.36 Nonoverlapping spatial multiplexing [after Collier and Pennington (37)].

dimensional object using an optical setup as shown in Fig.6.35.

E. Nonoverlapping Spatial Multiplexing

As described above, plane color holography of the spatial frequency multiplexing type and Lippmann-type color holography are both accompanied by several difficulties. Therefore, investigators began to seek a method for removing ghost images from the plane color hologram. The most immediate solution was a mosaic structure like color television kinescopes; that is, to record holographic fringe patterns for different colors at different places upon the hologram plate in a mosaic or striped pattern. This sort of color holography was first experimented with by Collier and Pennington (37) and was named the *nonoverlapping spatial multiplexing* method.

Figure 6.36 shows the optical arrangement used in their experiment. The object is a color transparency. The "mask" in the figure is a striped screen. When

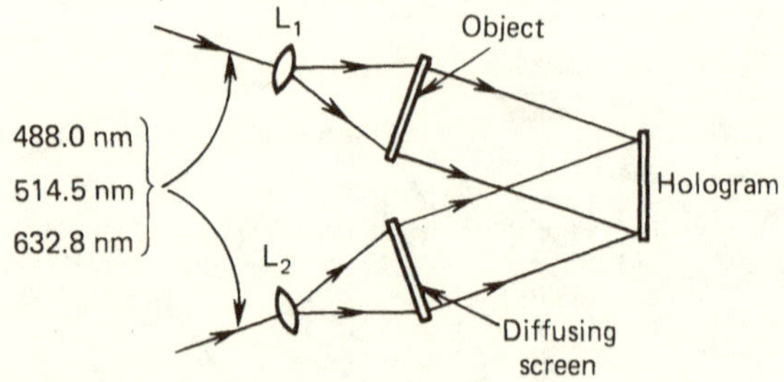

Fig.6.37 Coding of reference waves [after Collier and Pennington (37)].

either mirror M_1 or M_2 is tilted by a small, proper amount, the green and red reference beams illuminate the hologram plate in nonoverlapping stripes, and non-overlapping spatial multiplex recording can be achieved.

In reconstruction, an illuminating optical system identical to that for producing the reference beams must be used. Of course high accuracy is required in the optical arrangement to assure proper registration.

F. Coding of Reference Waves

In the same paper as described in the preceding item (37), Collier and Pennington proposed another method for removing ghost images. It features the use of reference waves each coded by different random spatial modulation signals given by a translucent diffuser, in practice a ground glass plate. In reconstruction, identically coded illuminations are used.

Figure 6.37 shows the optical setup of their experiment using three laser light sources. In this setup, three reference waves are once transmitted through a ground glass plate, which modulates the three waves in both amplitude and phase in complicated manners. Since the refractive index of the glass depends upon the wavelength and the surface roughness of the ground glass has spatial frequency components whose wavelengths are comparable to the reference beam wave-

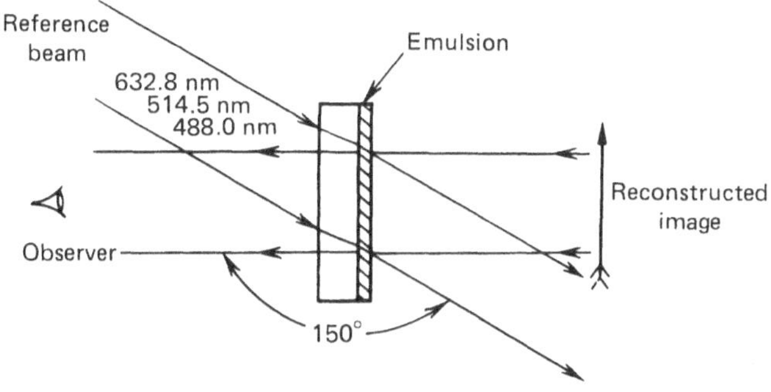

Fig.6.38 The optical setup used in the experiment of Lin and LoBianco (16).

lengths, the modulations given to the three wavelengths are different from one another. Therefore, the three wavefronts on the hologram plate are mutually different; in other words, they are "coded" differently by three spatial noiselike signals.

In reconstruction, illuminating beams having wavefronts identical to the reference beams are used. Generally, in a holographic process, reconstruction of the original image is accomplished by using an illuminating beam having a wavefront the same as the reference beam. This statement also holds for ghost images. Therefore, in the present scheme, no ghost image will be observed because the reference beams and illuminating beams are coded differently. A uniform background noise will be observed instead.

G. Prevention of Emulsion Shrinkage

In 1967, a method for preventing the shrinkage of the emulsion was proposed by Lin and LoBianco (16). In their demonstration, they used an ordinary optical system for Lippmann-type holography as shown in Fig. 6.38, but the emulsion was chemically "swollen" after the fixation process to compensate for the shrinkage. For that purpose the hologram plate was soaked in an

TABLE 6.3 Various Color-Holography Schemes Proposed

Authors	Reference and year	Figure	Object
K.S.Pennington L.H.Lin	(31) 1965	6.33	Two-dimensional transparency
L.H.Lin K.S.Pennington *et al.*	(34) 1966	6.34	Two-dimensional transparency
J.Upatnieks J.Marks R.J.Fedorowics	(35) 1966	--	Three-dimensional diffusing object
A.A.Friesem R.J.Fedorowics	(36) 1967	6.35	Three-dimensional diffusing object
R.J.Collier K.S.Pennington	(37) 1967	6.36	Two-dimensional transparency
R.J.Collier K.S.Pennington	(37) 1967	6.37	Two-dimensional transparency
L.H.Lin C.V.LoBianco	(16) 1967	6.38	Three-dimensional diffusing object

in 1965-1967

Type of hologram	Recording wavelengths (nm)	Reconstruction wavelengths (nm)	Comments
Bragg-effect	632.8 488.0	632.8 488.0	First experiment
Lippmann	632.8 488.0	White light	First white-light reconstruction
Lippmann	632.8 514.5 488.0	White light	First tricolor image
Various volume types	632.8 514.5 488.0	632.8 514.5 488.0	--
Plane	632.8 514.5	632.8 514.5	Nonoverlapping spatial multiplexing
Plane	632.8 514.5 488.0	632.8 514.5 488.0	Coding of reference beams
Lippmann	632.8 514.5 488.0	White light	Control of emulsion thickness

aqueous solution of triethanolamine $(CH_2OHCH_2)_3N$ and was then dried. Thus a proper amount of swell could be given to the emulsion to compensate for the shrinkage.

The optimum solution density and other soaking conditions depend upon the optical density of the hologram and characteristics of the hologram plate. To find these optimum conditions, we may use a trial-and-error approach because the triethanolamine could be washed away if the given swell is found excessive or insufficient.

6.13.5 SUMMARY OF THE EXPERIMENTS IN 1965-1967

The experiments in 1965-1967 described above are summarized in Table 6.3.

6.14 LIGHT SOURCES

6.14.1 REQUIREMENTS FOR LIGHT SOURCES (LASERS) USED IN HOLOGRAPHY

A. Temporal Coherence

Good temporal coherence is essential in a hologram recording process because fringe formation requires a stable phase relation between the object and reference waves. If the object has very small depth, we may reduce the requirements for coherence by arranging the system so that optical path lengths for the object and reference waves are nearly equal to each other. When the object is three-dimensional, however, such an arrangement is generally impossible; the object-wave path lengths cannot be equal to the reference-wave path length everywhere upon the object. Therefore, "temporal" coherence of the light source is needed because we must now make interference fringes of two waves originally emitted by the source at different times.

The degree of temporal coherence is represented by

$$\mu(\tau) = \frac{\langle A(t+\tau)A^*(t)\rangle}{\langle A(t)A^*(t)\rangle}, \tag{6.115}$$

where A(t) is the complex amplitude of the wave and the angle brackets < > indicate a time average. When the source is completely stable, A(t) is not a function of time but a constant. In the present case, however, it is assumed as fluctuating in both phase and amplitude. In Eq.(6.115), the degree of temporal coherence is given as a function of the time interval τ. If we define the optical path difference $\Delta L = c\tau$ where c is the velocity of light, we may rewrite $\mu(\tau)$ as $\mu(\Delta L)$.

A simpler expression of the degree of coherence is *coherence length*, defined as the optical length ΔL that gives $|\mu(\Delta L)| = 1/\sqrt{2}$. This gives a measure for the permissible optical path difference in holography, or in other words the permissible half-depth of the object.

When the oscillation line width of a laser is Δf, the coherence length

$$\Delta L = \alpha c/\Delta f \quad (\alpha = 0.1 \sim 0.3), \tag{6.116}$$

the factor α depending upon the type of discharge (7).

B. Spatial Coherence

The light emitted by a laser oscillating at a single transverse mode is spatially coherent. In holography, use of such a laser is recommended. Furthermore, the oscillating mode preferably should be the TEM mode which is the lowest mode; it gives more uniform and stable output than do the higher order modes.

When oscillation takes place in more than two

transverse modes, light from two spots on the output plane of the laser may not be coherent because outputs from different modes are usually incoherent. Such a source is said to be spatially incoherent. Generally, the degree of spatial coherence between two spots P_1 and P_2 upon the source is expressed, together with temporal coherence, as

$$\gamma_{12}(\tau) = \frac{\langle A_1(t+\tau)A_2^*(t)\rangle}{[\langle A_1(t)A_1^*(t)\rangle\langle A_2(t)A_2^*(t)\rangle]^{1/2}} \qquad (6.117)$$

where A_1 and A_2 are the complex amplitudes at P_1 and P_2, and asterisks denote complex conjugation. When $\tau = 0$, the above equation gives pure spatial coherence.

When a laser is oscillating in many modes but we hope to output only the TEM_{00} mode, we may use a simple spatial filter to remove spurious modes. The principle and construction of the spatial filter is rather simple; the laser output is once focused into a spot and transmitted through a pinhole having diameter close to the diffraction limit of the lowest mode. The transverse higher mode components cannot be focused within the same diameter and most of the energy associated with them is rejected. The smaller the diameter, the more complete the filtering but the greater the energy loss.

C. *Output Power*

We first consider the requirements for lasers used in recording holograms. In principle, the output-power requirement may not be demanding when the object is still because power can be traded with exposure time. However, in practice the power requirement is serious because vibration of the optical setup deteriorates the fringe pattern when exposure time is too long.

For most still objects, an output power of 100 mW is enough practically. However, this power level is not always easily obtainable. In many laboratory experiments, investigators use smaller lasers with outputs from a few through several tens of milliwatts.

When the object is subject to movement (a human being, for example) a pulse laser is needed (39). In this case a single-shot pulse energy of 0.05-0.5 J is required; this can be delivered by a ruby laser.

In reconstructing a three-dimensional image from a hologram, output power above 1 W is desirable for practical applications. However, in most laboratory experiments smaller lasers are usually used. When a small two-dimensional picture is reconstructed from a hologram as in holographic video recording systems, a small laser delivering 1 mW is sufficient.

6.14.2 LASERS USED IN HOLOGRAPHY

A. Helium-Neon Laser

The He-Ne laser became practical at the earliest stage (1963-1965) among all lasers; it is technologically the most stable and is the most reasonable in price. Its maximum power is about 50 mW which is sometimes insufficient, but He-Ne lasers are most commonly used in holography. Ordinarily, the oscillation at a wavelength of 632.8 nm is used. The spectral bandwidth is about 1 GHz, which gives a coherence length on an order of magnitude of 10 cm.

B. Argon Laser

Argon lasers deliver 1-2 W with a coherence length comparable to the He-Ne laser. This laser is technologically premature at present; performance is being improved steadily. By adjustment of the wavelength selector, any of the wavelengths 457.9 nm (purplish blue), 476.5 (blue), 488.0 (blue green), 496.5 (bluish green), 501.7 (green), and 514.5 (green) may be chosen. However, the strongest two wavelengths, 488.0 and 514.5 nm, are usually used. They deliver more than 500 mW each.

C. Krypton Laser

Krypton lasers deliver 1-2 W at various wavelengths ranging through almost the entire visible region. Simultaneous oscillations at three primary colors are even possible. However, the Kr laser is not technologically stable yet and less common in laboratories than the Ar laser.

D. Helium-Cadmium Laser

This delivers blue and ultraviolet light and is useful in making holograms upon dichromated gelatin films or photoresist materials.

E. Ruby Laser

This delivers very narrow (≃ 20 ns) pulsed output at 694.3 nm (red light). Many experiments using ruby lasers in holographic recording of animated objects have been reported because this is the only practical laser to be used in pulsed-laser holography.

6.15 HOLOGRAM RECORDING MATERIALS

6.15.1 REQUIREMENTS FOR PHOTOSENSITIVE MATERIALS USED IN HOLOGRAPHY

A. Sensitivity

High sensitivity is, of course, desirable. The sensitivity at 632.8 nm is especially important in most laboratory uses because He-Ne lasers are most commonly used.

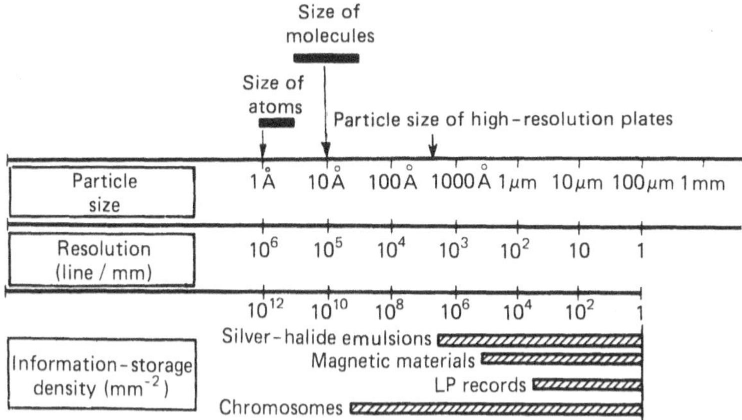

Fig.6.39 Particle size, resolution, and information storage density of various recording materials.

B. *Resolution*

High resolution is, of course, desirable. However, generally speaking, high sensitivity and high resolution are incompatible with each other. For example, for silver-halide emulsions, the following empirical relations are known:

$$\text{sensitivity} \propto d^2 \tag{6.118}$$

$$\text{resolution [lines/mm]} \simeq (1/10)d^{-1} \tag{6.119}$$

where d denotes the average diameter of the silver-halide particles.

Figure 6.39 shows the average particle size, resolution, and information storage density of various recording materials.

C. *Contrast*

In ordinary silver-halide photosensitive materials, the contrast is represented by a parameter γ (see

Section 6.15.2). A higher value of γ corresponds to a stronger contrast, that is, greater optical density variation for a given intensity variation.

In hologram recording, a greater γ gives greater fringe contrast; in the limit a sawtoothlike density variation may be achieved. This waveform is the best for obtaining a high diffraction efficiency as has been described in Section 6.7.1, but produces spurious higher order images.

D. Thickness of the Photosensitive Layer

In ordinary photography, the thickness of the photosensitive emulsion may be one of the causes of reduction in resolution. Therefore, in early stages of the development of photographic materials, it was hoped that the emulsion thickness could be reduced while retaining the necessary optical density.

On the contrary, in holographic processes we often take advantage of the emulsion thickness. In volume holography, a certain emulsion thickness is necessary. There are other cases in which emulsion thickness is needed. For example, in holographic digital memory devices, thick recording material stores information in a three-dimensional manner, and may give storage density higher than a theoretical limit computed upon a two-dimensional basis.

6.15.2 SILVER-HALIDE PHOTOGRAPHIC EMULSIONS

Most of the commercially available photographic films and plates consist of a silver-halide emulsion layer coated over a transparent substrate, a glass plate or an acetate film.

When the silver-halide emulsion is exposed to light, photons excite silver-halide particles to form a "latent preimage speck" which in the developing process is reduced to metallic silver particles as

$$Ag^+ + R \longrightarrow Ag + R^+,$$

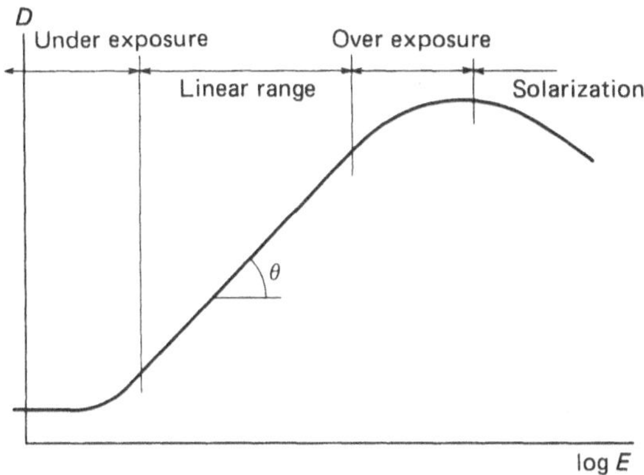

Fig.6.40 Exposure characteristics of silver-halide emulsions.

where R is the developing agent. The reduced silver particles increase the optical density of the plate. As to the mechanism of latent speck formation, many theories have been presented, by Garney and Mott (1938), Mitchell (1957), Trautweiler (1968), Hamilton (1968), and so forth. However, this mechanism has not been elucidated thoroughly.

To discuss the exposure characteristics of silver-halide emulsions, we first define optical density D and exposure E, as

$$D = \log_{10}(1/T) = \log_{10}(1/T_A^2), \tag{6.120}$$

$$E = I\tau, \tag{6.121}$$

where T and T_A are the intensity and amplitude transmittances obtained after exposure and development, respectively; I is the intensity of the incident light; and τ is the exposure time.

The exposure characteristics are usually represented by a curve on a log E-versus-D graph as shown in Fig.6.40. The contrast obtained with a silver-halide emulsion is characterized by a parameter γ, called simply gamma (or gamma value) or the *contrast*,

TABLE 6.4 Sensitivity and Resolution of Commercially Available Silver-Halide Photographic Emulsions

Photographic emulsions	Relative sensitivity at 632.8 nm (ASA value)	Resolution (lines/mm)
Neopan SSS (Fuji)	140	70
Neopan SS (Fuji)	57	80
Neopan F (Fuji)	14	120
Minicopy	1	320
Kodak 649F	0.03	2000
Scientia 10E70	0.5	3000

which is defined as

$$\gamma = \frac{dD}{d(\log_{10}E)} \tag{6.122}$$

in the linear range in the log E-D curve (tan θ in Fig.6.40). Usually γ is used to characterize the standard contrast of an emulsion.

In the following, characteristics of commercially available silver-halide emulsions will briefly be described.

A. *Sensitivity and Resolution*

Table 6.4 shows the sensitivity and resolution of photographic materials available commercially. For holography experiments, resolution better than 1000 lines/mm is usually needed. In the early stages of holography studies, Kodak 649F was widely used. However, since this was not originally intended for holographic uses, the spectral sensitivity is very wide (over the entire visible region) and the sensitivity is relatively low. Recently, Agfa-Gevaert, Inc. offered a new series of silver-halide emulsions which are particularly sensitized for laser light wavelengths;

Scientia 10E56 is specially sensitive to 514.5-nm green light and Scientia 10E70 to 632.8-nm red light. These are widely used in monochromatic hologram recording.

B. Contrast

The value of gamma is critically dependent upon the developing conditions. Hence, designation of its exact value is meaningless. Generally, it is lower for low-resolution, high-sensitivity emulsions (about 2) and higher for high-resolution, low-sensitivity emulsions.

C. Emulsion Thickness

A Kodak 649F plate has an emulsion layer of thickness 15 μm. Most of the other high-resolution emulsions (Kodak 649F film, Agfa-Gevaert 19E56, 10E70, 10E75, etc.) are about 6 μm thick.

6.15.3 OTHER PHOTOSENSITIVE MATERIALS

At present, only the silver-halide emulsion described above is a truly practical hologram recording material. However, there are other potentially practical recording materials which are described in the following:

A. Photoconductor-Thermoplastic Films

Photoconductor-thermoplastic film is sometimes called frost film, and a recording using this film is known as frost photography.

A photoconductive layer and a thermoplastic layer (electrically insulating) are formed on a glass substrate as shown in Fig.6.41. Before exposure the

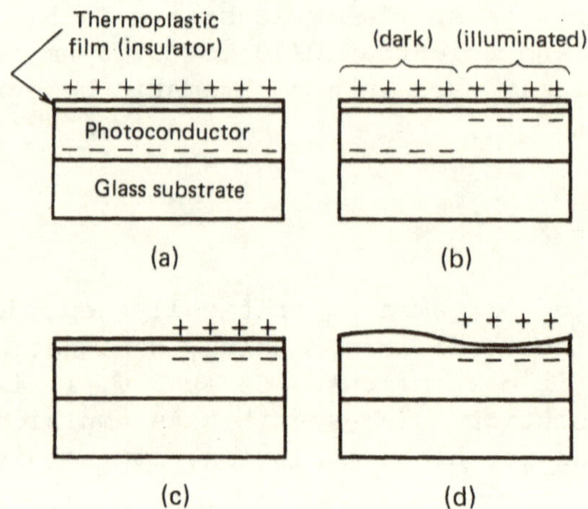

Fig.6.41 Photoconductor-thermoplastic film: (a) charging, (b) exposure, (c) second charging, (d) thermoplastic deformation.

surface of the thermoplastic layer is charged positively by a corona-discharge device. After exposure the negative charges are pulled through the photoconductive layer toward the rear surface of the thermoplastic layer, at those positions where the photoconductor was activated by photons. The film is then charged again by a corona-discharge device; this time the resultant surface-charge pattern is similar to the incident-light image. Thus an electrostatic latent image is obtained. In the development process, the thermoplastic layer is momentarily heated up to the softening temperature (60-100° C). The thermoplastic layer is then deformed; it becomes thinner where there had been strong incident light and thicker where dark.

The thickness modulation thus obtained is in practice sufficiently stable at room temperature. It can easily be erased, if necessary, by simply heating it to a temperature a little higher than the development temperature. In this step, the electrical conductivity of both the photoconductive and thermoplastic layers are increased by irradiation of light, and the surface tension of the thermoplastic film makes itself flat again.

Photoconductor-thermoplastic film has a peculiar spatial frequency response; it shows a broad peak typically at a spatial frequency of 1000 lines/mm. The bandwidth is ordinarily half of the center frequency. This property may readily be understood; when the spatial frequency is too low, in the development process the thermoplastic material has difficulty moving to places where it was less irradiated.

When used in holography, photoconductor-thermoplastic film may form a reversible, phase hologram. It features relatively high sensitivity (ASA 0.3), relatively high diffraction efficiency (12-34%), and relatively high resolution (1000-1200 lines/mm).

B. *Photochromic Materials*

Materials whose optical transmittance varies under irradiation by light are called photochromic materials, and the phenomenon itself is called photochromism. In many cases, photochromic darkening takes place under irradiation at a specific wavelength, whereas transmittance can be restored (the material can be bleached) at another wavelength. Hence the photochromic material can be used as a reversible holographic recording material (40). This reversibility is its greatest advantage. Other advantages are the lack of granularity and that no development process is needed. The disadvantage is its low sensitivity (several joules per square centimeter). When used as a hologram recording material, it forms an amplitude (absorption-type) volume hologram. The efficiency is not high, usually less than 1%.

At present, the following photochromic materials are being experimented with in laboratories.

(1) *Photochromic glass.* Glass containing very fine particles of silver-halide photosensitive materials is called silver-halide photochromic glass or

simply photochromic glass. The darkening and bleaching take place at wavelengths of 488.0 and 632.8 nm, respectively. It has a rather low sensitivity, but is chemically and mechanically stable.

(2) *Photochromic crystals.* Several kinds of crystalline photochromic materials are known: (a) CaF_2 crystals doped with La, Ce, or Tb; (b) $SrTiO_3$ doped with Fe-Mo or Ni-Mo; and (c) alkali halides such as KBr or KI. These materials are stable and have relatively high sensitivity, but are difficult to form into large crystals.

(3) *Organic photochromic materials.* A famous example is the spiropyrane derivative. The sensitivity is high, but chemical stability is low. After long use, the photochromism deteriorates. It is usually suspended in transparent plastic films for experimental use.

C. Dichromated Gelatin Films

Gelatin films containing a small amount of dichromate become insoluble in water after irradiation by light, whereas the unilluminated portion can be washed away in water. This material is said to be an almost ideal hologram recording material because a volume phase hologram with very little absorption can be made with it (7).

This material is not commercially available at present, but can be prepared in laboratories. [Refer to Ref. (7) for the detailed process and also for related literature.] The resolution is very high (up to 3000 lines/mm); successful formation of a diffraction grating with a pitch of 270 nm has been reported. The disadvantage is that it is insensitive to He-Ne laser light, and has very low sensitivity (1/100 as compared with Kodak 649F) even for a wavelength of 488.0 nm.

D. *Photopolymers*

When photopolymer materials are illuminated, polymerization of molecules takes place and a light-scattering polymer is produced in the form of a coloid. The fixation can be carried out by irradiation with an ultraviolet light.

For photographic or holographic uses, the photo-polymer material is sandwiched by two flat glass plates with a spacing of 10-20 μm. It features a relatively high resolution (above 1000 lines/mm) and high diffraction efficiency (45%), but is an irreversible recording material.

E. *Miscellaneous*

At present, also being investigated or experimented with are: the use of the refractive-index change of ferroelectric materials ($LiNaO_3$, $BaTiO_3$, and so forth) due to optical damage; evaporation of metallic films; and demagnetization of magnetooptic materials by strong ruby-laser light.

6.16 SUMMARY

In this chapter, the principle and technological aspects of holography have been described. Emphasis has been placed upon its features as a three-dimensional imaging technique. Some topics, such as the quantity of information recorded, reduction techniques, and special holography techniques (holographic screen or projection-type holography) have been omitted; these will be described separately in Chapters 7 and 8.

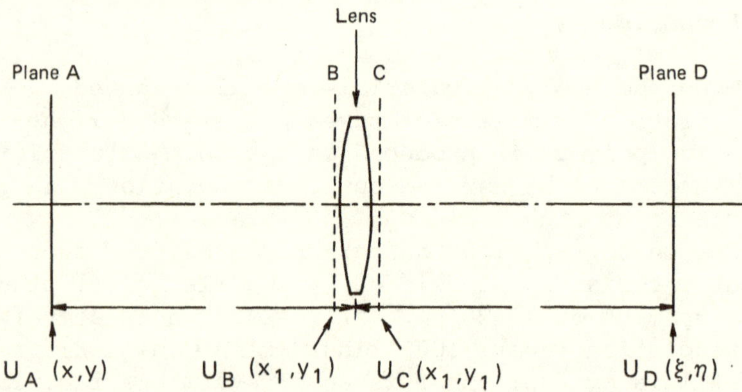

Fig.A.1 An optical system for producing a Fourier-transform hologram.

APPENDIX I ANALYSIS OF A FOURIER-TRANSFORM HOLOGRAM WITH LENS

It is stated in Section 6.3.3 that two image planes separated from a convex lens (focal length f) by distance f each (see Fig.6.3b) are related to each other through Fourier transforms. It is also stated that, therefore, a hologram constructed in an arrangement like Fig.6.3b is called a Fourier-transform hologram or a Fraunhofer hologram. In this appendix, a mathematical proof of the first statement will be given.

We consider an optical system as shown in Fig.A.1. Suppose an image $u_A(x,y)$ is placed at plane A and is transmitted to plane D. Planes A and D are separated from lens L by distances d and f, respectively, where f is the focal length of the lens.

First, the amplitude distribution upon plane B is considered. It is given as a Fresnel diffraction of the image on A. From a relation similar to Eq. (6.13) of the text, it is given as

$$u_B(x_1,y_1) \propto \exp(-jkd) \iint_{-\infty}^{\infty} u_A(x,y)$$

$$\exp\left(-jk\frac{(x-x_1)^2+(y-y_1)^2}{2d}\right)dx\,dy, \tag{A.1}$$

where an approximation similar to Eq.(6.27)

$$\sqrt{d^2+(x-x_1)^2+(y-y_1)^2} \simeq d + \frac{(x-x_1)^2+(y-y_1)^2}{2d} \tag{A.2}$$

is used. An integral of the form in Eq.(A.1), which gives a Fresnel diffraction pattern of a plane object, is called a Fresnel-Kirchhoff integral.

The convex lens L has the property of converting a plane wave to a converging spherical wave, by giving a phase shift expressed as $\exp[jk(x_1^2+y_1^2)/2f]$. Therefore, the amplitude distribution upon plane C is given as

$$u_C(x_1,y_1) \propto \exp[jk(x_1^2+y_1^2)/2f]u_B(x_1,y_1). \tag{A.3}$$

Finally, the amplitude distribution upon D is again given as the Fresnel diffraction of $u_C(x_1,y_1)$, so that

$$u_D(\xi,\eta) \propto \iint_{-\infty}^{\infty} u_C(x_1,y_1)\ \exp\left(-jk\frac{(x_1-\xi)^2+(y_1-\eta)^2}{2f}\right)dx_1\,dy_1$$

$$\propto \exp\left(-jk\frac{\xi^2+\eta^2}{2f}\right)\iint_{-\infty}^{\infty} u_B(x_1,y_1)\exp\left(jk\frac{x_1\xi+y_1\eta}{f}\right)dx_1\,dy_1. \tag{A.4}$$

Equations (A.1) and (A.4) relate $u_A(x,y)$ and u_D. However, it is rather tedious to combine these equations straightforwardly. We notice that the integral in Eq.(A.4) has the form of a Fourier transform. If we define the Fourier transform of $u_B(x,y)$ as

$$B(\omega_x,\omega_y) = \frac{1}{2\pi}\iint_{-\infty}^{\infty} u_B(x_1,y_1)\ \exp[j2\pi(x_1\omega_x+y_1\omega_y)]dx_1dy_1, \tag{A.5}$$

then Eq.(A.4) is rewritten as

$$u_D(\xi,\eta) \propto \exp\left(-jk\frac{\xi^2+\eta^2}{2f}\right)B\left(\frac{\xi}{\lambda f},\frac{\eta}{\lambda f}\right). \qquad (A.6)$$

Next, we take the Fourier transforms of both sides of Eq.(A.1). Using the theory that the Fourier transform of a convolution of two functions is given as the product of the Fourier transforms of each function, we may write

$$B(\omega_x,\omega_y) \propto A(\omega_x,\omega_y)\ \exp[j\pi\lambda d(\omega_x^2+\omega_y^2)] \qquad (A.7)$$

where $A(\omega_x,\omega_y)$ denotes the Fourier transform of $u_A(x,y)$, and a relation

$$\mathcal{F}[\exp(-ax^2)] = (1/2a)\ \exp(-\pi^2\omega^2/a)$$

has been used, where $\mathcal{F}$ denotes Fourier transform. If we set

$$\omega_x = \xi/\lambda f, \quad \omega_y = \eta/\lambda f, \qquad (A.8)$$

Eqs.(A.6) and (A.7) directly lead to

$$u_D(\xi,\eta) \propto \exp\left[-j\frac{\pi}{\lambda}\,\frac{\xi^2+\eta^2}{f}\,(1-d/f)\right]A\left(\frac{\xi}{\lambda f},\frac{\eta}{\lambda f}\right). \qquad (A.9)$$

In particular, when d = f, the above equation is simplified to

$$u_D(\xi,\eta) \propto A(\xi/\lambda f,\eta/\lambda f), \qquad (A.10)$$

which is the relation to be proved. Even when $d \neq f$, u_D is given as the Fourier transform of $u_A(x,y)$ when the phase difference is neglected.

Optical systems as described above are often employed in information processing schemes.

APPENDIX II ANALYSIS OF THE RESOLUTION OF AN IMAGE RECONSTRUCTED FROM A FINITE HOLOGRAM

The purpose of this appendix is to derive Eq. (6.85) in the text. We first put $z = 0$ into the right-hand side of Eq.(6.74) and equate it to Eq.(6.84), so that

$$C_0 T(x) = \int_{-\pi/2}^{+\pi/2} p(\theta)\ \exp(-jkx\sin\theta)\,d\theta. \tag{A.11}$$

Under the paraxial approximation $\theta \ll \pi$, the above equation is approximated as

$$C_0 T(x) = \int_{-\infty}^{\infty} p(\theta)\ \exp(-j2\pi x\theta/\lambda)\,d\theta, \tag{A.12}$$

where $\lambda = 2\pi/k$ has been used. The above equation tells us that $T(x/\lambda)$ and $p(\theta)$ are related as

$$\mathcal{F}^{-1}[p(\theta)] = C_0 T(x/\lambda) \tag{A.13}$$

where $\mathcal{F}^{-1}$ denotes the inverse Fourier transform. Therefore, $p(\theta)$ should be obtained simply as the Fourier transform of the transmittance distribution, Eq.(6.84).

Equation (6.84) is given as a product of the sinusoidal transmittance distribution identical to Eq.(6.71):

$$T_1(x) = \left(1 - \frac{m}{2}\right) + \frac{m}{2}\cos\left(\frac{2\pi x}{d}\right), \tag{A.14}$$

and a rectangular window function

$$T_2(x) = \begin{cases} 1 & \left(-\dfrac{D}{2} < x < +\dfrac{D}{2}\right) \\ 0 & \left(x < -\dfrac{D}{2},\ \dfrac{D}{2} < x\right). \end{cases} \tag{A.15}$$

We note that the Fourier transform of the product of two functions is given as the convolution of the Fourier transforms of the two functions. The Fourier transform of $T_1(x/\lambda)$ is known [see Eq.(6.76)], whereas that of $T_2(x/\lambda)$ is given as

$$\mathcal{F}[T_2(x/\lambda)] = \frac{\sin(\pi D\theta/\lambda)}{\pi D\theta/\lambda}. \qquad (A.16)$$

Equation (6.85) is obtained as the convolution of Eq. (6.76) and Eq.(A.16).

APPENDIX III DERIVATION OF SPECKLE-NOISE FORMULA

The derivation of Eq.(6.91) will be described. Putting Eq.(6.89) into Eq.(6.90), we obtain

$$F(X) = \frac{\sum_n \sum_m I_n I_m (1+g_{nm}) \sin c^2[(k_n-k_m)X/2]}{\left(\sum_n I_n\right)^2} \qquad (A.17)$$

where

$$I_n = A_n A_n^*, \quad I_m = A_m A_m^*, \qquad (A.18)$$

and g_{nm} is a quantity defined as

$$g_{nm} = A_n A_m^* \sum_{n'} \sum_{m'} A_{n'} A_{m'}/I_{nm} \qquad (A.19)$$

for all sets of n, m, n', m' satisfying

$$k_n - k_m = k_{n'} - k_{m'}, \quad (n,m) \neq (n',m'). \qquad (A.20)$$

In computing Eq.(A.17), the relation

$$\langle \exp[-j(k_n-k_m-k_{n'}+k_{m'})x_0 + j(\phi_n-\phi_m-\phi_{n'}+\phi_{m'})]\rangle_{x_0}$$

$$= \delta(k_n-k_m-k_{n'}+k_{m'}) \tag{A.21}$$

is used, where δ denotes a delta function. In actual cases, the directions of the incident light are distributed in a two-dimensional manner so that

$$\begin{aligned} k_{nx} - k_{mx} &= k_{n'x} - k_{m'x}, \\ k_{ny} - k_{my} &= k_{n'y} - k_{m'y} \end{aligned} \tag{A.22}$$

are required. Therefore, approximating as $g_{nm} = 0$ brings forth little error.

Next, we modify Eq. (A.17) to an integral form assuming that the sources are distributed continuously. If the wave number k is distributed between 0 and K, Eq. (A.17) may be rewritten as

$$F(x) = \frac{\int_0^K \int_0^K \sin c^2[(k_n-k_m)X/2]dk_n\, dk_m}{\int_0^K \int_0^K dk_n\, dk_m} \tag{A.23}$$

$$= \left(\frac{1}{2}KX\right)^{-2}\left[(KX)\,\mathrm{Si}(KX) - \mathrm{Cin}(KX) - 2\sin^2\left(\frac{1}{2}KX\right)\right], \tag{A.24}$$

where

$$\mathrm{Si}(z) = \int_0^z (\sin t/t)dt, \quad \mathrm{Cin}(z) = \int_0^z [(1-\cos t)/t]dt. \tag{A.25}$$

References

(1) Latta, J. N., The Bleaching of Holographic Diffraction Gratings for Maximum Efficiency, *Appl.*

Opt. 2, No. 12, 2409-2418 (Dec. 1968).
(2) Sheridon, N. K., Producing of Blazed Holography, *Appl. Phys. Lett. 12,* 316-318 (May 1968).
(3) Meier, R. W., Cardinal Points and the Novel Imaging Properties of a Holographic System, *J. Opt. Soc. Amer. 56,* No. 2, 219-223 (Feb. 1966).
(4) Meier, R. W., Magnification and Third-Order Aberrations in Holography, *J. Opt. Soc. Amer. 55,* No. 8, 987-992 (Aug. 1965).
(5) Smith, H. M., "Principles of Holography." Wiley (Interscience), New York, 1969.
(6) Kogelnik, H., Coupled Wave Theory for Thick Hologram Gratings, *Bell Syst. Tech. J. 48,* No. 9, 2902-2947 (Nov. 1969).
(7) Collier, R. J., Burckhardt, C. B., and Lin, L. H., "Optical Holography." Academic Press, New York, 1971.
(8) Lee, W. H., and Greer, M. O., Noise Characteristics of Photographic Emulsions Used for Holography, *J. Opt. Soc. Amer. 61,* No. 3, 402-409 (March 1971).
(9) Amodei, J. J., and Mezrich, R. S., Holograms in Thin Bismuth Films, *Appl. Phys. Lett. 15,* No. 2, 45-46 (July 1969).
(10) Nomura, H., Projection-Type Holography, M.S. Dissertation, Department of Electronic Eng., Univ. of Tokyo, March 1972.
(11) Pennington, K. S., and Harper, J. S., Techniques for Producing Low-Noise, Improved Efficiency Holograms, *Appl. Opt. 9,* No. 7, 1643-1650 (July 1970).
(12) Takeda, Y., Nonogami, S., and Miyamura, Y., Bleached Phase Hologram, Oyo Butsuri *(J. Appl. Phys. Jap.)* 40, No. 1, 41-48 (Jan. 1971).
(13) Bryngdahl, O., Formation of Blazed Gratings, *J. Opt. Soc. Amer. 60,* No. 1, 140-141 (Jan. 1970).
(14) George, N., and Mathews, J. W., Holographic Diffraction Gratings, *Appl. Phys. Lett. 9,* No. 5, 212-215 (Sept. 1966).
(15) Lin, L. H., Hologram Formation in Hardened Dichromated Gelatin Films, *Appl. Opt. 8,* No. 5, 963-966 (May 1969).

(16) Lin, L. H., and LoBianco, C. V., Experimental Techniques in Making Multicolor White-Light Reconstructed Holograms, *Appl. Opt. 6,* No. 7, 1255-1258 (July 1967).
(17) Colburn, W. S., and Haines, K. A., Volume Hologram Formation in Photopolymer Materials, *Appl. Opt. 10,* No. 7, 1636-1641 (July 1971).
(18) Gerritsen, H. J., Hannan, W. J., and Ramberg, E. G., Elimination of Speckle Noise in Hologram With Redundancy, *Appl. Opt. 7,* No. 11, 2301-2311 (Nov. 1968).
(19) Gabor, D., Laser Speckle and Its Elimination, *IBM J. Res. Develop.,* 509-514 (Sept. 1970).
(20) Reinheimer, C. J., and Copple, J. D., Experimental Fourier Analysis of Speckle Patterns, *Appl. Opt. 10,* No. 6, 1456-1458 (June 1971).
(21) Pole, R. V., Three-Dimensional Imagery and Holograms of Objects Illuminated in White Light, *Appl. Phys. Lett. 10,* No. 1, 20-22 (Jan. 1, 1967).
(22) McCrickerd, J. T., and George N., Holographic Stereogram From Sequential Component Photographs, *Appl. Phys. Lett. 12,* No. 1, 10-12 (Jan. 1, 1968).
(23) Kasahara, T., Kimura, Y., and Kawai, M., Three-Dimensional Construction of Imaginary Objects by the Method of Holographic Stereogram, A paper read at the US-Japan Holography Seminar, Tokyo, Oct. 1969.
(24) Yano, A., Improved Method of Making Holographic Stereograms Using Stereoscreens, *Opt. Comm. 2,* No. 5, 209-211 (Oct. 1970).
(25) Burckhardt, C. B., Display of Holograms in White Light, *Bell Syst. Tech. J. 45,* No. 12, 1841-1844 (Dec. 1966).
(26) DeBitetto, D. J., White-Light Viewing of Surface Holograms by Simple Dispersion Compensation, *Appl. Phys. Lett. 9,* No. 12, 417-418 (Dec. 15, 1966).
(27) Stroke, G. W., White-Light Reconstruction of Holographic Images Using Transmission Holograms Recorded With Conventionally Focused Images and In-Line Background, *Phys. Lett. 23,* No. 5, 325-327 (Oct. 31, 1966).

(28) Rosen, L., Focused-Image Holography With Extended Sources, *Appl. Phys. Lett. 9,* No. 9, 337-339 (Nov. 1, 1966).

(29) Leith, E. N., and Upatnieks, J., Wavefront Reconstruction With Diffused Illumination and Three-Dimensional Objects, *J. Opt. Soc. Amer. 54,* No. 11, 1295-1301 (Nov. 1964).

(30) Lohmann, A. W., Reconstruction of Vectorial Wavefronts, *Appl. Opt. 4,* No. 12, 1667-1668 (Dec. 1965).

(31) Pennington, K. S., and Lin, L. H., Multicolor Wavefront Reconstruction, *Appl. Phys. Lett. 7,* No. 3, 56 (Aug. 1, 1965).

(32) Mandel, L., Color Imagery by Wavefront Reconstruction, *J. Opt. Soc. Amer. 55,* No. 12, 1697-1698 (Dec. 1965).

(33) Marom, E., Color Imagery by Wavefront Reconstruction, *J. Opt. Soc. Amer. 57,* No. 1, 101-102 (Jan. 1967).

(34) Lin, L. H., Pennington, K. S., Stroke, G. W., and Labeyrie, A. E., Multicolor Holographic Image Reconstruction With White-Light Illumination, *Bell Syst. Tech. J. 45,* No. 4, 659-660 (April 1966).

(35) Upatnieks, J., Marks, J., and Fedorowics, R. J., Color Holograms for White-Light Reconstruction, *Appl. Phys. Lett. 8,* No. 11, 286-287 (June 1, 1966).

(36) Friesem, A. A., and Fedorowics, R. J., Multicolor Wavefront Reconstruction, *Appl. Opt. 6,* No. 3, 529-536 (March 1967).

(37) Collier, R. J., and Pennington, K. S., Multicolor Imaging From Holograms Formed on Two-Dimensional Media, *Appl. Opt. 6,* No. 6, 1091-1096 (June 1967).

(38) Vilkomerson, D. H. R., and Bostwick, D., Some Effects of Emulsion Shrinkage on a Hologram's Image Space, *Appl. Opt. 6,* No. 7, 1270-1272 (July 1967).

(39) Siebert, L. D., Front-Lighted Pulse Laser Holography, *Appl. Phys. Lett. 11,* No. 10, 326-328 (Nov. 15, 1967).

(40) Leith, E. N., Upatnieks, J., Hildebrand, B. P., and Haines, K., Requirements for Wavefront Reconstruction Television Facsimile System, *J. SMPTE 74,* No. 10, 893-896 (Oct. 1965).

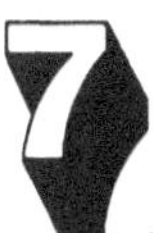

Information Recorded in Three-Dimensional Images and Its Reduction

The purpose of this chapter is to consider the quantity of information recorded in various three-dimensional images: holography, integral photography, and multiple photography. To date, the theories on information quantity have not been completed. Described in this chapter, therefore, are only analyses of the number of sampling points required to determine uniquely a three-dimensional image. The results obtained are summarized in a short table.

In the latter half of this chapter, methods of reducing the quantity of information are described. It is emphasized that the most effective way of reducing information is to give up vertical parallax information.

7.1 INTRODUCTION

7.1.1 TECHNICAL BACKGROUND

The amount of information recorded in a three-dimensional image is usually tremendous. This fact offers three engineering tasks. The first is to compute the real information content, that is, to make clear what part of that tremendous amount of information is really necessary for reconstruction of the image, and what part is to be attributed to redundancy of information and hence could be removed. The second task is to develop information-reduction techniques, with which we can reduce the redundancy of information and record the minimum necessary information. If a transmission technique for three-dimensional images is established in future, it will inevitably employ some such information-reduction techniques. The third task is to develop optical memory devices having a tremendous information capacity, on the basis of the same principle as three-dimensional imaging techniques. An example of such a memory device is a holographic memory; it features an associated readout function and the potential for gigantic memory capacity in future.

In this chapter, recent efforts toward the first task (computation of the information content) and the second (information-reduction techniques) will be described. At present, the theory of the information recorded in a three-dimensional image has not been completed. Only available are analyses on the number of sampling points required to "determine uniquely" a three-dimensional image. This sort of computation is the first necessary step for computing the true information content. These theories for determining the sampling point number have appeared quite recently; they have not been stated in a systematic manner. In this chapter, therefore, the author tries to systematize the available theories on various three-dimensional imaging techniques, and also to clarify the mutual relationships among these theories. (The symbols used are unified, and hence might sometimes be different from the original papers.) It was found, however, that the presently available theories (including the author's) are not sufficient to allow construction of a complete, systematic set of theories. Some new theories, therefore, have been developed and added in this book. It will be seen that the framework of the theories on the sampling point number have almost been completed.

The true information content is obtained as the product of the sampling point number and the information (bit) per sampling point. In computing the latter quantity, the S/N (single-to-noise) ratio of the recording medium is used; in other words, turbidity of the photographic material as well as speckle noise have to be taken into account. It should be noted that such a S/N ratio theory associated with three-dimensional imaging has not been presented; this is truly a challenging task for the near future.

In the latter half of this chapter, several proposals for reducing hologram information will be described. As to integral photography, two practical methods of information reduction (LS pictures and multiple-picture projection-type three-dimensional images) have been discussed in detail in Chapters 4 and 5. These will, therefore, be rediscussed only rather briefly in Section 7.7.

7.1.2 HISTORICAL BACKGROUND

To begin, theories on the information content recorded in three-dimensional images and its reduction will be mentioned briefly in chronological order. (However, such an historical review might sometimes be troublesome for those readers who do not work in this specific field. In case the reader really feels so, he should skip this subsection and proceed to Section 7.2. He may return to this subsection later, if necessary.)

First, works on information or sampling point number of three-dimensional images have been presented in the past 10 years as follows:

1962 Leith and Upatnieks (1) discussed the holography technique in connection with information theory.

1965 Leith *et al.* (2) computed the sampling point number of a hologram recording two-dimensional picture information.

1968 Burckhardt computed the sampling point number of an integral photograph (3), and also of a hologram

TABLE 7.1 Six Categories of Computation of Sampling Point Number in Three-Dimensional Imaging

(A) Computation of sampling point number upon the recording plane	(1) Integral photography
	(2) Multiple photography
(B) Computation of resolvable spot number of the reconstructed image	(3) Holography

recording three-dimensional picture information (4).

1969 Saito and Nagata (5) computed the "spatial resolvable spot number" of a three-dimensional spatial image reconstructed from a hologram.

1970 Okoshi (6) computed the "depth resolvable spot number" of a lens-sheet three-dimensional picture, and also of a projection-type three-dimensional image.

We can classify the above theories (except for the first, which is of a fundamental nature) according to a classifying principle as shown in Table 7.1 into six categories as follows:

(A-1) computation by Burckhardt (3),

(A-2) computation described in Section 7.5.1,

(A-3) computation by Leith (two-dimensional object) (2) and computation by Burckhardt (three-dimensional object) (4),

(B-1) computation described in Section 7.4.3,

(B-2) computation by Okoshi (6) and computation described in Section 7.5.2,

(B-3) computation by Saito and Nagata (5).

Note that some computations (A-2, B-1, B-2) have recently been performed to complete the list of analyses. In the first half of this chapter, the above six analyses are described; in the latter half, the following six techniques for reducing hologram information will be described:

1968 experiment of sampled hologram by Burckhardt (4),

1968 experiment of sampled Fourier-transform hologram by Lin (7),

1968 proposal of information reduction using a diffuser by Haines and Brumm (8),

1969 experiment of unidirectional holography by Fritzler and Marom (9),

1970 experiment of double-reflection holography by Hamasaki and Nagata (10),

1970 experiment of projection-type holography by Okoshi and Yano (11).

These experiments and proposals will be introduced in the above chronological order.

7.2 BASIC PROPERTIES OF FOCUSING OF A LIGHT BEAM IN IMAGE SPACE

7.2.1 SIMILARITY AND DIFFERENCE BETWEEN LENS AND HOLOGRAPHIC SYSTEMS

As stated in Chapter 6, the equations describing focusing phenomena in a lens system and in a holographic system are similar to each other, as far as a single image point is concerned. However, in most holographic processes, the image to be reconstructed is not a single image point, but an ensemble of a large number of image points; this is in practice the biggest difference between the above two cases. In other words, a lens collimates a beam into a spot, whereas a hologram collimates it into many spots scattered in image space as shown in Fig. 7.1.

7.2.2 LATERAL AND DEPTH RESOLUTION LIMITS

We consider lens-type and holographic optical systems as shown in Fig.7.1. Both of them are assumed to

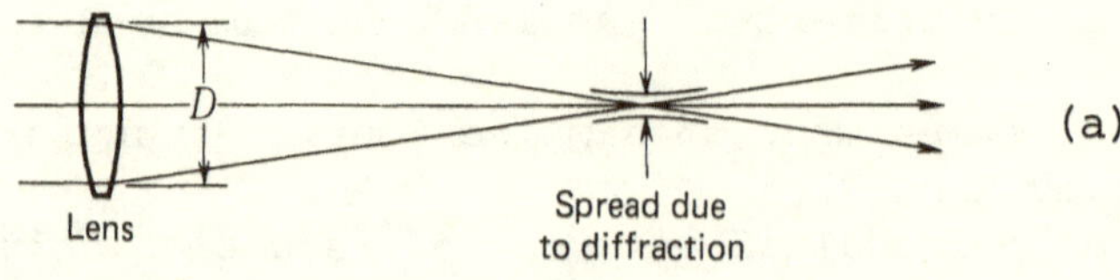

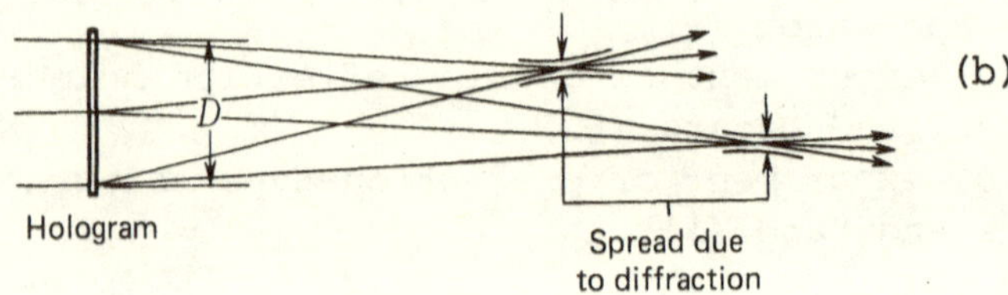

Fig.7.1 (a) Focusing in a lens system. (b) Focusing in a holographic system.

be free from aberrations, and are of square shape with dimensions D × D. In both systems, the focusing can never be complete, since a certain spread of the light beam due to diffraction is always present; the width of the spread is determined by D.

We use symbols as defined in Fig.7.2 to derive the amplitude fluctuation in the space around the spot image. The spread of the light beam due to diffraction δ_1 as defined in Fig.7.2a is given, when so-called Fraunhofer diffraction is assumed, as

$$\delta_1(z) = 2\lambda z/D, \tag{7.1}$$

where λ is the wavelength. On the other hand, when $z \neq a$, spread of the beam due to defocus is also present. It is given from Fig.7.2b as

$$\delta_2(z) = |z-a|D/a. \tag{7.2}$$

If we make the approximation that the spread for any z is given as $\delta(z) = \{\delta_1^2(z) + \delta_2^2(z)\}^{1/2}$, we obtain

$$\delta^2(z) = \left(\frac{z-a}{a} D\right)^2 + \left(\frac{2\lambda z}{D}\right)^2. \tag{7.3}$$

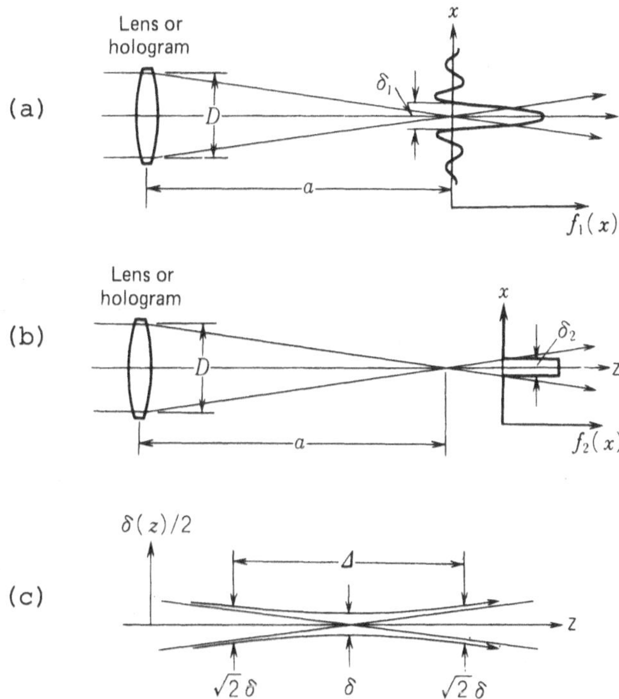

Fig.7.2 (a) Blurring due to diffraction. (b) Blurring due to defocus. (c) Definitions of lateral resolution limit and depth resolution limit.

The spread given by Eq.(7.3) implies an envelope of the light beam as illustrated in Fig.7.2c.

We define, referring to Fig.7.2c, the lateral resolution limit δ as the resolution limit $\delta(z)$ of the point image at its minimum (waist) value. We also define the depth resolution limit Δ as the distance between two points upon the z axis where the light intensity is reduced by 1/2, *i.e.*, where $\delta^2(z) = 2\delta^2$.

Then the lateral resolution limit is obtained, from Eq.(7.3), as

$$\delta = \{\delta(z)\}_{min} = (2\lambda a/D)[1 + (2\lambda a/D^2)^2]^{-1/2}, \qquad (7.4)$$

for a distance given by

$$z = a[1 + (2\lambda a/D^2)^2]^{-1}. \tag{7.5}$$

However, in most practical cases $2\lambda a/D^2 \ll 1$ holds. [This means that the spread due to diffraction is much smaller than the lens (hologram) size; that is, $2\lambda a/D \ll D$.] Hence, the best focus is obtained at $z \doteqdot a$ and

$$\delta \doteqdot 2\lambda a/D. \tag{7.6}$$

The depth resolution limit Δ is obtained, after some computations, as

$$\Delta = (4\lambda a^2/D^2)\{1 + (2\lambda a/D^2)^2\}^{-1}, \tag{7.7}$$

which may also be simplified, under the above assumption, as

$$\Delta \doteqdot 4\lambda a^2/D^2. \tag{7.8}$$

In the following considerations Eqs.(7.6) and (7.8) will be used throughout.

7.3 SAMPLING POINT NUMBER OF A HOLOGRAM

7.3.1 CASE OF A TWO-DIMENSIONAL OBJECT

To facilitate later discussions, the sampling point number of a hologram recording a two-dimensional object will be considered first.

A. Spatial Frequency Distribution of the Hologram Pattern

The picture information of a two-dimensional object with dimensions $w \times w$ is recorded in a hologram having dimensions $D \times D$, placed at a distance a from the object, as shown in Fig.7.3a. A plane reference wave with an incident angle θ_R is assumed. The angle under which the hologram sees the object is denoted by ψ.

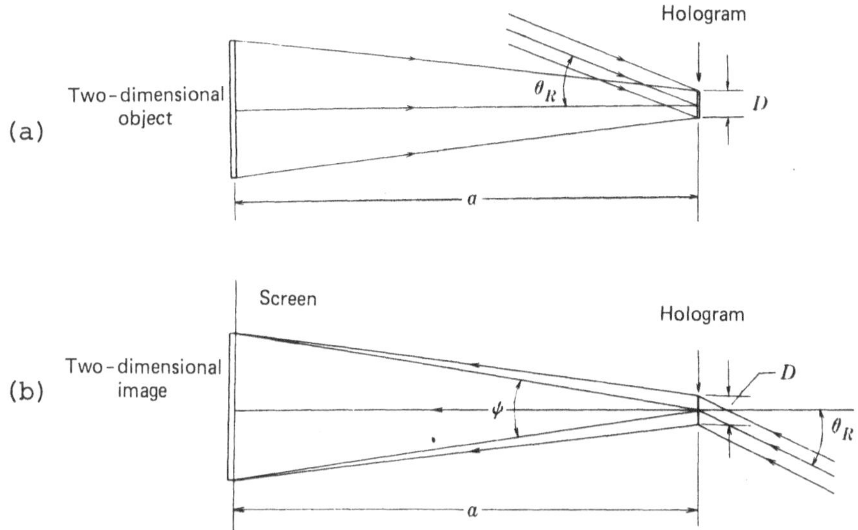

Fig.7.3 Hologram of a two-dimensional object. (a) Recording. (b) Reconstruction.

In the reconstruction process, the hologram might reproduce either a real or a virtual image, but for simplicity we tentatively assume that a real, two-dimensional image is reconstructed upon a flat projection screen as shown in Fig.7.3b.

In the hologram-recording process shown in Fig. 7.3a, the angular distribution of the light incident upon the hologram will be something like that shown in Fig.7.4. The object-beam energy spreads between $-\psi/2$ and $+\psi/2$, whereas the reference-beam energy is concentrated at $\theta = \theta_R$. When a hologram is made under irradiation by two beams as described above, what will be the spatial frequency spectra of the recorded fringes? We first make the rather crude approximations

$$a \gg w \gg D \quad \text{and} \quad \theta_R \ll (\pi/2) . \tag{7.9}$$

Then, since $\sin \theta_R \simeq \theta_R$ and $\sin \theta_0 \simeq \theta_0$ hold in Eq. (6.62), from that equation

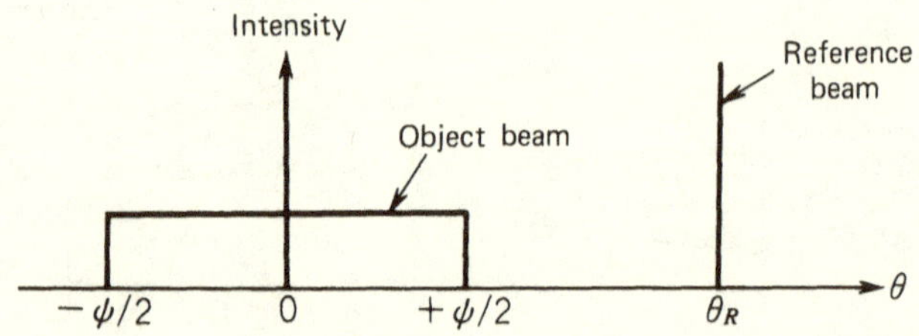

Fig.7.4 Angular distribution of light intensity in the optical system shown in Fig.7.3a.

$$f(\text{spatial frequency}) = \Delta x^{-1} \simeq |\theta_0 - \theta_R|/\lambda \tag{7.10}$$

is obtained for fringes produced by the interaction between the object and reference beams. The spectrum as expressed by Eq.(7.10) will be something like the square-shaped one in Fig.7.5; it is derived simply by inserting the distributions of θ_0 and θ_R shown in Fig. 7.4 into Eq.(7.10).

In addition to the above square-shaped spectrum which is the very component that should be recorded upon the hologram, another triangular, lower frequency spectrum is found in Fig.7.5. This component is produced by interactions between different angular components in the object beam ranging between $-\psi/2$ and $+\psi/2$. We should note that a relation similar to Eq.(7.10), $f = |\theta_1 - \theta_2|/\lambda$, holds for any two angular components θ_1 and θ_2 in the object beam.

B. *Choice of the Reference-Beam Angle θ_R*

We consider the desirable reference-beam angle on the basis of the preceding discussion. To assure that the triangular spectrum produced by the intermodulation of the object beam does not overlap the square-shaped spectrum, an inequality

$$\theta_R \gtrsim (3/2)\psi \tag{7.11}$$

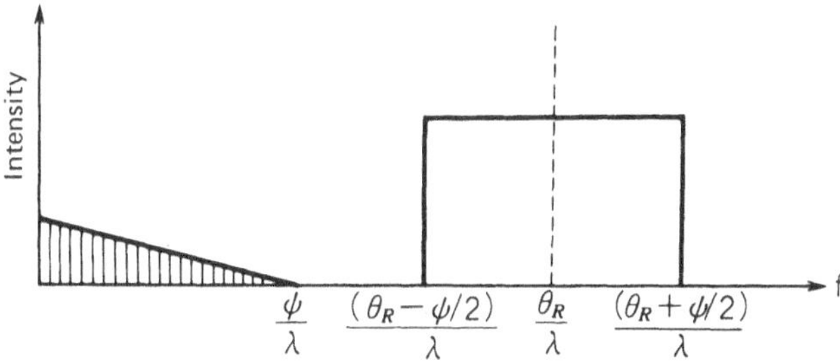

Fig.7.5 Spatial frequency spectra of the hologram pattern.

must hold, as is obviously derivable from Fig.7.5. This is the safest choice for the reference-beam angle which is illustrated in Fig.7.6a.

If the maximum spatial frequency of the object beam measured along the hologram plane is denoted by W, it is given in general as

$$W = \sin(\psi/2)/\lambda \doteqdot \psi/2\lambda. \tag{7.12}$$

Therefore, referring to Fig.7.6a and using the above W, we may say that the maximum spatial frequency of the hologram pattern is equal to 4W when $\theta_R = 3\psi/2$ is used. This last condition implies that point B in Fig.7.6a approaches point A.

In many practical cases, however, the object beam is much weaker than the reference beam. In such a case the disturbance given by the triangular spectrum can be neglected, and the reference-beam angle may be reduced so that

$$\theta_R/\lambda \simeq W, \quad \text{hence} \quad \theta_R \simeq \psi/2, \tag{7.13}$$

as illustrated in Fig.7.6b. The above condition gives, as seen in Fig.7.3, the minimum reference-beam angle with which we can separate the direct image, the conjugate image, and the undiffracted component of the illuminating beam. Obviously, in such a case the

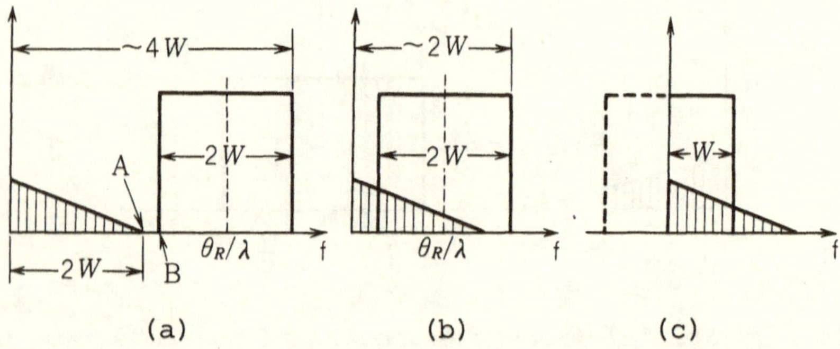

Fig.7.6 Spatial frequency spectra of the hologram pattern for three cases: (a) when $\theta_R \sim (3/2)\psi$; *(b) when* $\theta_R \sim (1/2)\psi$; *(c) when* $\theta_R = 0$.

maximum spatial frequency of the recorded hologram pattern is 2W.

In some extreme special cases, we dare to make $\theta_R = 0$ and let the three components mentioned above be superposed upon each other. This was done by Gabor in his first experiment with holography (see Section 2.6); such a scheme is now called the *on-axis reference-beam* method. (In contrast to this, the ordinary method in which a tilted reference beam is used is called the *off-axis reference-beam* method.) More up-to-date examples of the on-axis reference-beam scheme will be described in Section 8.9 in connection with the transmission of holograms via television; in these cases the disturbance due to the superposition of spectra is eliminated by special techniques. Anyway, the maximum spatial frequency of the hologram pattern is W when an on-axis reference beam is used, as illustrated in Fig. 7.6c.

Before proceeding into the next step of the discussion, we derive the relation between W and the resolvable line number of the reconstructed image N. Using the lateral resolution limit δ and the width of the image w, we can write, from Eq.(7.6),

$$w/N = \delta \doteqdot 2\lambda a/D. \tag{7.14}$$

Therefore, from Eqs.(7.12), (7.14), and $\psi \doteqdot w/a$,

$$WD = N \tag{7.15}$$

is obtained.

C. *Necessary Sampling Point Number on a Hologram*

In the information theory of continuous signals, we first consider a problem: How many points are needed in order that the whole signal be determined uniquely by giving values of the signal only on these "sampling" points? The number of such sampling points is called the *sampling point number*. As described earlier, computing the sampling point number is the first step in computing the quantity of information.

The underlying idea is Shannon's sampling theorem, which says that when the maximum frequency of a (time-dependent) signal is f_{max}, the whole signal (waveform) can be determined uniquely if the instantaneous values of the signal at discrete, sampling instants separated by $\Delta T = 1/2f_{max}$ are given.

It is not difficult to extend the above theorem to the case of a two-dimensional signal such as a hologram pattern. For example, when a hologram is of on-axis reference-beam type and has a maximum spatial frequency W in lateral directions, from the above statement and Eq.(7.15), the sampling point number is given as

$$P = (2WD)^2 = 4N^2. \tag{7.16}$$

Similarly, when a hologram is of on-axis reference-beam type in one lateral direction but $\theta_R = \psi/2$ for the other lateral direction,

$$P = (2WD)(4WD) = 8N^2 \tag{7.17}$$

holds. When $\theta_R = \psi/2$ for both (horizontal and vertical) lateral directions,

$$P = (4WD)^2 = 16N^2. \tag{7.18}$$

Finally, when the reference-beam angle is chosen in the safest way and $\theta_R = 3\psi/2$ holds for both lateral directions,

$$P = (8WD)^2 = 64N^2. \tag{7.19}$$

The above equations indicate that the necessary sampling point number is greatly dependent upon the choice of the reference-beam angle. The required sampling point number is smallest in the on-axis reference-beam case [Eq.(7.16)], which number is equal to that of an ordinary photograph recording the image of the object directly.

7.3.2 DISCUSSIONS ON THE TWO-DIMENSIONAL OBJECT CASE

It is well known that a hologram has a tremendous amount of information. However, looking at Eqs.(7.16)-(7.19), some readers might think that as far as the object is two-dimensional, the recorded information is far from tremendous; the required sampling point number is at most only 16 times greater than the truly necessary sampling point number $4N^2$.

The above statement is correct in some special cases, but not correct in many practical cases. In many cases the factor of redundancy is very much larger than 16. This is because the hologram size is rarely as small as the minimum size giving the necessary resolution ($D = 2\lambda a/\delta$) but much bigger than that in most cases. Especially when a three-dimensional virtual image is reconstructed and observed "through" the hologram as in most ordinary hologram experiments, D is usually as large as 10 cm, or even larger than that. We should note that in such cases extremely fine lateral resolution is realized due to the hologram's large size.

However, practically such a resolution is too high to be observed directly by human eyes. (The case of such a three-dimensional virtual image will be discussed later in Sections 7.3.3.B. and 7.3.3.C.)

As an example, values indicated by Leith *et al.* in their early paper (2) will be cited. When $\lambda = 500$ nm, $D = 10$ cm, $a = 9$ m, and $w = 6$ m, Eq. (7.14) gives $\delta = 90$ μm and $N = 6.5 \times 10^6$. If we further use Eq. (7.16) we obtain $P = 1.7 \times 10^{10}$. (Leith *et al.* indicate that if we transmit such a signal via television having a frame frequency of 30 Hz, the required bandwidth is as large as 250 GHz!) Such a tremendous sampling point number can be attributed solely to the fact that from the above hologram size an extremely high resolution (90 μm at a distance of 9 m) is obtained, which cannot be recognized by human eyes anyway. In other words, it is because the hologram size mentioned above is much greater than the pupil diameter of human eyes (refer to the discussion in Section 7.3.3.C). We should note that the resolution of a healthy eye is approximately equal to the diffraction limitation of the pupil diameter.

If we reduce the resolvable line number to commercial TV quality ($N = 500$), $P = 4N^2 = 10^6$ is obtained. If we further assume the object size and viewing distance both equal to commercial TV standard values ($w = 50$ cm, $a = 3$ m), from Eq. (7.14) we may conclude that the necessary hologram size is about 3 mm, which is comparable to the pupil diameter of a human eye.

There are two methods for observing a two-dimensional image reconstructed from such a tiny hologram. The first one is a method which should be called eye-piece holography. In this method a virtual image is observed through a tiny hologram, having a diameter ranging from several to ten millimeters, and placed just in front of the human eye. Such a method may not have been discussed elsewhere and thus be newly proposed in this book. The second method is to project a real image from a tiny hologram upon a white screen. This method is often employed in reproducing two-dimensional images; well-known examples are page-type holographic

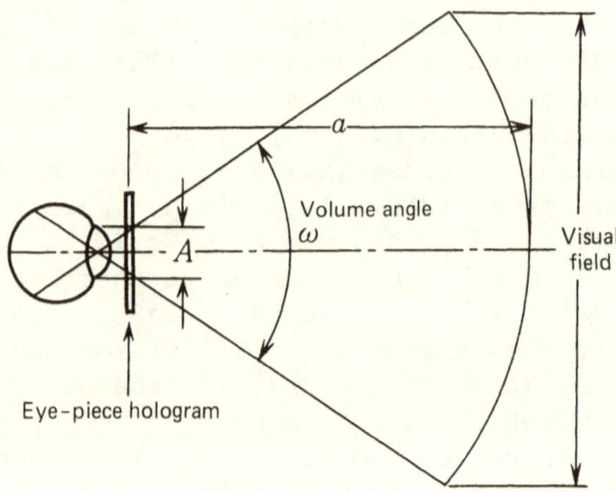

Fig.7.7 Eye-piece holography.

digital memory devices, and so-called Selectavision (12). (Among projection-type reconstructions of holographic images, there is another important category, that is, the reconstruction of three-dimensional picture information. This will also be discussed in Section 7.8.6.)

7.3.3 *CASE OF A THREE-DIMENSIONAL OBJECT*

A. *Eye-Piece Hologram*

We consider the case in which the reconstructed image is observed by two human eyes through two tiny eye-piece holograms as shown in Fig.7.7. Since the viewing position is fixed, the necessary sampling point number does not depend essentially upon whether the image is two or three dimensional.

First, let us compute the necessary sampling point number of the hologram, supposing that we observe the image with a single eye. If the reconstructed image is three dimensional, only accommodation is effective for depth perception. We assume that the angular resolution β of the eye is given in terms of the diffraction limitation of the pupil diameter, that is,

$$\beta = 2.44\lambda/A. \tag{7.20}$$

The volume angle of a cone having vertex angle β is given as

$$\Delta\omega = \pi(\beta/2)^2 = \pi(1.22\lambda/A)^2. \tag{7.21}$$

Next, we tentatively assume that the volume angle ω representing the whole visual field of an eye is approximately half a hemisphere, that is, $\omega = \pi$.* If the reference beam is of on-axis type in the horizontal direction and of off-axis type ($\theta_R = \psi/2$) in the vertical direction, N^2 in Eq.(7.17) may be replaced by ($\omega/\Delta\omega$) to obtain

$$P = 8(\omega/\Delta\omega) \simeq 5.5(A/\lambda)^2. \tag{7.22}$$

In obtaining the above equation, Eq.(7.21) has been used. Assuming A = 3 mm and λ = 500 nm leads to $P = 2 \times 10^8$.

For a fictitious binocular eye-piece holography device, $P = 4 \times 10^8$. If we transmit this information via television with a frame frequency of 30 Hz, the required bandwidth will be 6 GHz.

B. Wide Viewing Zone Hologram (Case I. When the Information Is Not Reduced)

The concept of the eye-piece hologram described above is not conventional. Practical holograms for

---

* We should note that in reality the visual field provided with good resolution as given by Eq.(7.20) is much narrower than π. The above computation, therefore, assumes a fictitious device with which we can "look around" in a wide visual field with high resolution given by Eq.(7.20).

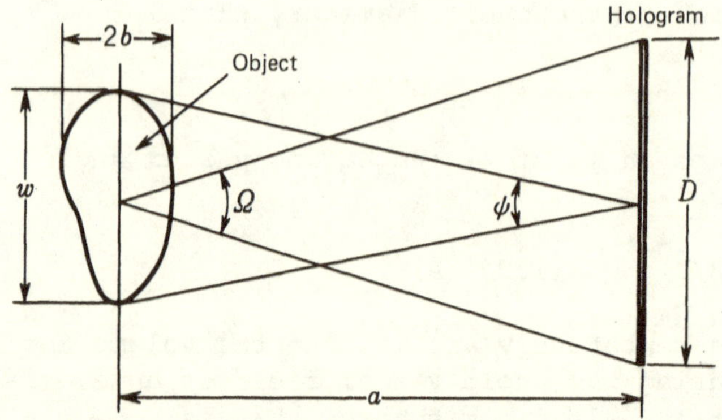

Fig.7.8 Symbols used in the computation of the sampling point number of a hologram recording a three-dimensional object.

reconstructing a three-dimensional virtual image are usually made in an appreciable size. When the virtual image is observed through the hologram, the hologram size gives the viewing zone. Hence a certain size is necessary to assure a reasonable viewing zone. Therefore, in most holograms the tremendous amount of information is stored not to provide adequate resolution but to provide a sufficient viewing zone. As a result, they usually give too good, useless resolution which can never be resolved by human eyes.

The necessary sampling point number for such a case can be obtained from Eqs.(7.16)-(7.19). However, as described above, a certain sampling point number P is required not to give N but to give D. Therefore, formulas using N in Eqs.(7.16)-(7.19) are rather meaningless; we ought better rewrite these equations in terms of D and w. If we consider Eq.(7.18) as a typical example, we may rewrite it, using Eq.(7.14), as

$$P \simeq 16N^2 = 4(D\omega/\lambda a)^2. \tag{7.23}$$

We may further rewrite the above equation using quantities illustrated in Fig.7.8, $\psi \sim w/a$ and $\Omega \sim D/a$, as

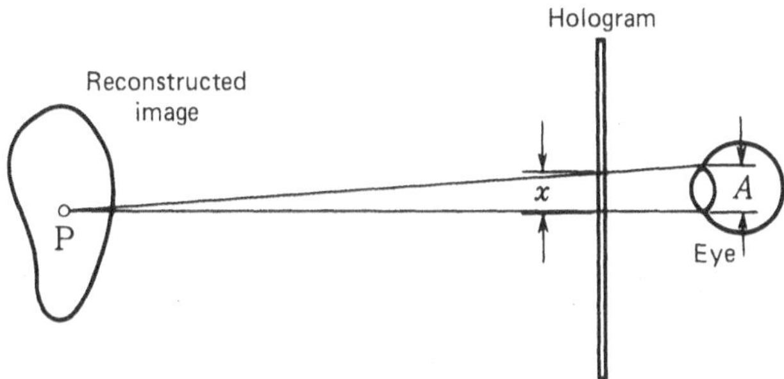

Fig.7.9 An eye looking at a fixed point in the image.

$$P = 4\Omega^2\psi^2(a/\lambda)^2. \tag{7.24}$$

For example, when $\Omega = 0.5$, $\psi = 0.25$, $a = 20$ cm, and $\lambda = 500$ nm, we obtain $P = 1 \times 10^{10}$. This value is a typical one for holograms usually made in laboratories to record a three-dimensional object.

C. *Wide Viewing Zone Hologram (Case II. When the Information Is Reduced)*

The wide viewing zone hologram as described above has an extremely high resolving power. However, when a virtual image is reconstructed and a point P on it is observed through the hologram as shown in Fig.7.9, what the observer's eye perceives is the component of light diffracted on the hologram within width x, which is even smaller than the pupil diameter A. Therefore, the resolution is reduced automatically to the human eye's diffraction limitation, which is discussed in the last part of Section 6.9. Thus the extremely high resolution inherent in a wide viewing zone hologram is actually useless when a virtual image is observed directly by human eyes.

The redundancy of holograms described above is, however, a necessary redundancy to assure an appreciable viewing zone, as far as we hope that the resolution is

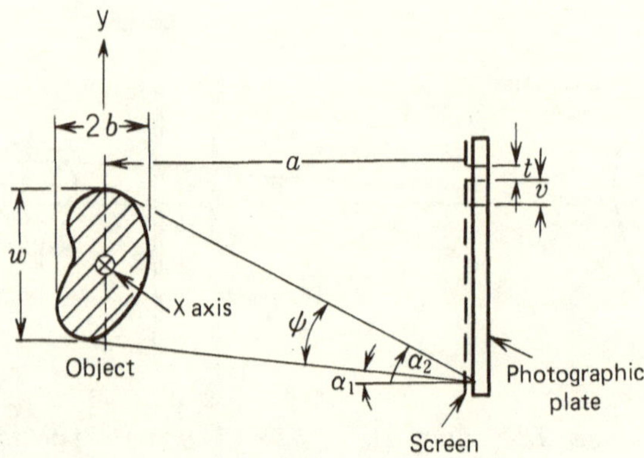

Fig.7.10 Recording of a reduced-information hologram [after Burckhardt (4)].

not reduced by the holographic process but limited only by the human eye's diffraction. Burckhardt (4) considered how much information reduction could be achieved if we dared reduce the resolution below the human eye's limitation by recording only discretely sampled portions of the hologram. The first part of his discussion describes a proposal for an information-reduction technique, and hence should be described in the latter half of this chapter; however, in the following discussions, as a preliminary stage for a discussion of the intrinsic limitations on information reduction, his information-reduction technique itself has to be described first.

When the hologram is recorded, a stripelike "sampling mask" is placed just in front of the hologram plate as shown in Fig.7.10, so that some parts of the plate with width t are exposed, whereas other parts with width v remain unexposed. If a virtual image is observed through such a stripelike hologram, the observer will feel as if the image is reconstructed behind a stripelike screen. Burckhardt called this effect the *screen effect*.

He proposed the following method for minimizing this screen effect. A secondary hologram (H2) is made

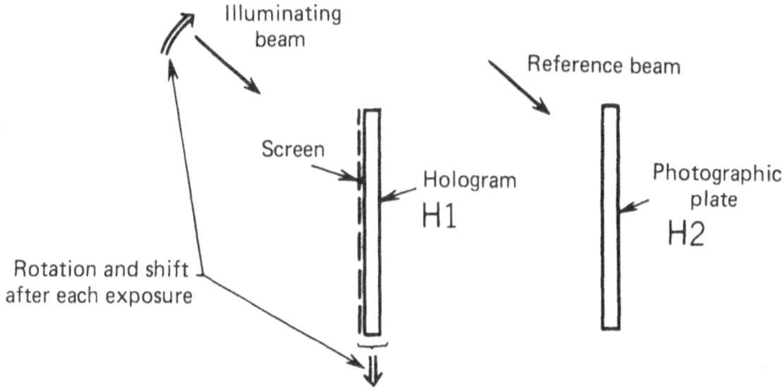

Fig.7.11 Removal of the screen effect (secondary hologram recording).

from the first one (H1), using a setup as shown in Fig.7.11. The first hologram is shifted laterally stepwise with a shift of t, and after each shift the image reconstructed by H1 is recorded on H2. If we write

$$k = v/t, \qquad (7.25)$$

the number of exposures required to cover the entire unexposed region of H1 is k, and hence k+1 exposures are needed in all. If we use, instead of the stripe-like mask, a mask provided with many rectangular apertures of dimensions $t \times t$, and transfer the information of H1 onto H2 by shifting H1 in both (x and y) lateral directions, the number of exposures is $(k+1)^2$. We consider hereafter the latter case.

If, after developing H2, we observe the virtual image reconstructed by H2 from the right, we will see a small flipping for each t and a larger flipping for each t+v. To minimize these flippings, Burckhardt proposed rotating the direction of the light beam illuminating H1, little by little at each exposure of H2, so that the flipping is entirely canceled at least at the median plane of the reconstructed image.

In Fig.7.12, points P_1, P_2, and P_3 are all located

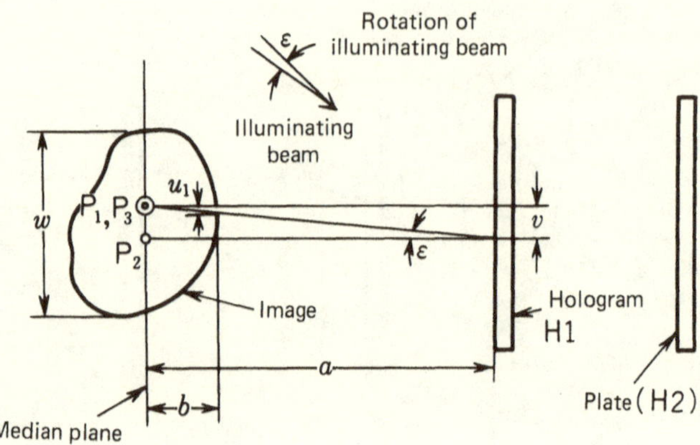

Fig.7.12 Arrangement in secondary hologram recording.

upon the median plane. Point P_1 is an original image point. The illuminating beam is assumed to be a plane wave. When the first hologram H1 has been shifted by v (= kt) while the illuminating beam angle remains fixed, the image point moves to P_2 through a distance v. We may bring back that image point to P_3, the same point as P_1, by properly rotating the illuminating beam. If we use the paraxial approximation, the necessary rotating angle ε is simply given as

$$\varepsilon \simeq kt/a. \tag{7.26}$$

The necessary rotation for each step t, therefore, is given by t/a.

We can thus cancel the flipping completely for image points on the median plane. However, for other image points reconstructed off the median plane, the cancellation is not complete. The magnitude of the residual flipping for a shift of v, as seen in Fig.7.12, is proportional to the distance b between the point of interest and the median plane, and is given as

$$u_1 = v(b/a) = ktb/a. \tag{7.27}$$

There is another cause of deterioration of image

quality: the blurring of the image due to diffraction of the slits (or apertures) having width t; the amplitude distribution function representing this blurring is given as

$$y = K \frac{\sin(\pi x t/\lambda a)}{\pi x t/\lambda a}, \tag{7.28}$$

where x denotes the lateral displacement around the image point. If we define the width of the blurring u_2 as the distance between the two smallest zeros of Eq. (7.28),

$$u_2 = 2\lambda a/t. \tag{7.29}$$

We should note that the observer's pupil also gives diffraction, which is given as [see Eq. (7.20)]

$$u_3 = 2.44\lambda c/A, \tag{7.30}$$

where c is the distance between the cye and the image, and A is the diameter of a pupil. The image quality is limited by the above three kinds of blurring u_1, u_2, and u_3.

To facilitate the discussion, we approximate the amplitude distribution function of a point image being subjected to the three blurrings, as

$$y = K \exp(-4x^2/u^2) \tag{7.31}$$

where

$$\begin{aligned} u^2 &= u_1^2 + u_2^2 + u_3^3 \\ &= (ktb/a)^2 + (2\lambda a/t)^2 + (2.44\lambda c/A)^2. \end{aligned} \tag{7.32}$$

The optimum t that minimizes Eq. (7.32) can easily be obtained as

$$t_{opt} = a(2\lambda/kb)^{1/2}, \tag{7.33}$$

which gives

$$u_{min} = [4kb\lambda + (2.44\lambda c/A)^2]^{1/2}. \tag{7.34}$$

Under the above conditions the resolvable line number N is given by

$$N = w/u_{min} = w/\{4kb\lambda + (2.44\lambda c/A)^2\}^{1/2}. \tag{7.35}$$

Rewriting the above equation in a form that gives k explicity and using the inherent resolution of the human eye N' ($= w/u_3 = wA/2.44\lambda c$), leads to

$$k = [w^2 - N^2(2.44\lambda c/A)^2]/4\lambda bN^2$$

$$= (w^2/4\lambda b)(1/N_r^2) \tag{7.36}$$

$$[\text{where } 1/N_r^2 = (1/N^2) - (1/N'^2)].$$

The necessary sampling point number of the hologram recorded in the above manner is reduced, like its total area, by a factor $1/(k+1)^2$ as compared with Eq.(7.24). If we assume for simplicity that $N^2 \gg N'^2$ and $k^2 \gg 1$, the reduced sampling point number P is obtained, from Eqs.(7.24) and (7.36), as

$$P = 4\Omega^2\psi^2(a/\lambda)^2/(1+k)^2$$

$$= [64\Omega^2b^2N_r^4/w^2][k^2/(1+k)^2] \tag{7.37}$$

$$\doteqdot 64\Omega^2b^2N^4/w^2. \tag{7.38}$$

Equation (7.38) has a more meaningful form than Eq. (7.24) does. In the former, P is given as a function of the required N, whereas in the latter, the required resolution has, apparently, nothing to do with P.

Numerical examples mentioned by Burckhardt are as follows. When w = 20 cm, b = 5 cm, λ = 632.8 nm, A = 3 mm, and a = c = 1 m, then N' = 389. If we reduce the requirement of N well below N', N = 200 for example, we get from Eq.(7.36) k = 5.80, and P can be reduced by a factor of 1/46.3. Assuming D = 10 cm leads to $P = 8.56 \times 10^7$, which would require a bandwidth of 1.28 GHz if transmitted via television having a frame frequency of 30 Hz. The last value is comparable to the value in the eye-piece hologram, and is well below the original requirement.

D. Computation of Spatial Resolution

Another approach to the present problem is computation of the spatial resolution, in other words, the number of resolvable spots in the image space, or the number of three-dimensional "picture elements" in the image space. The discussion in this section will follow the computation by Saito and Nagata (5), but some symbols and coefficients might be rewritten for consistency in this book. The result obtained finally, therefore, is a little different from that in Ref.(5) in coefficients, but essentially identical.

We consider a hologram having size $D \times D$. The maximum spatial frequency of the object wave measured along the hologram plate is denoted by W. (We assume that spatial frequencies above W cannot be recorded.) We define on the z axis as shown in Fig.7.13 a point P, whose position is determined by the angle ψ (= $2\lambda W$) shown in the figure. Next we draw two straight lines (actually, two surfaces) mm' and nn' that pass through P and the lower and upper edges of the hologram, respectively.

If the paraxial approximation is used, an object-wave component whose incident angle is greater than W (= $\psi/2$) can never be recorded, since the spatial frequency of the object wave along the hologram cannot exceed W. Therefore, the region in the object space from which any beam of the object wave arriving at the hologram plate never produces a spatial frequency along the hologram greater than W is the region below line mm' and above nn'. It is shown in Fig.7.13 as a hatched region V and will hereafter be called the *completely reconstructed region*. In the vicinity of V, there is

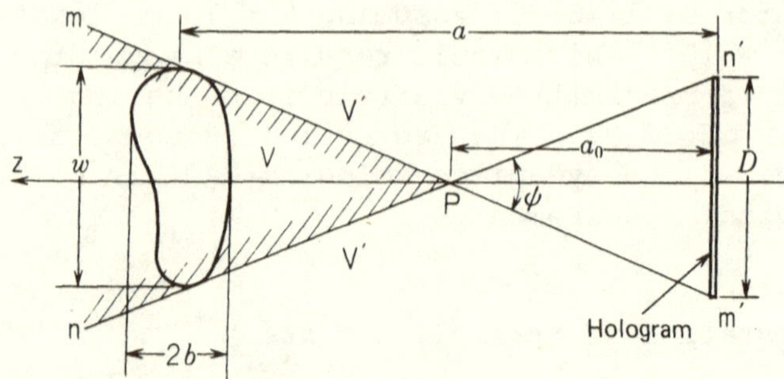

Fig.7.13 Symbols used in the computation of the spatial resolution of a hologram.

a degraded region V' where recording (hence, also reconstruction) is possible in some part of the hologram, but impossible in the other part. Such a region will be called an *incompletely reconstructed region.*

As a measure of spatial resolution, let us compute the resolvable spot number in the completely reconstructed region. From Eqs.(7.6) and (7.8), the lateral resolution limit (in the x and y directions) and the depth resolution limit (in the z direction) are given, respectively, as

$$\delta = 2\lambda z/D\,, \tag{7.39}$$

$$\Delta = 4\lambda z^2/D^2\,. \tag{7.40}$$

Therefore, the spatial resolution in V will be

$$M = \int_V (\delta^2\Delta)^{-1}dv$$

$$= \frac{D^4}{16\lambda^3}\int_{a_0}^{\infty} z^{-4}dz\int_{-D(z-a_0)/2a_0}^{+D(z-a_0)/2a_0} dx\int_{-D(z-a_0)/za_0}^{+D(z-a_0)/2a_0} dy$$

$$= D^6/48\lambda^3a_0^3\,. \tag{7.41}$$

Since $a_0 \simeq D/\psi = D/2\lambda W$ holds, the above equation can be rewritten as

$$M = D^3W^3/6. \tag{7.42}$$

On the other hand, if we assume a reference wave tilted by $\theta_R = \psi/2$ in both the x and y directions, the sampling point number upon the hologram is given, from Eq.(7.18), as

$$P = (4WD)^2. \tag{7.43}$$

Combining Eqs.(7.42) and (7.43) leads to

$$M = P^{3/2}/384. \tag{7.44}$$

Equation (7.44) tells us that M is not proportional to P, and when P (or M) exceeds 384^2 ($\simeq$ 150,000), M is greater than P. This statement is equivalent to claiming that if we construct a fictitious three-dimensional sensor system having M elements, we might be able to read out information with spatial resolution M which is greater than P, originally recorded in the two-dimensional plate. This is apparently a contradiction. Probably we should draw the following conclusion from the above statement: the information of M spots cannot always be read out from a single hologram because when too much information is to be read out in parallel, disturbance given by the light bundles for reproducing other spots becomes intolerably great, and the signal-to-noise ratio is deteriorated. Therefore, M is merely the total number of picture elements whose size is computed by assuming that a relatively small number of spots are recorded in the hologram.

It is also interesting to compute the number of depth resolvable steps from $z = a_0$ through $z = \infty$. It is given as

$$N_D = \int_{a_0}^{\infty} \frac{dz}{\Delta} = \frac{D^2}{4\lambda a_0} = \frac{1}{2} WD. \tag{7.45}$$

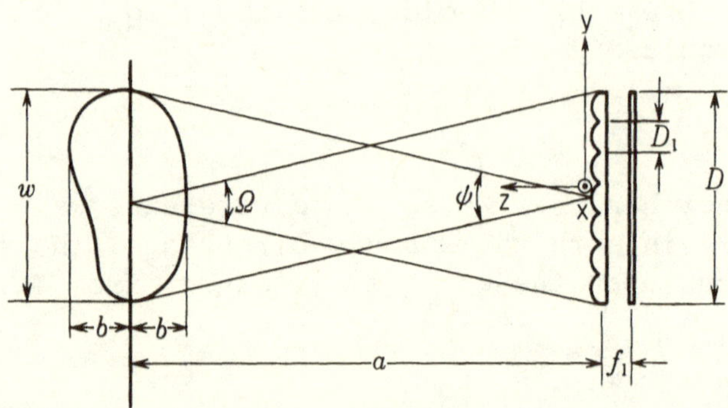

Fig.7.14 Symbols used in the computation of the sampling point number of an integral photograph.

Therefore, from Eqs.(7.42) and (7.45),

$$M = D^3W^3/6 = \frac{1}{3} N_D N^2 \tag{7.46}$$

is obtained. The factor 1/3 in the last formula roughly gives the ratio of information used in the completely reconstructed region to the entire information reproduced.

7.4 SAMPLING POINT NUMBER IN INTEGRAL PHOTOGRAPHY

7.4.1 CASE OF A THREE-DIMENSIONAL OBJECT

Following Burckhardt's considerations (3), let us first show that an equation similar to Eq.(7.38) holds for integral photography (IP). The IP to be discussed here is IP of remote-imaging type, as defined in Section 4.4.3. The symbols used in the analysis are shown in Fig.7.14. The following discussion starts from some conclusions of Section 4.4.3.

The spread of the point image upon the photographic

plate of the secondary integral photograph is given by Eq.(4.39) in Chapter 4, and the spread of the point image in the final image space is given by Eq.(4.42):

$$u_2 = \sqrt{3}\, u_1, \tag{7.47}$$

$$\begin{aligned} u_{final} &= \sqrt{4}(a/f_1)u_1 \\ &= [(1.66b^2D_1^2/a^2) + (3.93\lambda^2a^2/D_1^2)]^{1/2} \end{aligned} \tag{7.48}$$

[In the above equations some symbols different from Eqs.(4.39) and (4.40) are used; the spread w in Eqs. (4.39) and (4.40) are replaced here to prevent confusion with the width of image, and the lens diameter D is replaced by D_1 to prevent confusion with the hologram size.] From Eq.(7.47), the necessary sampling point number upon the photographic plate of the secondary integral photograph is given as

$$P = D^2/u_2^2 = D^2/3u_1^2. \tag{7.49}$$

On the other hand, the resolvable line number is given in terms of the image width w, as

$$N = w/u_{final} = wf_1/2au_1. \tag{7.50}$$

Combining Eqs.(7.49) and (7.50), we obtain

$$\begin{aligned} P &= 4a^2D^2N^2/3f_1^2w^2 \\ &= (4a^2D^2N^4/3f_1^2w^4)u_{final}^2. \end{aligned} \tag{7.51}$$

When we use the entire area of the photographic plate (without gaps), $a/f_1 = w/D_1$ holds, and hence from Eq.(7.48)

$$P = \frac{2.21D^2N^4b^2}{a^2w^2}\left(1 + \frac{2.37\lambda^2a^4}{D_1^4b^2}\right). \tag{7.52}$$

If we use Ω ($\simeq D/a$) shown in Fig.7.14, and note that the second term in parentheses of Eq.(7.52) is unity when $D_l = D_{l\ opt}$ (see Section 4.4.3), we obtain

$$P = \frac{4.42\Omega^2 b^2 N^4}{w^2}. \tag{7.53}$$

This equation is identical with Eq.(7.38) except for the difference between the coefficients.

When the resolution requirement is low, that is, $N \ll N_{max}$, we may set $D_l \ll D_{l\ opt}$, and the second term in parentheses of Eq.(7.52) can be neglected to give

$$P = \frac{2.21\Omega^2 b^2 N^4}{w^2}. \tag{7.54}$$

From the above discussions and Eq.(7.38), we may conclude that the relation $P \propto \Omega^2 b^2 N^4/w^2$ is essential in reconstruction of three-dimensional images provided with both horizontal and vertical parallax information. The fact that P is proportional to N^4 suggests that a little variation in the resolution requirement greatly affects the sampling point number. Therefore, determination of the resolution requirement upon psychological and physiological bases is an important task for the future.

It is difficult to tell rigorously to what the difference between coefficients in Eqs.(7.38) and (7.53) should be attributed. If we are satisfied by qualitative explanations, we may mention two facts: (1) defocus is not present in holography; and (2) the amplitude and phase of the object wave are recorded in the hologram, whereas only intensity is recorded in IP.

7.4.2 CASE OF A TWO-DIMENSIONAL OBJECT

The computation described in this section should be regarded as merely a check of the validity of the above computations.

Let $b \to 0$ in Eq.(7.52); this gives the relation between P and N for a two-dimensional object. At such a limit, the first term in parentheses in Eq.(7.52) can be neglected, and hence

$$P_{(b=0)} = 5.24\Omega^2 N^4 \lambda^2 a^4 / w^2 D_1^4. \qquad (7.55)$$

When a two-dimensional object is recorded, defocus never takes place. Therefore, the best choice of D_1 is to make it as large as possible, that is, $D_1 = D$. (This does not mean IP but ordinary photography with a single lens.) Therefore,

$$P_{(b=0)} = 5.24\Omega^2 N^4 \lambda^2 a^4 / w^2 D^4. \qquad (7.56)$$

On the other hand, when $b \to 0$ in Eq.(7.48),

$$u^2_{final(b=0)} = 3.93\lambda^2 a^2 / D_1^2 = 3.93\lambda^2 a^2 / D^2. \qquad (7.57)$$

Comparing Eqs.(7.50), (7.56), and (7.57), we obtain

$$P_{(b=0)} = (4/3)N^2. \qquad (7.58)$$

The factor 4/3 in the above equation represents the deterioration of the picture quality taking place during secondary IP recording and the final recording. [Compare Eqs.(7.47) and (7.48).]

7.4.3 COMPUTATION OF SPATIAL RESOLUTION

Next, let us compute the spatial resolution of IP and compare it with the sampling point number P. The computation will be carried out on rather crude assumptions since the result is not particularly significant.

The symbols used in the analysis are shown in Fig. 7.15. The lateral resolution limit δ is given in terms of the lateral resolvable line number N as

$$\delta = w/N. \qquad (7.59)$$

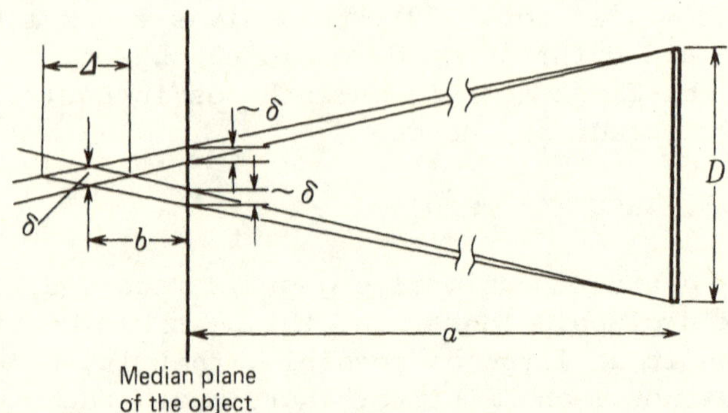

Fig.7.15 Relation between lateral and depth resolution limits in an integral photograph.

Under the assumption that $a \gg b$, the depth resolution limit Δ is given in terms of the size of the integral photograph, from geometrical considerations in Fig.7.15, as

$$\Delta \simeq 2a\delta/D. \tag{7.60}$$

Therefore, the spatial resolution M is given in terms of the volume of the image space, $w \times w \times 2b$:

$$M = 2bw^2/\delta^2\Delta = bN^3D/aw = \Omega bN^3/w. \tag{7.61}$$

It is difficult to discuss this equation in connection with Eq.(7.53) describing sampling point number P. The reason is that $P \propto N^4$ in Eq.(7.53), whereas $M \propto N^3$ in Eq.(7.61). Such a nonlinear relation between P and M was also discovered in holography. Comparing the two equations, we find that an inequality $P > M$ holds for

$$N > (w/4.42\Omega b). \tag{7.62}$$

The above inequality is satisfied in most practical integral photography; it means that IP is generally an uneconomical recording medium for three-dimensional imaging.

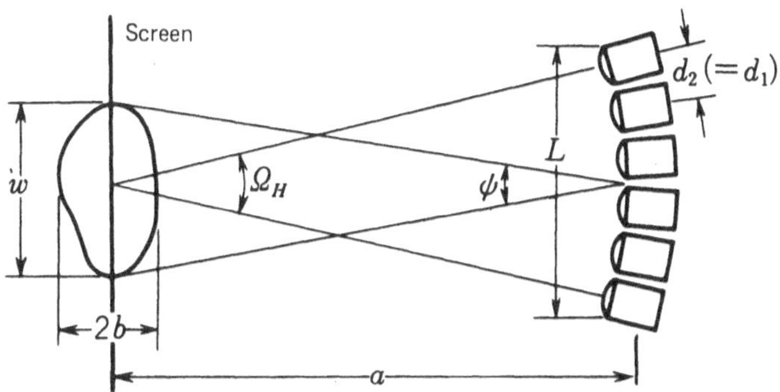

Fig.7.16 Symbols used in the computation of the sampling point number of a multiple photograph.

7.5 SAMPLING POINT NUMBER IN MULTIPLE PHOTOGRAPHY

We first note that in this book the term *multiple photography* is used generically for lens-sheet three-dimensional pictures of close-reconstruction type and projection-type three-dimensional displays. In Section 4.4.3, it was shown that when the image is closely reconstructed upon the second stage integral photograph, the optimum lens size of the first stage becomes much greater than that of the second integral photograph. In other words, close-reconstruction-type IP is always a kind of multiple photography.

In the following discussion, we shall further restrict our considerations to multiple photography having only unidirectional (horizontal) parallax. It will be found that in such cases a linear relationship between P and M is obtained; it is a relation simpler than those obtained for ordinary IP and holography.

The following computation will intentionally be based upon rather crude assumptions. This is merely because the crude assumptions make the process and results of the computation more comprehensible. A projection-type display will be kept in mind throughout; however, lens-sheet three-dimensional pictures allow an identical analysis.

7.5.1 SAMPLING POINT NUMBER OF THE PRIMARY PICTURES IN PROJECTORS

This computation is rather simple. If resolutions in the horizontal lateral direction and vertical lateral direction are denoted by N_H and N_V, then the total number of sampling points in all n* projectors is given as

$$P = n(2N_H)(2N_V). \tag{7.63}$$

Using the symbols shown in Fig.7.16 and assuming that $n \gg 1$, we may write $n \doteqdot L/d_2 \doteqdot \Omega_H a/d_2$, and hence

$$P = (\Omega_H a/d_2)(2N_H)(2N_V). \tag{7.64}$$

As is shown later in Eq.(7.72), the above equation can be rewritten in a form more convenient for comparison with equations for holography.

7.5.2 COMPUTATION OF SPATIAL RESOLUTION

Next, let us compute the spatial resolution and compare it with Eq.(7.64). To facilitate discussion, we assume the following conditions:

(1) $\alpha = \beta = \gamma = 1$. [Refer to Eqs.(4.1)-(4.3) in Chapter 4.)

(2) The screen is an ideal unidirectionally direction-selective screen, having ideal directivity (zero angular spread).

(3) The image is reconstructed around a screen located at distance a with a depth range $\pm b$.

* In Chapter 3, the projector number was denoted by N to avoid possible confusion with the refractive index. In this chapter, however, the lower case letter n is used for the projector number to avoid confusion with the resolvable line number N.

(4) Flipping is masked by blurring. In particular, if we use the symbols defined in Section 4.3.3, the factor k_1 is chosen as

$$k_1 \ (= D_1/d_1) = 1. \tag{7.65}$$

(5) To make the sampling point number as small as possible, the horizontal resolution N_H is reduced to the value determined by blurring.

We note that the blurring S is given for a spot at depth b, as derived in Eq.(4.11), as

$$S = d_2 b/(a+b) \doteqdot d_2 b/a. \tag{7.66}$$

If we reduce N_H to a value determined by the value of S given above,

$$N_H = w/S = wa/d_2 b. \tag{7.67}$$

Next we compute the depth resolution, or in other words, resolvable step number in the depth direction. We save labor by starting at Eq.(4.27), although this equation does not completely fit the present case. In the present model to be considered, blurring S is present upon the screen, the screen is ideal (zero spread), and the width of the visual field is denoted by L $(= \Omega_H a)$. Therefore, we let

$$\beta \to 0, \quad P \to L, \quad p \to S, \quad a_2 \to a, \quad B_2 \to b/a$$

in Eq.(4.27), and further double the depth resolution to take into account the fact that the image is reconstructed behind and in front of z = a. Thus we obtain

$$N_D \doteqdot \frac{L}{S}\frac{b}{a+b} \doteqdot \frac{LN_H}{w}\frac{b}{a} = \frac{\Omega_H N_H b}{w}. \tag{7.68}$$

[Note that the above result agrees with that given by another approach using Eq.(7.60) which was obtained for IP.] If we further use Eq.(7.67), we obtain

$$N_D = (\Omega_H b/w)(wa/d_2 b) = \Omega_H a/d_2. \tag{7.69}$$

Comparing Eqs.(7.64) and (7.69) leads to

$$P = N_D(2N_H)(2N_V). \tag{7.70}$$

Therefore, the spatial resolution M is

$$M = N_D N_H N_V = (1/4)P. \tag{7.71}$$

On the other hand, using Eqs.(7.67), (7.69), and (7.70), and letting $N_H = N_V = N$, we may write

$$P = 4\Omega_H b N^3/w. \tag{7.72}$$

The final equation has a form convenient for comparison with Eq.(7.38) for a hologram and Eq.(7.53) for IP.

7.6 SUMMARY OF COMPUTATIONS OF P AND M

The results of the computations of P and M in the preceding three sections are summarized in Table 7.2. The last item, unidirectional holography, has not been discussed so far. It will be described later in Section 7.8.6.

To facilitate a more intuitive understanding, the bandwidths required to transmit the various three-dimensional images via television channels are listed in Table 7.3. The following conditions are assumed: λ = 500 nm, Ω = 0.2, a = 2 m, b = 20 cm, w = 40 cm, and f_F (frame frequency) = 30 Hz.

7.7 INFORMATION REDUCTION IN INTEGRAL PHOTOGRAPHY

The reduced-information version of integral photography is the lenticular-sheet three-dimensional

TABLE 7.2 Summary of Computations of the Sampling Point Number P and Spatial Resolution M for Various Three-Dimensional Imaging Techniques

	P	M
Holography		
Two dimensional		
On-axis reference	$(2N)^2$	---
Off-axis reference	$(4N)^2$	---
Three dimensional		
Eye-piece type	$11(A/\lambda)^2$	---
Wide viewing zone	$4\Omega^2\psi^2(a/\lambda)^2$	$(1/3)N_DN^2$
Reduced-information	$64\Omega^2b^2N^4/w^2$	---
Integral photography	$4\Omega^2b^2N^4/w^2$	$\Omega bN^3/w(=N_DN^2)$
Multiple photography (unidirectional)	$4\Omega_HbN^3/w$	$(1/4)P(=N_DN_HN_V)$
Unidirectional holography	$16\Omega_HbN^3/w$	---

TABLE 7.3 Bandwidths Required to Transmit Three-Dimensional Images Via Television[a]

	B (MHz)	
Holography		
Two dimensional		
On-axis reference	30	(N=500)
Off-axis reference	120	(N=500)
Three dimensional		
Eye-piece type	6000	
Wide viewing zone	1,500,000	
Reduced-information	600,000	(N=500)
Integral photography	42,000	(N=500)
Multiple photography (unidirectional)	750	(N=500)
Unidirectional holography	3000	(N=500)

[a] f_F = 30 pictures/s, λ = 500 nm, Ω = 0.2, a = 200 cm, b = 20 cm, and w = 40 cm.

picture. The sampling point number for lenticular-sheet three-dimensional pictures has been computed, as an example of multiple photography, in Section 7.5. Technical matters were discussed in Chapter 4; they will not be repeated in this section. Only the degree of information reduction will be discussed.

As seen in Table 7.1, no essential difference is found between IP and multiple photography as to spatial resolution (resolvable spot number) M. On the other hand, from the formulas in the table,

$$\frac{P\ (\text{IP})}{P\ (\text{multiple photography})} = \Omega bN/w. \tag{7.73}$$

If we assume typical values $\Omega = 0.4$, $b/w = 0.3$, $N = 500$, then we obtain a factor of 60.

7.8 INFORMATION REDUCTION IN HOLOGRAPHY

In this section, five methods for reducing the information of a hologram are described in chronological order.

7.8.1 SAMPLING METHOD [BURCKHARDT (4)]

This has been described in Section 7.3.3.C in connection with the computation of the sampling point number of a hologram. In Ref.(4), photographs of the reconstructed image are shown.

7.8.2 SAMPLING OF A LENSLESS FOURIER-TRANSFORM HOLOGRAM [LIN (7)]

The sampling method proposed by Burckhardt is ingenious. The experimental process, however, is rather cumbersome. Lin showed, almost at the same time as Burckhardt's proposal, that a sampling process similar

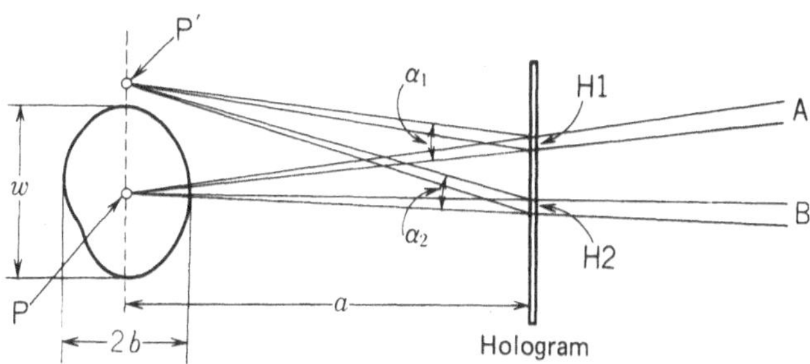

Fig.7.17 Principle of Lin's method. In a lensless Fourier-transform hologram, the position of a reconstructed point image P remains fixed when a minute hologram H1 moves to position H2.

to Burckhardt's could be made in a much easier way simply by making the hologram of Fourier-transform type.

The experiment could be performed conveniently with a setup for a "lensless" Fourier-transform hologram. As shown in Fig.7.17, in recording the hologram a spherically diverging reference beam is used, whose point source P' is at the same distance (a) as the median plane of the object from the photographic plate. [It was stated in Section 6.3.3 that in such a setup the recorded hologram is, approximately, equivalent to a Fourier-transform hologram; but the proof was not given. Readers who are interested in a mathematical proof should refer to, for example, Ref.(13). In the present discussion, however, that proof is not necessary.]

We assume that a minute hologram H1 shown in Fig. 7.17 is shifted to the position H2, and duplicated. If the shift distance and the distance PP' are both much shorter than a, the paraxial approximation $\alpha_1 = \alpha_2$ holds. Hence observers located at points A and B will see the spot P at approximately the same position. This is the principle of the simple sampling/duplicating method proposed by Lin.

Although he showed neither the rate of information

reduction brought about by this method nor the optimum sampling parameters, they should evidently be identical to Burckhardt's.

In Ref.(7), Lin showed photographs of the reconstructed image obtained from a sampled/duplicated hologram with dimensions 13 × 10 cm. The principal parameters were PP' = 8 cm, w = 8 cm, a = 22 cm, and 2b = 3 cm. Three photographs are shown for the following cases:

(1) Ten strip-shaped holograms with 1-cm height and 13-cm width are duplicated vertically to form a 13 × 10 cm hologram. (Reduction factor = 1/10.)

(2) A hundred strip-shaped holograms with 1-mm height and 13-cm width are duplicated vertically to form a 13 × 10 cm hologram. (Reduction factor = 1/100.)

(3) The hologram as described above (2) is further sampled horizontally by a factor of 1/10 with 6-mm pitch. Hence the original dimensions of the unit holograms are 1 mm (height) by 0.6 mm (width); they are prepared with 5.4-mm separations, and then duplicated horizontally. (Reduction factor = 1/1000.)

It is noteworthy that by Lin's method vertical parallax information can easily be removed from the hologram.

7.8.3 INFORMATION REDUCTION BY USING A DIFFUSER [HAINES AND BRUMM (8)]

In the two sampling methods described above, though the total information is reduced, the sampled information is recorded densely on the hologram in its segmented portions which have very small area. Hence the maximum spatial frequency recorded upon the hologram is never reduced.

Haines and Brumm devised a method with which maximum spatial frequency, as well as total information, could be reduced. Their method requires only the addition of a diffuser plate to an ordinary setup for holography.

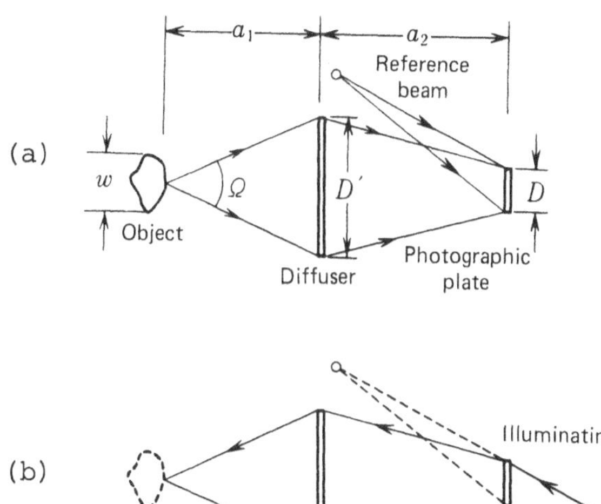

Fig.7.18 Reduction of hologram size by using a diffuser. (a) Recording. (b) Reconstruction.

The recording and reconstruction setups are shown in Figs.7.18a and 7.18b, respectively. In recording the hologram, a relatively small photographic plate is used. Instead, a sheet of diffuser (ground glass, for example), larger than the plate, is inserted between the object and the hologram plate. In such a recording setup, information from the entire object, observed from a wide viewing angle Ω is carried via diffuser to the hologram plate and recorded.

In the reconstruction process, the developed hologram plate is placed in the same position as the original relative to the diffuser. If a diverging spherical wave has been used as the reference beam in the recording, the conjugate, converging wave is used as the reconstruction beam. In such a case, a pseudoscopic real image will be observed at the original position if the diffuser and hologram are exactly positioned with respect to each other and the diffuser is free from attenuation. (When an amplitude-type diffuser is used, the image is not well reconstructed. A phase-type diffuser such as a ground glass plate is desirable.)

Let us consider the reduction factors of the maximum spatial frequency upon the hologram W and the sampling point number. For simplicity we assume that $w \ll D'$ and $D \ll D'$ in Fig.7.18a. We consider that in the "original" setup (before information reduction), a hologram plate having size $D' \times D'$ was placed at the position of the diffuser. In that case the maximum spatial frequency is given as

$$W' \simeq \Omega/2\lambda \simeq D'/2\lambda a_1. \tag{7.74}$$

In the reduced-information setup as shown in Fig.7.18a, from the assumption $D \ll D'$, we obtain

$$W \simeq D'/2\lambda a_2. \tag{7.75}$$

Therefore, the reduction factor for the spatial frequency is given as

$$W/W' \simeq a_1/a_2. \tag{7.76}$$

The above relations directly give the reduction factor for the sampling point number as

$$R = WD/W'D' \simeq a_1 D/a_2 D'. \tag{7.77}$$

In practice, the above reduction is obtained for both lateral (horizontal and vertical) directions, and a reduction of R^2 is achieved.

Following the proposal of Haines and Brumm, Hildebrand (14) generalized their method and discussed a method using a moving diffuser.

7.8.4 LENS-TYPE UNIDIRECTIONAL HOLOGRAPHY [FRITZLER AND MAROM (9)]

As has been emphasized, the most effective way to reduce information is to give up vertical parallax.

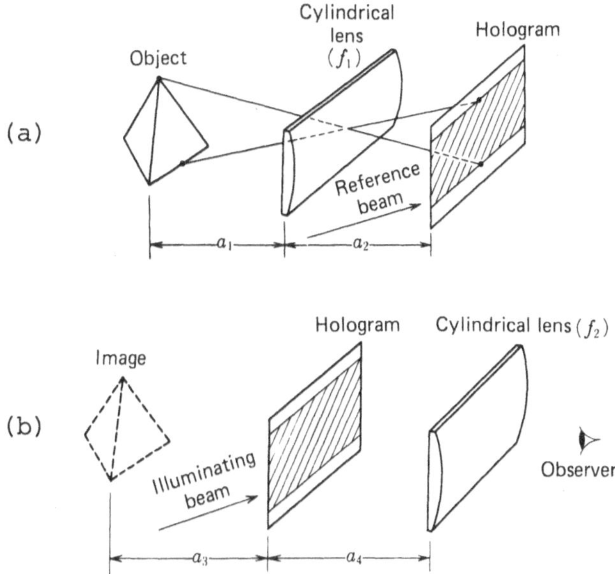

Fig.7.19 Unidirectional holography using a cylindrical lens. (a) Recording. (b) Reconstruction.

Fritzler and Marom proposed a method in which an anamorphic optical system is used to achieve the above purpose; it is an optical system that is holographic horizontally but of lens type vertically.

The recording and reconstruction setups are shown in Fig.7.19a and 7.19b. A cylindrical lens having focal length f_1 is used to make a real image of the object upon the plate with respect to the vertical direction. The distances in Fig.7.19 are chosen so as to satisfy

$$(1/a_1) + (1/a_2) = 1/f_1. \tag{7.78}$$

With respect to the horizontal direction, the recording process is entirely holographic. A hologram thus obtained is called a *unidirectional hologram*.

In reconstruction, a cylindrical lens (focal length f_2) is again used as shown in Fig.7.19b. The position of the reconstructed image is given, in terms of a_3 and a_4 shown in the figure, as

$$(1/a_4) - [1/(a_3+a_4)] = 1/f_2. \tag{7.79}$$

The condition for distortion-free reconstruction (equal magnifications in the horizontal and vertical directions) is given as

$$(a_2/a_1)[(a_3+a_4)/a_4] = a_3/(a_1+a_2). \tag{7.80}$$

In Ref.(9) a photograph of the reconstructed image is shown.

7.8.5 *DOUBLE-REFLECTION-TYPE HOLOGRAPHY [HAMASAKI AND NAGATA (10)]*

As described in Section 7.3.3.B, the resolution inherent in an ordinary wide-viewing-zone hologram is determined by the width of the viewing zone. The resolution thus determined is usually very high; it is too high to be perceived by human eyes. This fact is the main cause of the tremendous quantity of information in an ordinary hologram.

The sampling point number of a wide-viewing-zone hologram without information reduction is given, by Eq.(7.24), as

$$P = 4\Omega^2\psi^2(a/\lambda)^2. \tag{7.81}$$

This equation tells us that if the wavelength is enlarged or "equivalently enlarged" in some way, information could be reduced without reduction of the viewing zone.

If human eyes had a response up to the wavelengths of the far infrared, 100 μm for example, and good photosensitive materials were available at such infrared wavelengths, an ordinary holographic process would suffice. However, in reality both of the above statements are untrue.

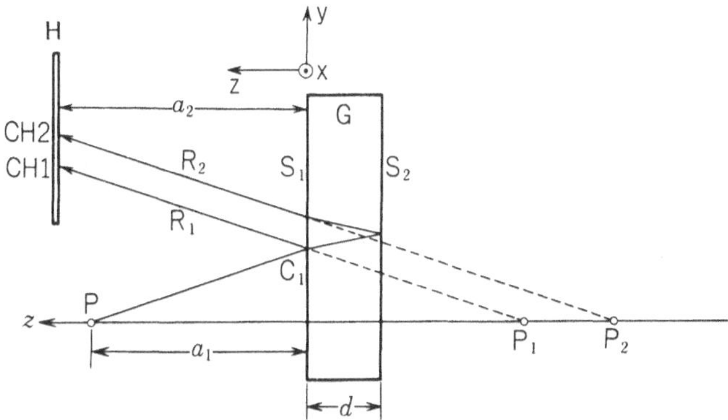

Fig.7.20 Transformation of equivalent wavelength using double reflection.

Hamasaki and Nagata proposed enlarging the recording wavelength equivalently by making a kind of wavelength transformation in the recording process. An optical system as shown in Fig.7.20 is used. In this figure H is the hologram plate, G is a homogeneous glass plate having optically flat, parallel surfaces S_1 and S_2. We set coordinates x, y, z as shown in the figure, and consider that point P represents the entire object. The ray of light emitted at P constitutes a component R_1 reflected at surface S_1, another component R_2 reflected at S_2, and other components of multiple reflections. The component R_1 has the same wavefront as that emitted at P_1, the mirror image of P with respect to surface S_1; its amplitude distribution upon the hologram plane is given as

$$A_1 = A_0 \exp\left\{-j\,\frac{\pi(x^2+y^2)}{\lambda(a_1+a_2)} - j\,\frac{2\pi(a_1+a_2)}{\lambda}\right\}, \qquad (7.82)$$

where a relation $x^2+y^2 \ll (a_1+a_2)^2$ is assumed.

As to the component R_2, we should first note that the equivalent source P_2 is located right of P_1 by a distance

$$d_e = 2d/n, \tag{7.83}$$

where n is the refractive index of the glass plate G. On the other hand, the glass plate also gives a phase delay corresponding to a path length 2nd. Therefore, the amplitude distribution of R_2 upon the hologram is written as

$$A_2 = A_0 \exp\left\{-j \frac{\pi(x^2+y^2)}{\lambda(a_1+a_2+d_e)} - j \frac{2\pi(a_1+a_2+2nd)}{\lambda}\right\}. \tag{7.84}$$

[We should note that the factor A_0 in Eqs.(7.82) and (7.84) are not exactly equal. However, equating these factors to one another leads to no essential problem.] Therefore, the fringe pattern formed by the interference between R_1 and R_2 upon the hologram is expressed as

$$A_1A_2^* + A_1^*A_2 = |A_0|\exp\left\{-j \frac{\pi(x^2+y^2)}{\Lambda(a_1+a_2)} + j \frac{4\pi nd}{\lambda}\right\} + |A_0|\exp\{c.c.\}, \tag{7.85}$$

where c.c. denotes a complex conjugate term, and

$$\Lambda = \lambda(a_1+a_2+d_e)/d_e. \tag{7.86}$$

Note that the first term in Eq.(7.85) has a form identical to the amplitude distribution of a spherical wave diverging from point P, but having a wavelength of Λ. Therefore, a transformation of the wavelength to a longer one can be achieved; the parameter Λ as defined in Eq.(7.86) is the equivalent, enlarged wavelength.

The experiment was performed for a simple two-dimensional transparent object, a letter S. The parameters are $a_1+a_2 = 1.59$ m, $\lambda = 632.8$ nm, and $\Lambda = 51$ μm. The obtained long-wavelength hologram was reduced by

1/10 and printed on a Kodak 649F photographic plate. The image was reconstructed using a He-Ne laser beam.

This method has several practical disadvantages such as distortion and aberration in the reconstructed image, reduction of the image size, and that a spatially incoherent illumination of the object is needed to avoid interference between beams from different points on the object. To meet this requirement, in the experiment described above, a high-speed rotating diffuser disk is inserted between the illuminating laser source and the object. However, the basic idea of wavelength transformation by double reflection is an interesting one, and might be applied in various systems.

7.8.6 PROJECTION-TYPE UNIDIRECTIONAL HOLOGRAPHY [OKOSHI AND YANO (11)]

A. Technical Background

As has been stated elsewhere, the most effective method of reducing hologram information is to give up vertical parallax information, in other words, to make the hologram unidirectional. For this purpose Lin used an array of strip-shaped Fourier-transform holograms and obtained reduction by 1/10 or 1/100, and a further reduction by 1/1000 by employing the sampling principle with respect to the horizontal direction (see Section 7.8.2). Fritzler and Marom used an anamorphic (cylindrical) lens system for the above purpose (see Section 7.8.4). However, both of the above methods have practical disadvantages; the former requires multiple exposures in the recording process, and the latter requires a lens system in the reconstruction.

We should note that the use of a unidirectionally direction-selective screen offers a much simpler method of observing a three-dimensional image reconstructed from a unidirectional hologram; we may simply project a real image from a strip-shaped hologram onto a unidirectionally (only horizontally) direction-selective screen. In this section, this method will be described

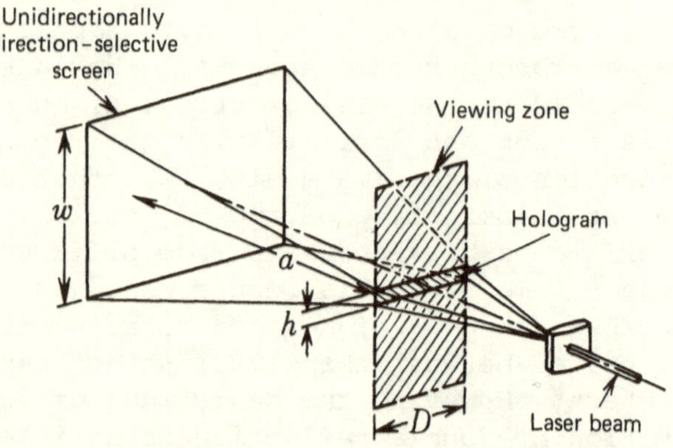

Fig.7.21 Principle of the projection-type, unidirectional holography.

as an example of unidirectional holography, and the sampling point number inherent to this sort of imaging will be derived.

B. Principle of the Method

The principle of projection-type, unidirectional holography is shown in Fig.7.21. (In this figure a strip-shaped hologram is shown. However, in later discussions, a horizontally sampled hologram will also be considered.)

A laser beam is spread horizontally by a cylindrical lens and illuminates the strip-shaped hologram. Such a hologram contains information from the object viewed from various horizontal positions but from a single vertical position. The real, pseudoscopic three-dimensional image is projected from the strip-shaped hologram onto the unidirectionally direction-selective screen. The reconstructed image is observed from the hologram side as an orthoscopic one. In such a setup the viewing zone is expanded in the vertical direction to the height determined by the diffusing angle of the screen.

As will be seen later, the hologram width D is determined by the required width of the viewing zone, whereas its height h is determined by the requirements for vertical resolution of the reconstructed image.

C. Limitation of Information Reduction in the Vertical Direction

To obtain the rate of information reduction with respect to the vertical direction, we first consider the minimum hologram height. If we use symbols as defined in Fig.7.21 and denote the wavelength by λ, the vertically resolvable line number on the screen is given, from Eq.(7.14), as

$$N_V = wh/2\lambda a. \tag{7.87}$$

For example, when $w = a/5$, $\lambda = 500$ nm, and $N_V = 400$ are required, then the necessary hologram height h is 2 mm. On the other hand, the maximum vertical spatial frequency component W_V of the hologram pattern is given, when a vertically on-axis reference beam is used, as

$$W_V = w/2\lambda a. \tag{7.88}$$

Hence, the relation $N = hW_V$ holds. This means no redundancy is present with respect to the vertical direction.

The sampling point number of the hologram in the vertical direction P_V is obtained in the same way as Eq.(7.16). For an on-axis reference beam,

$$P_V = 2N_V. \tag{7.89}$$

When the reference beam is vertically off-axis so that disturbance by direct (undiffracted) light can be removed,

$$P_V = 4N_V. \tag{7.90}$$

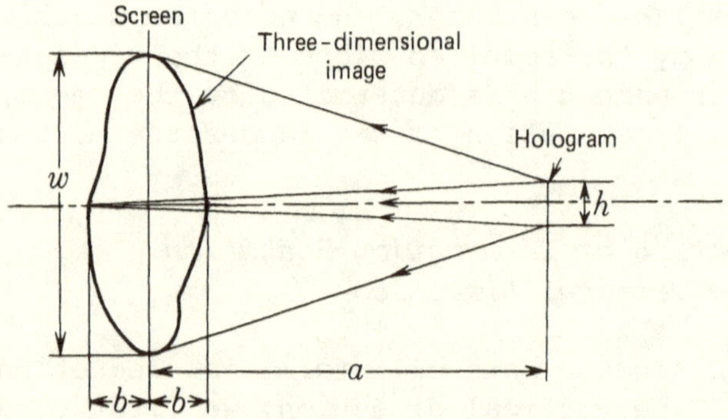

Fig.7.22 Symbols used in the analysis.

D. Optimum Hologram Height

The hologram height h should not be too large. Since the screen has a diffusing property in the vertical direction, a vertical "blurring" will take place when h is too large and the reconstructed spatial image has a certain depth. When the depth of the image is denoted by 2b as shown in Fig.7.22, it is easily seen that the width of the vertical blurring upon the screen is expressed as

$$\delta_1 \doteqdot h(b/a). \qquad (7.91)$$

Note that this is proportional to h.

On the other hand, as stated before, h should not be too small. The vertical resolution limit due to diffraction measured upon the screen is given directly from Eq.(7.87), as

$$\delta_2 = w/N_V = 2\lambda a/h. \qquad (7.92)$$

Note that this is inversely proportional to h.

Parameters δ_1 and δ_2 are shown as functions of h in Fig.7.23. To get the optimum hologram height, we assume that the overall blurring is expressed as

$$\delta = \sqrt{\delta_1^2+\delta_2^2}. \qquad (7.93)$$

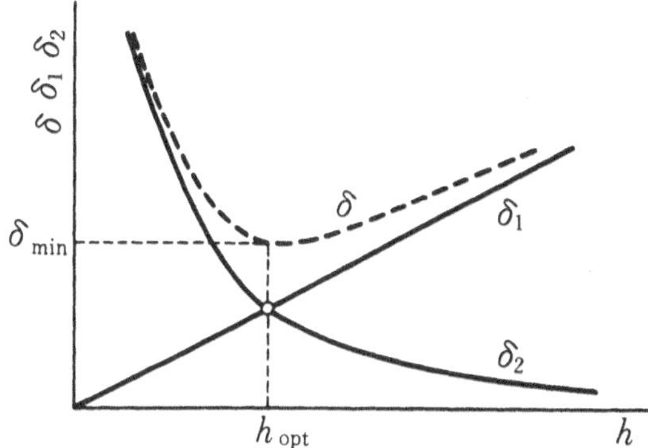

Fig.7.23 Lateral resolution limits as functions of the hologram height.

From the above three equations,

$$\delta_{min} = 2\sqrt{\lambda b} \tag{7.94}$$

and

$$h_{opt} = a\sqrt{2\lambda/b} \tag{7.95}$$

are obtained. For example, when b = 10 cm, λ = 500 nm, and a = 50 cm, then we obtain δ_{min} = 0.5 mm and h_{opt} = 3.2 mm.

E. Limitation of Information Reduction in the Horizontal Direction

The strip-shaped hologram shown in Fig.7.21 may further be sampled in the horizontal direction; in that case a real image is projected from a number of tiny, windowlike holograms. In such an ultimately reduced-information version, the sampling point number in the horizontal direction P_H is equal to the square root of Eq.(7.38). Hence, if the screen size is w × w,

$$P_H = 8\Omega_H b N_H^2/w = 8DbN_H^2/wa. \tag{7.96}$$

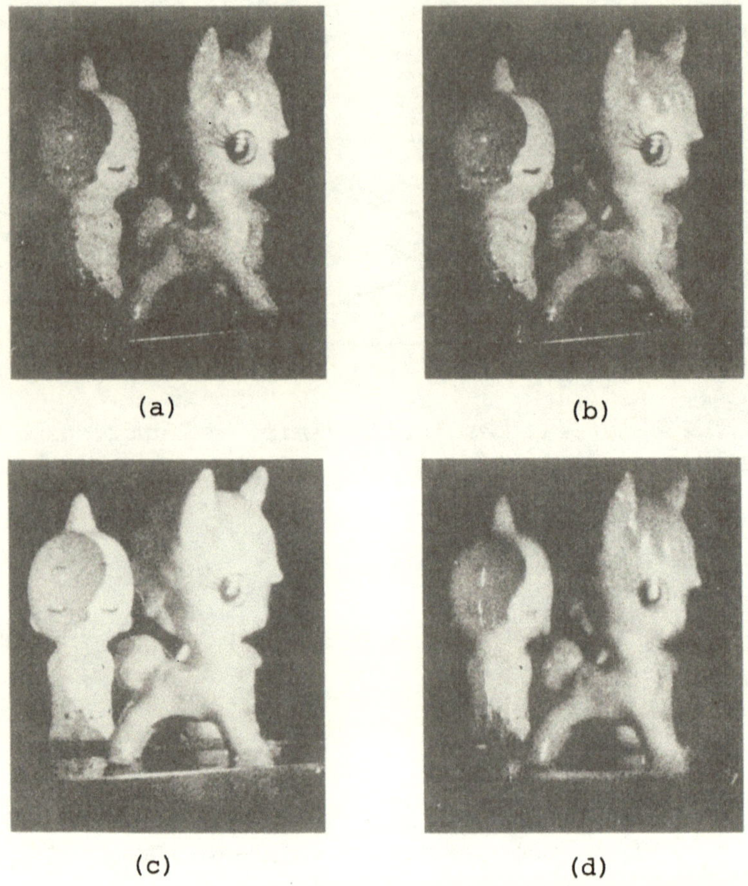

Fig.7.24 Images obtained by the projection-type holography for (a) h = 1 mm, (b) h = 2 mm, (c) h = 5 mm, and (d) h = 40 mm.

If we use Eq.(7.89), the total sampling point number is given by

$$P = P_V P_H = 16DbN_V N_H^2/wa. \tag{7.97}$$

For example, if we consider a low-resolution case similar to Section 7.3.3.C, w = 20 cm, D = 10 cm, b = 5 cm, a = 1 m, N_V = 200, we obtain P = 3.2 × 10^6. A high-resolution example, D = w = 20 cm, b = 5 cm, a = 40 cm, $N_V = N_H = 500$, gives $P = 5 \times 10^8$.

F. Experiments

Experiments were performed using both a lenticular-sheet stereoscreen and a rear-projection-type stereoscreen consisting of a Fresnel-lens sheet and a cylindrical-lens sheet (11). Photographs in Fig.7.24 show the images obtained upon the latter screen for various hologram heights. A He-Ne laser and an Ar laser (green: 514.5 nm) were used in taking the hologram and in the projection, respectively. Photographs were taken from a distance of 900 mm. A camera with an F/8 lens was used. The dimensions of the image are height = 12 cm, width = 10 cm, depth (2b) = 15 cm, and a = 50 cm. Therefore, Eq.(7.95) gives h_{opt} = 1.8 mm.

Photograph (b) for h = 2 mm shows the finest resolution, although the difference between (a) and (b) is not great. Hologram heights less than 1 mm could not be tested because of the difficulty in collimating the laser beam.

Experiments with projection-type holography using a horizontally sampled hologram were reported by Okoshi and Nomura (15). The resolution of the reconstructed image measured upon a standard test chart showed agreement with theoretical prediction within ±20%.

7.9 SUMMARY

The most important implication of Table 7.2 is as follows. For a three-dimensional image provided with both horizontal and vertical parallaxes, P is proportional to N^4. For those provided with only horizontal parallax, P is proportional to N^3. In the former, N is not proportional to M; whereas in the latter, proportionality holds. This means that the most effective and efficient method of reducing information in reconstructing a three-dimensional spatial image is to give up vertical parallax information.

References

(1) Leith, E. N., and Upatnieks, J., Reconstructed Wavefronts and Communication Theory, *J. Opt. Soc. Amer. 52*, No. 10, 1123-1130 (Oct. 1962).

(2) Leith, E. N., Upatnieks, J., Hildebrand, B. P., and Haines, K., Requirements for Wavefront-Reconstruction Television Facsimile Systems, *SMPTE 74*, No. 10, 893-896 (Oct. 1965).

(3) Burckhardt, C. B., Optimum Parameters and Resolution Limitation of Integral Photography, *J. Opt. Soc. Amer. 58*, No. 1, 71-76 (Jan. 1968).

(4) Burckhardt, C. B., Information Reduction in Holograms for Visual Display, *J. Opt. Soc. Amer. 58*, No. 2, 241-246 (Feb. 1968).

(5) Saito, S., and Nagata, Y., Resolvable Spot Number of a Three-Dimensional Spatial Image Reconstructed From a Hologram (in Japanese), 1969 National Convention of Japan Institution of Electronics and Communication Engineers, Paper No. 553.

(6) Okoshi, T., Optimum Design and Depth Resolution of Lens-Sheet and Projection-Type Three-Dimensional Displays, *Appl. Opt. 10*, No. 10, 2284-2291 (Oct. 1971).

(7) Lin, L. H., A Method of Hologram Information Reduction by Spatial Frequency Sampling, *Appl. Opt. 7*, No. 3, 545-548 (March 1968).

(8) Haines, K. A., and Brumm, D. B., Holographic Data Reduction, *Appl. Opt. 7*, No. 6, 1185-1189 (June 1968).

(9) Fritzler, D., and Marom, E., Reduction of Bandwidth Required for High Resolution Hologram Transmission, *Appl. Opt. 8*, No. 6, 1241-1243 (June 1969).

(10) Hamasaki, J., and Nagata, Y., Double-Reflection Holography--A Method of Reducing Hologram Information (in Japanese), 1971 Joint National Convention of Four Institutions on Electrical Engineering, Paper No. 236.

(11) Okoshi, T., and Yano, A., Reduced-Information Projection-Type Holography Using a Horizontally Direction-Selective Stereoscreen, *Opt. Commun. 3*, No. 2, 85-88 (April 1971).

(12) Bartolini, R. *et al.*, Embossed Hologram Motion Pictures for Television Playback, *Appl. Opt. 9*, No. 10, 2283-2290 (Oct. 1970).
(13) Collier, R. J., Burckhardt, C. B., and Lin, L. H., "Optical Holography." Academic Press, New York, 1971.
(14) Hildebrand, B. P., Hologram Bandwidth Reduction by Space-Time Multiplexing, *J. Opt. Soc. Amer. 60*, No. 2, 259-264 (Feb. 1970).
(15) Okoshi, T., and Nomura, H., Three-Dimensional Imaging by Projection-Type Holography (in Japanese), Proceedings of the 3rd Conference on Imaging Techniques, Paper No. 2-8, Nov. 21-24, 1972.

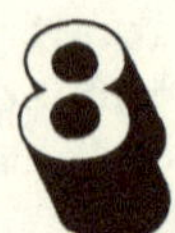

Miscellaneous Three-Dimensional Imaging Techniques

Miscellaneous three-dimensional imaging techniques including computer-generated holograms, computer-generated stereoscopic drawings, varifocal mirror techniques, three-dimensional x-ray imaging, holographic screens, and three-dimensional television are described.

8.1 INTRODUCTION

In this chapter, miscellaneous three-dimensional imaging techniques other than lens-sheet pictures, projection-type displays, and holography, which have been discussed in detail in the preceding three chapters, will be described. Among these, the computer-generated hologram technique constitutes a substantial field for image engineering; it might deserve an independent book. What can be presented concerning it in this book, therefore, is only an introductory description of that field.

8.2 COMPUTER-GENERATED HOLOGRAMS

8.2.1 HISTORICAL BACKGROUND

Proposals for making holograms using a computer appeared first in the literature in 1966-1967 (1),(2). The number of papers in this field became tremendous in 1968-1971; most dealt with two-dimensional imaging, and rather few of them with three-dimensional images. However, it seems to the author that the investigators finally aim to realize a three-dimensional, on-line computer output display in the distant future. This field has already been established and comprehensive review papers are available in technical journals. In this section, detailed mathematical discussions will be omitted; a few experiments particularly oriented toward three-dimensional image reconstruction will be described.

8.2.2 CLASSIFICATION OF COMPUTER-GENERATED HOLOGRAMS

When an image is reconstructed from a hologram, the illuminating beam will have, after being transmitted through the hologram, a wavefront expressed in general as

$$A(x,y) = A_0(x,y)\exp[j\phi(x,y)], \tag{8.1}$$

including both phase and amplitude variations. A possible method for reconstructing the same wavefront as Eq.(8.1) from a computer-generated hologram will be as follows. We prepare a translucent plate giving the amplitude variation $A_0(x,y)$ and a phase plate giving the phase variation $\phi(x,y)$, pile them up, and illuminate them from the back by a normally incident plane wave. However, practically such a scheme is rather tedious. In most computer-generated hologram schemes, therefore, either of the following two methods is employed:

(1) A gratinglike transmittance variation is given to form both $A_0(x,y)$ and $\phi(x,y)$. In other words, an amplitude-type hologram is generated. Appropriate reference and illuminating wavefronts are assumed, and the required phase variation is generated by diffraction upon the hologram pattern.

(2) Only the phase variation $\phi(x,y)$ is given by a finely shaped phase plate. The transmittance variation is neglected because in most holograms (except for image holograms), the transmittance $A_0(x,y)$ is a slowly varying function and may be approximated by a constant.

In method (1), a significant difference between the computer-generated hologram and an ordinary optically prepared hologram is that in the former, a delicate optical density variation cannot easily be provided. In most computer-generated holograms, therefore, efforts are made to prepare equivalently the wavefront of Eq.(8.1) by using just "binary" density modulation, that is, by giving transmittance of 1 or 0 as a function of the coordinates (x,y). Moreover, to facilitate the computation and preparation of the hologram pattern, the density modulation is given in most cases in relatively simplified patterns such as arrays of circular or rectangular apertures whose sizes and positions are modulated. However, recent progress with peripheral computer equipment is making half-tone plotters available, *i.e.*, plotters capable of providing a gray scale. The number of reports on experiments using such plotters is increasing.

An example of method (2) is the Kinoform to be described in the following subsection.

On the other hand, computer-generated holograms are classified into two groups according to the image to be reconstructed, one group aiming at two-dimensional image reconstruction and another three-dimensional. Most of the past reports dealt with the former, whereas reports of three-dimensional imaging occur with increasing frequency these days.

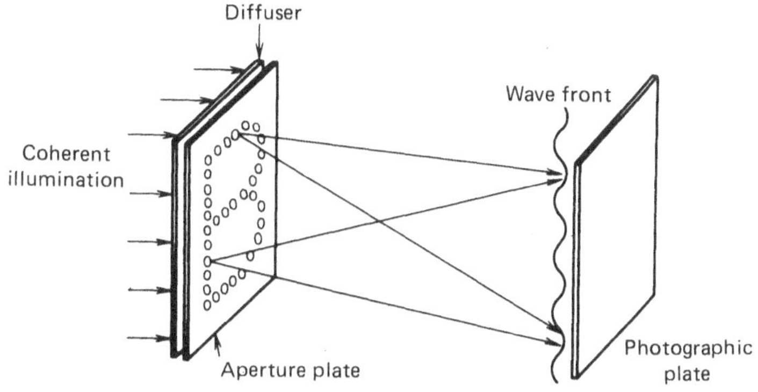

Fig.8.1 Principle of Kinoform.

8.2.3 KINOFORM

Kinoform was proposed by Lesem *et al.* of IBM in 1969 (3). It was asserted that this was not a hologram but a new wavefront reconstruction technique. However, it seems to the author that it may still be considered as a kind of hologram.

Kinoform is the best example of method (2) mentioned in the preceding subsection. The amplitude transmittance variation $A_0(x,y)$ is entirely neglected, whereas the computed phase variation $\phi(x,y)$ required to reconstruct an image is given by a phase plate deliberately prepared through a photographic process. Suppose for simplicity that we want to prepare a Kinoform recording a fictitious, simple two-dimensional object as shown in Fig.8.1. It is not difficult to compute the phase of the wavefront upon the photographic plate if we assume that each spotlike light source constituting the letter B emits light of equal amplitude and the same phase. The computed phase distribution is converted into a chart provided with a gray scale using a half-tone plotter. The chart is then reduced in size photographically, developed, bleached to give the phase variation, and a Kinoform is completed.

An experiment was performed with respect to a

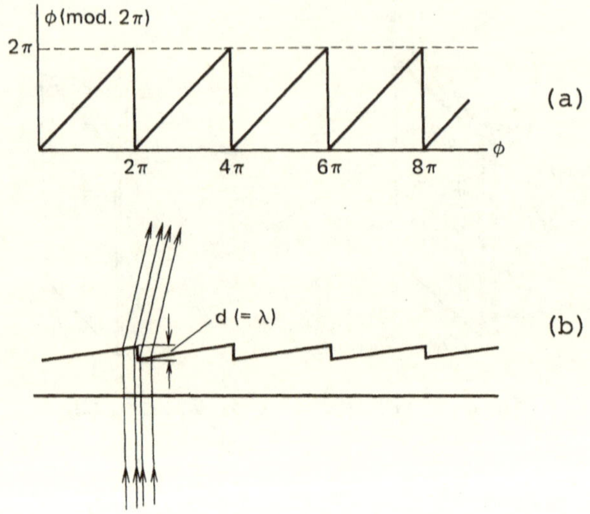

Fig.8.2 (a) The function ϕ(mod 2π). (b) The phase-matching condition of a phase hologram.

simple, fictitious three-dimensional object and a two-dimensional object (letter B). The object is first represented by sampling points with a maximum number of 64 × 64, and the phase variation $\phi(x,y)$ is computed. A half-tone plotter capable of generating 32-step gray scale, resolvable spot number of 12,000 × 8000 in a 1.5 × 1.0 m^2 space, draws the original Kinoform pattern in which

$$\text{optical density} \propto \phi(x,y)\ (\text{mod } 2\pi), \tag{8.2}$$

where the symbol (mod 2π) represents a sawtoothlike function as shown in Fig.8.2a. The obtained pattern is then reduced in size photographically to 18 × 12 cm^2, and the photographic emulsion is bleached to form a transmission-type phase hologram.

Figure 8.2b shows schematically the variation of the equivalent optical thickness of the emulsion. The bleaching process should be controlled carefully so that the height of the step d shown in Fig.8.2b provides

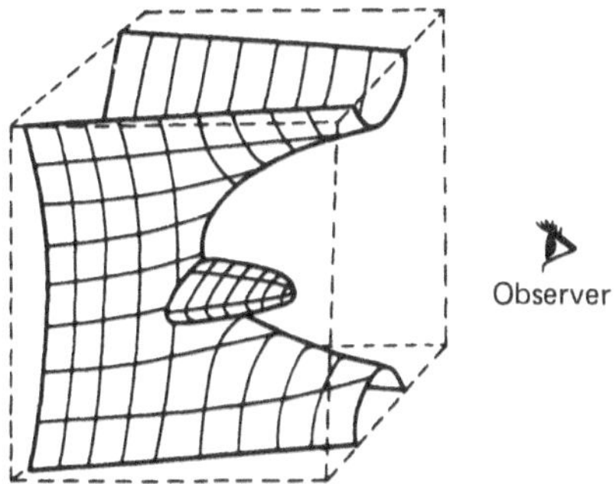

Fig.8.3 A three-dimensional pattern used in the experiment of the computer-generated integral photography.

an optical path difference of one wavelength, in other words, a phase difference of 2π. This condition is called the *phase-matching* condition. Careful bleaching control for this condition may bring forth a diffraction efficiency as high as 90%.

In Ref.(3), photographs of two-dimensional and three-dimensional reconstructed images are shown.

8.2.4 COMPUTER-GENERATED HOLOGRAPHIC STEREOGRAM

As stated before, the computer-generated hologram technique seems to aim finally at on-line three-dimensional display of computer output. However, at present real-time display is almost impossible. To be practical, simplified approaches are investigated.

King *et al.* of Bell Telephone Laboratories proposed and experimented with a practical method in which the images of a fictitious three-dimensional object viewed from various directions are first computed and displayed, and then a holographic stereogram giving these discrete images is prepared in a photographic manner (4).

In the experiment reported, an ensemble of 40

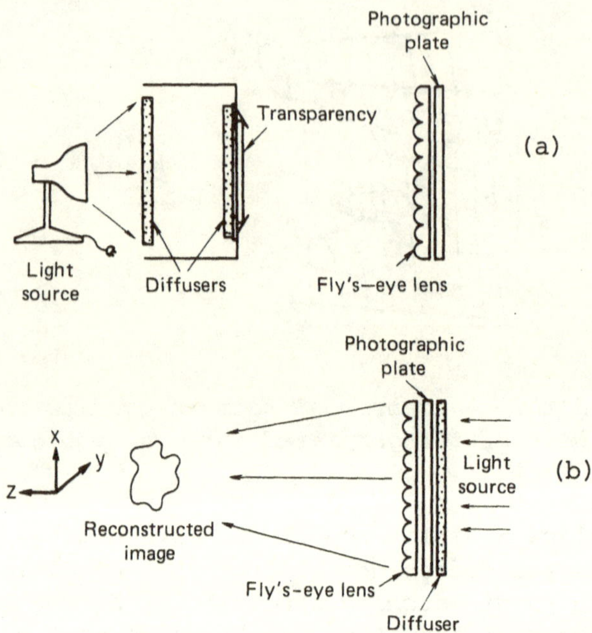

Fig.8.4 Preparation of a computer-generated integral photograph.

fictitious bars spatially intersecting with each other was assumed to be the object. The planar images of the object seen from 193 directions distributed within 59° were computed and recorded on microfilm using a computer-controlled microfilm plotter. After being developed and fixed, the images in the microfilm were synthesized to a unidirectional holographic stereogram using a setup similar to Fig.6.25 in Chapter 6. The required computer time was typically 20 s.

8.3 COMPUTER-GENERATED INTEGRAL PHOTOGRAPH

A trial of synthesizing an integral photograph with the aid of a computer was reported by Chutjian and Collier of Bell Laboratories in 1968 (5). The object recorded was a curved surface as shown in Fig. 8.3. Cross sections of such a spatial pattern at its various depths are generated by the computer, and they are recorded in an integral photograph through multiple exposures, using a setup as shown in Fig.8.4a. The reconstructed image is observed in a setup shown in Fig.8.4b. In Ref.(5) photographs of the reconstructed images are shown.

8.4 BINOCULAR STEREOSCOPIC DISPLAY OF COMPUTER OUTPUTS

Three-dimensional display of computer-output information, in particular computer-processed three-dimensional radar information, is sometimes desired; such demand will further increase in future. For this purpose, computer-generated hologram and integral photograph techniques are not practical at present; simpler techniques such as binocular display or a varifocal mirror method (see the next section) are employed (6).

Binocular display of computer output, or stereoscopic drawing by computer, has been investigated in the past several years. In early experiments, figures were drawn by a plotter and then observed using a stereoscopic viewer. In some recent works, the images are displayed on a cathode-ray tube to facilitate real-time, on-line operation (7). The most difficult problem with these techniques is the "hidden-line problem," that is, to distinguish and erase those image lines that are hidden behind the image itself. This requires a sophisticated algorithm.

Figure 8.5 shows an example of binocular display of a computer-simulated electron trajectory in an electronic device (a special-purpose electron-beam collector to be used in microwave traveling-wave tubes), obtained in the author's laboratory (8). Such drawings were put into binocular viewers specially designed for this purpose, and were found useful in improving the arrangement of the electrodes on a trial-and-error basis to optimize the electron trajectory.

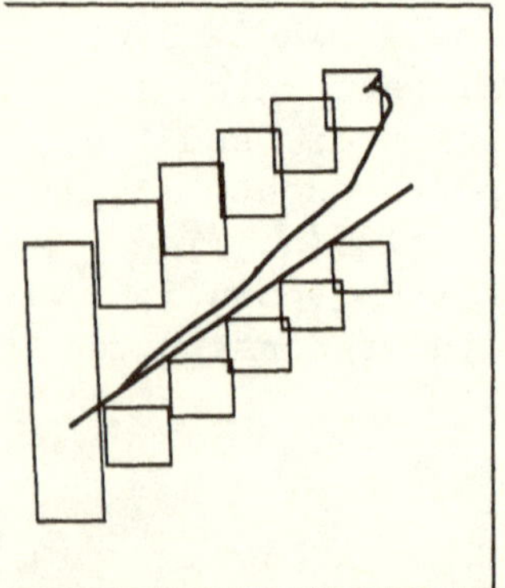

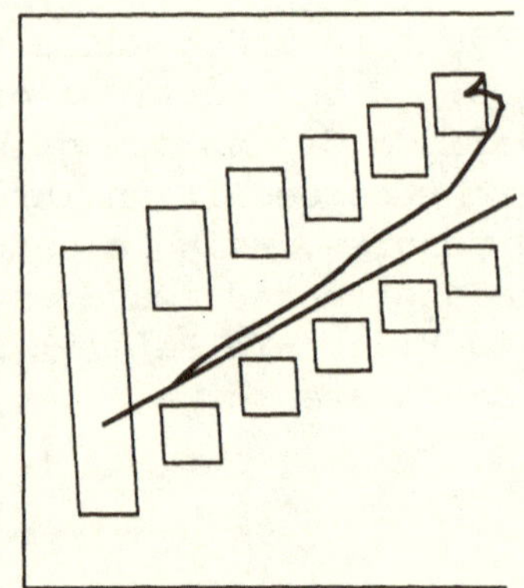

Fig.8.5 An example of the binocular stereoscopic display of computer output.

8.5 VARIFOCAL MIRRORS

A varifocal mirror is a vibrating-membrane mirror devised for three-dimensional imaging. It was invented and developed by A. C. Traub of the Mitre Corporation, who aimed at a practical three-dimensional display for airport traffic control (9)-(11). Later, this technique was further investigated by Rawson, King, and Berry of Bell Laboratories for application to three-dimensional computer-generated movies (12) and three-dimensional television (13).

8.5.1 PRINCIPLE OF A VARIFOCAL MIRROR

A thin membranelike mirror is put on the frame of a speaker and is vibrated by a pressure drive on its back surface. A cathode-ray tube (CRT) display is observed through such a mirror as shown in Fig.8.6. Cross sections of the three-dimensional image to be reconstructed are displayed successively on the CRT. In such an arrangement, if the vibrating frequency is much higher than the maximum response frequency of the eye, a three-dimensional sensation is obtained because these successive images are observed at different distances.

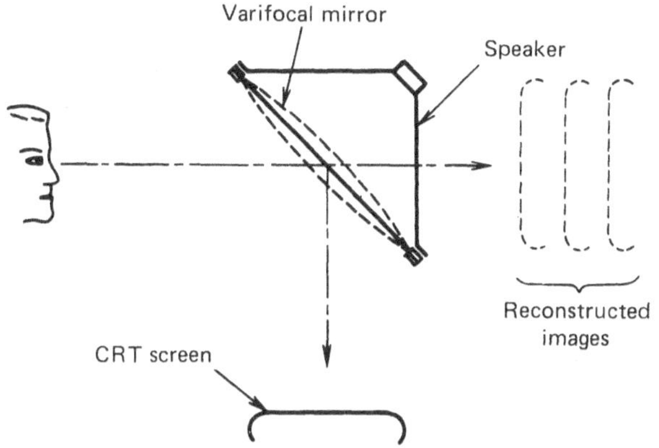

Fig.8.6 Three-dimensional imaging using a varifocal mirror.

In his experiment, Traub used uniform polyester films with thicknesses of 6.35 and 12.7 μm, upon one surface of which aluminum was sputtered. The vibrating frequency was 30-60 Hz.

A disadvantage of this system is that the more distant the image appears, the bigger it looks, and vice versa, if the cross sections are displayed on the CRT with constant magnification. This is because such a mirror focusing system has larger magnification for a more distant image. Therefore, when used in computer-output display or in computer-processed information display, the distortion described ought to be precompensated in the computer.

8.5.2 APPLICATION TO THREE-DIMENSIONAL MOVIES

Rawson proposed and experimented with the application of the varifocal mirror technique to three-dimensional movies (12). In his experiment, the images were generated in a computer and recorded successively on a 16-mm movie roll film. It was then played back synchronously with the vibration of the varifocal mirror.

The difficult problem was how to obtain uniform depth resolution. When the varifocal mirror device is driven with a sinusoidal waveform, the density of the reconstructed image in the depth direction is far from uniform; it is dense at the front and rear ends and less at the center. In Rawson's experiment, therefore, a sawtoothlike waveform was used in driving the speaker to attain a uniform image density over 80% of a cycle.

8.5.3 APPLICATION TO THREE-DIMENSIONAL TELEVISION

This will be described separately in Section 8.9.7 (13).

8.6 THREE-DIMENSIONAL X-RAY IMAGING

Probably the first experiment with three-dimensional x-ray imaging was reported by Kasahara *et al.* (14). They took a number of x-ray photographs of an object from various directions and synthesized them into a holographic stereogram. X-ray imaging using the holographic stereogram technique was further investigated by Groh and Kock (15).

On the other hand, binocular x-ray imaging has been tried for many years. Many medical doctors have devised individually stereoscopic viewers or special viewing techniques for their own use.

A new topic in this field is the use of a lenticular sheet for binocular x-ray pictures, proposed and experimented with by C. Yamada of Toppan Printing Company and E. Kaku of the Medical School of the University of Tokyo (16). They developed a quick preparation system of lenticular-sheet binocular x-ray pictures, to make available in surgical operations on the brain detailed information on the configuration of fine vessels. The original picture is recorded several hours before

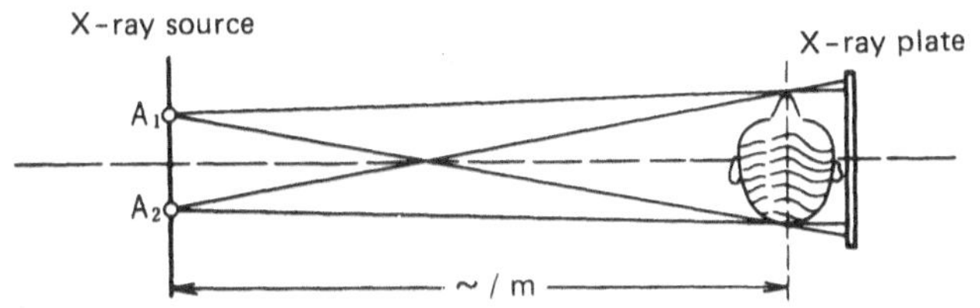

Fig.8.7 Recording of binocular x-ray pictures.

the start of the operation, quickly duplicated, and distributed to the staff to provide the latest information on the vessels.

Figure 8.7 shows the arrangement for x-ray image recording. Two x-ray pictures of the brain are taken on the right, first with the x-ray source at A_1, and second at A_2. These pictures are then synthesized into a lenticular-sheet binocular picture. The separation between A_1 and A_2 is usually 12-13 cm. A specially designed x-ray tube having very small source area must be used to assure proper depth resolution. The source size in ordinary x-ray tubes is 0.3-1.0 mm in diameter, whereas in their experiment it was typically 0.05 mm. Just in front of the photographic plate or film, a fine venetian-blind-like structure made of lead is placed, which passes only the directly transmitted component of the x ray and rejects the scattered components. This scheme is necessary to assure proper contrast.

The stereoscopic pictures obtained proved themselves useful for positioning the fine vessels in the brain.

8.7 HOLOGRAPHIC SCREEN

The first proposal to make in a holographic manner a screen to be used in a projection-type display was made by Gabor (17). This sort of screen might provide a technical breakthrough in projection-type display techniques.

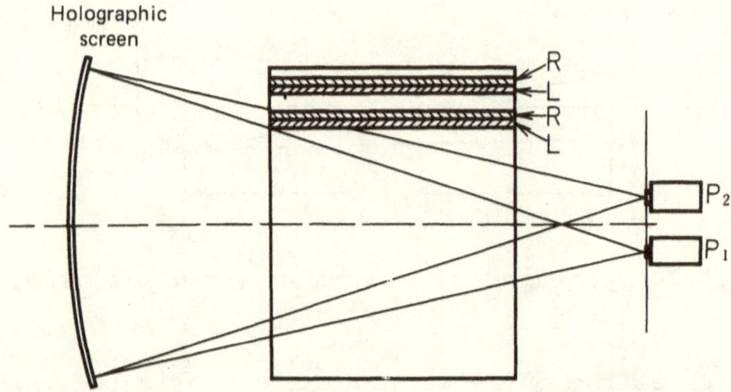

Fig.8.8 Projection-type binocular display using a holographic screen.

8.7.1 GABOR'S PROPOSAL

Gabor conceived an idea for binocular stereoscopic movie theaters using gigantic Lippmann-type volume holograms as the projection screens (17). The basic concept is illustrated in Fig.8.8. The screen is a Lippmann hologram which reflects light from projector P_1 toward the right-eye positions of the viewers sitting in the seats (hatched area R), and light from P_2 toward the left-eye positions (hatched area L). Theoretically, there is no essential difficulty in making such a Lippmann hologram. However, of course, the realization of such an idea would encounter many practical difficulties.

8.7.2 UNIDIRECTIONALLY DIFFUSING HOLOGRAPHIC SCREEN

Taking advantage of the interference in coherent light, we may make various screens for use in three-dimensional imaging. An example of this is the unidirectional diffusing screen made by Nomura in the author's laboratory (18).

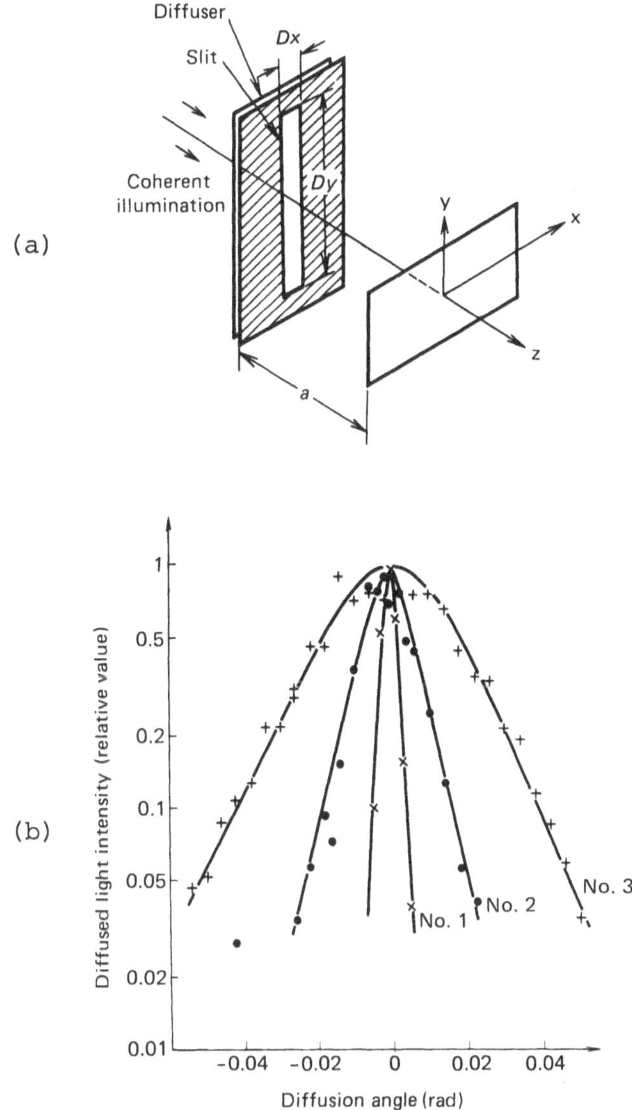

Fig.8.9 The unidirectionally diffusing holographic screen. (a) Optical setup for the preparation. (b) Horizontal diffusing property. x, *sample 1;* •, *sample 2;* +, *sample 3.*

The unidirectionally diffusing screen is a screen that is transparent in one tangential direction (horizontally, for example) but is diffusing like a ground glass plate in another tangential direction (vertically, for example). An example of such a screen is a fine lenticular sheet, which is often employed to give a vertically diffusing property in composite-type horizontally direction-selective screens (see Section 5.3).

Figure 8.9a shows a method for producing a unidirectionally diffusing screen using the interference phenomenon of coherent light. At distance a from a photographic plate is placed a vertical slit with dimensions $D_x \times D_y$, illuminated from the back through a diffusing (actually, ground glass) plate. After exposure, the photographic plate is developed, and bleached to form a kind of phase hologram.

The phase hologram thus prepared shows a vertically diffusing property because it is an in-line hologram which records a diffusing, narrow vertical slit. Figure 8.9b shows the horizontal diffusing properties of three samples, each made with conditions of a = 200 mm, D_y = 60 mm, and D_x = 0.5, 2, and 5 mm, respectively. The transmission efficiency of the diffused component is above 80%, whereas direct transmission was hardly observable.

8.8 BINOCULAR STEREOSCOPIC TELEVISION

8.8.1 INTRODUCTION

Binocular stereoscopic television schemes have a long history of development. Many test models or test systems have been developed for individual purposes, and finally disappeared.

On the other hand, today people are talking about fantastic three-dimensional television in the distant future. It seems that most people are thinking implicitly of genuinely three-dimensional, in other words *autostereoscopic,* television, when they talk about the future three-dimensional television.

The author believes, as do most people, that if three-dimensional television becomes common in future society, it will inevitably be autostereoscopic as well as of very high resolution. It is widely accepted that high-quality, superwide screen television systems will appear within 10 years, prior to the age of three-dimensional television. Any three-dimensional television scheme will not be able to supersede the superwide screen television unless the former is superior to the latter also in resolution and screen width.

However, binocular systems are still of special value for some specific purposes, especially in the educational and medical fields. Therefore, prior to the discussion of genuinely three-dimensional television in the following section, we review the past development of binocular stereoscopic television.

8.8.2 BINOCULAR TELEVISION SYSTEMS PROPOSED IN THE PAST

A. Direct-View Systems

(1) *Two-color method.* The right-eye and left-eye images are displayed on a color cathode-ray tube with two different colors, for example, the former with green and the latter, red. The viewer observes the image with a pair of two-color glasses, the green glass on the right and red on the left. In the literature, for example, Mengel's experiment (19) is found. This method, however, is not promising because of eye fatigue.

(2) *Polarization method.* An example is shown in Fig.8.10. Two cathode-ray tubes display the right-eye and left-eye images, and they are synthesized by a half mirror through polaroid filters giving polarizations orthogonal to one another. The viewer observes the images with a pair of polaroid glasses, and a depth sensation is attained. Appearing in the literature are a three-dimensional visual aid for an isotope manipulator developed by the Argonne Laboratory of the AEC (Atomic Energy Commission, USA) (20), and a commercially available system developed in the United Kingdom (21).

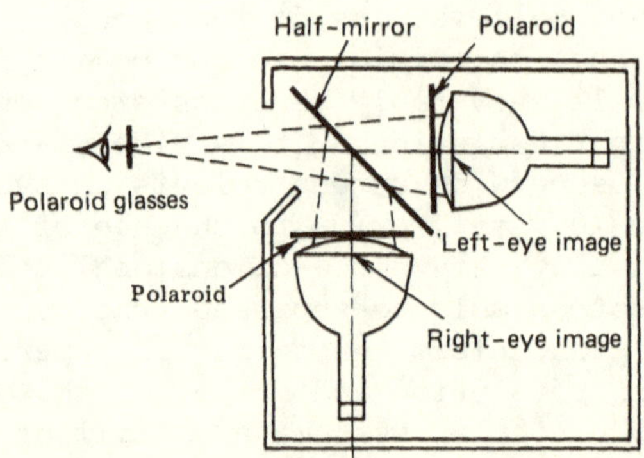

Fig.8.10 A binocular television system using two cathode-ray tubes.

(3) *Time-division method.* In this method two images are given to the two eyes alternately. For example, in a system developed by ATI (American Television, Inc.), the observer holds a pair of simple shutter mechanisms in front of his eyes, which shields each eye alternately, whereas the right-eye and left-eye images appear on the cathode-ray tube alternately and synchronously (22). A possible technical drawback is that the two images may not be completely separated because of the "afterimage" effect of the cathode-ray tube.

(4) *Double-slit parallax barrier method.* If we use a construction similar to Fig.8.10, but replace the polaroid filters by double-slit parallax barriers as shown in Fig.8.11, we may realize a more or less "freely viewable" stereoscopic television system without glasses. Such a system was investigated late in the 1950s in NHK (Japan Broadcasting Corporation) Research Laboratories (23).

Such a system might be realized most easily among all the "glasses-free" methods. However, it is not considered promising because of the technical limitations of the parallax barriers, which were described in Section 2.3.5.

(5) *Lens-and-slit parallax barrier method.* The drawback of the double-slit parallax barrier method

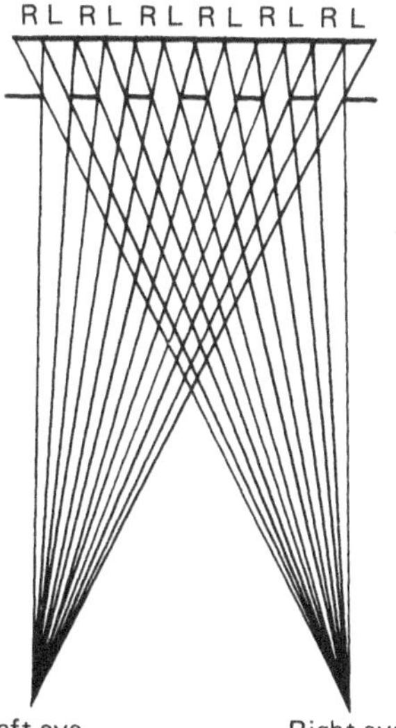

Fig.8.11 A double-slit parallax barrier. Apertures are made on R for the right-eye image and on L for the left-eye image.

described in the preceding item may be overcome by this method. In this method, a lenticular-sheet-and-slit structure as shown in Fig.8.12 is inserted in place of the polaroid filters in Fig.8.10 (24). Such a lens-and-slit structure is less visible than the double-slit structure in the above method because the light spreads in the entire width of each cylindrical lenslet. This method might be one of the most practical methods among the direct-view systems.

(6) *Vibrating-slit method.* If one plate of the double-slit structure vibrates with an amplitude of half the slit pitch, and the right-eye and left-eye images are displayed on a cathode-ray tube alternately and synchronously with the slit vibration, a single-tube parallax barrier system can be realized (24). Other versions include vibrating the lenticular sheet in the lens-and-slit parallax barrier system and electrical switching of slits using electrooptic effects (24).

(7) *Use of special-purpose cathode-ray tubes.* If the use of a specially designed display tube is permitted, many other methods are possible. An example

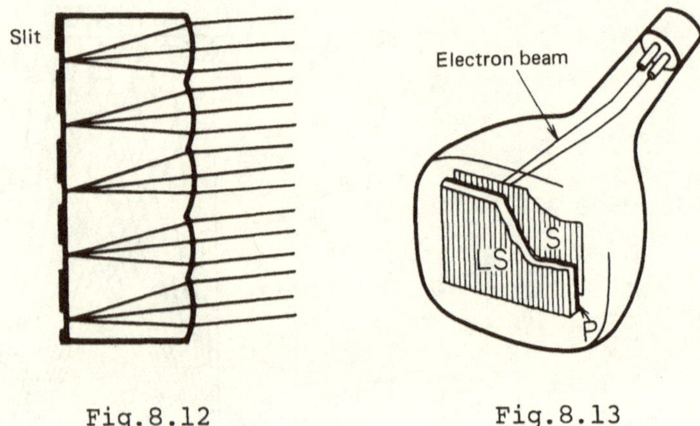

Fig.8.12 A lens-and-slit parallax barrier.
Fig.8.13 A special cathode-ray tube for stereoscopic display.

is to use a cathode-ray tube in which a shadow mask or line grid S is placed just behind the fluorescent screen, whereas a slit sheet or a lenticular sheet is placed in front of the screen as shown in Fig.8.13. This scheme was first proposed by Yamaguchi and Fukushima in 1961 (23), but has not been experimented with so far.

B. *Projection-Type Systems*

In projection-type systems, the right-eye and left-eye images are once displayed on projection kinescopes (or a kinescope), and they are projected onto a specially designed screen. Therefore, the inherent technical problems are almost identical to those of projection-type three-dimensional still pictures and three-dimensional movies.

(1) *Two-color method.* Differently colored right-eye and left-eye images are projected onto an ordinary screen, and observed through a pair of colored glasses. This method is sometimes used in stereoscopic movies, but is not promising because of eye fatigue [see Section 8.8.2.A (1)].

(2) *Polarization method.* Orthogonally polarized right-eye and left-eye images are projected from cathode-ray tubes onto a screen, and observed through a pair of orthogonal polaroid glasses. This method is practical and promising. The surface-coating material of the screen must be chosen carefully to minimize the rotation and spread of the polarization.

A single-tube version of the above method designed for medical use (x-ray television) is described in the following subsection (8.8.3).

(3) *Time-division method.* This is essentially identical to the time-division direct-view system described in the preceding section, except that animated images are projected from a cathode-ray tube onto a screen. If very lightweight, all-electronic "switching glasses" are developed in the future, this method might become the most practical one.

(4) *Use of a direction-selective screen.* The principle of this method is identical to the binocular, projection-type still display, except that animated images are projected from cathode-ray tubes. Most of the technical problems are held in common with still displays. This method also seems promising (24).

8.8.3 BINOCULAR X-RAY TELEVISION FOR MEDICAL PURPOSES

There have been many proposals and experiments in this field, but few papers can be found reporting design and performance of practical systems. In this subsection, the system developed by Noda *et al.* (25) of Hitachi Denshi Company will be described briefly.

Noda *et al.* have developed the following three systems for use in the medical education:

(1) Direct-view binocular x-ray television system.

A block diagram of the system is shown in Fig. 8.14. The binocular images are displayed on a single 5-in. cathode-ray tube half-and-half, and observed through convex lenses (f = 60 mm) at a distance of

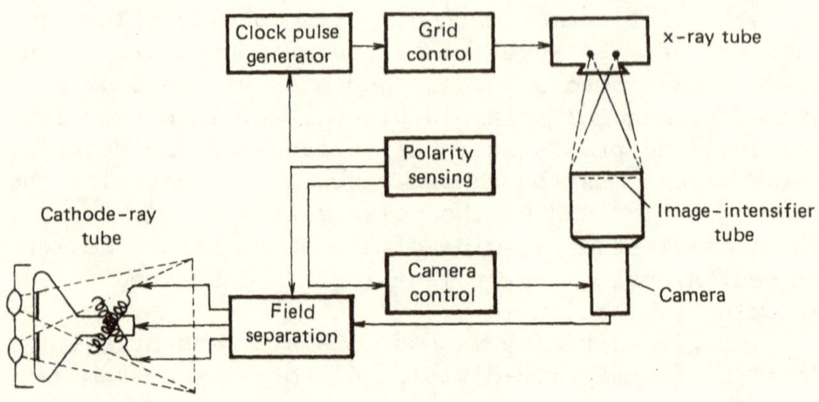

Fig.8.14 A direct-view binocular x-ray television system [after Noda et al. (25)].

50 mm, so that the virtual image of the cathode-ray-tube screen is focused at 300 mm.

(2) Direct-view polarization-method binocular x-ray television system.

This system has overall construction similar to Fig.8.14, and a display system like that in Fig.8.10.

(3) Projection-type polarization-method binocular x-ray television system.

This system aims at simultaneous observation by many students. A block diagram of the entire system is shown in Fig.8.15. The right-eye and left-eye images are displayed on an Eidophor tube (26) alternately, in front of which a polaroid disk as shown in the figure rotates synchronously. The observer watches the projected image through a pair of orthogonal polaroid glasses. The dimensions of the projected image are about 600 × 600 mm^2.

The basic idea underlying the experiments of Noda *et al.* is that methods without glasses such as projection systems using direction-selective screens, although

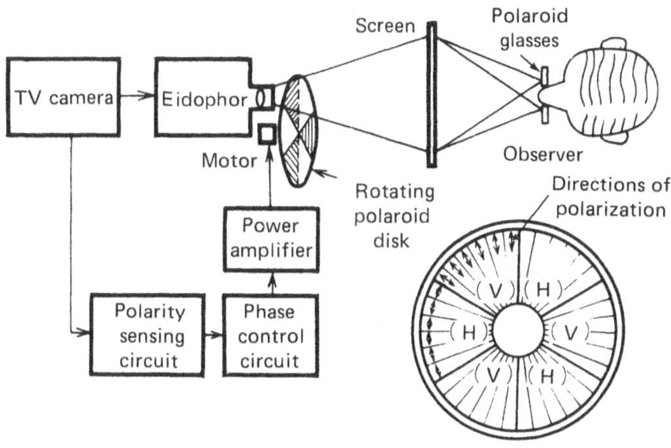

Fig.8.15 A projection-type binocular x-ray television system [after Noda et al. (25)].

they might look smart, are not practical at present, and that for the time being we should prefer the practicality of methods with glasses. Almost simultaneously with Noda *et al.*, Herman discussed the advantages and disadvantages of various binocular stereoscopic television systems and stated a conclusion identical to that of Noda (27).

8.9 AUTOSTEREOSCOPIC THREE-DIMENSIONAL TELEVISION

8.9.1 INTRODUCTION

As described in Section 8.8.1, in future society people might enjoy three-dimensional television of an autostereoscopic type. However, at present nobody can tell what kind of technique will be used in that future system because probably we shall experience many technical innovations before that time. Thus the discussion on autostereoscopic television systems is apt to become a mere dream story. In this section, therefore, we shall restrict ourselves to presently predictable subjects concerning three-dimensional television; these are the problem of the quantity of information to be transmitted and displayed, some transmission experiments with holograms and varifocal-mirror images, and the psychological background of the art.

8.9.2 TWO MAJOR DIFFICULTIES IN REALIZING AUTOSTEREOSCOPIC TELEVISION

At present there are two major difficulties to be overcome before we realize autostereoscopic television.

The first difficulty is that we do not yet have any satisfactory display device to be used in it. This difficulty must be overcome in two steps. In the first place, three-dimensional still image display techniques must be developed to a practical level because these are presently at an immature state. Still-image techniques are a prerequisite to animated-image techniques. The second step is to animate the image; this step is left for our efforts in the future.

The second difficulty is that the amount of information to be transmitted is tremendous. This problem was discussed in Chapter 7. It was shown that if we transmit a complete autostereoscopic television signal having resolution and frame frequency both comparable to the ratings of today's commercial television, the required bandwidth will be as high as several 10 GHz or more. Therefore, development of information-reduction techniques will be indispensable.

It is almost impossible for the author to state how the first difficulty will be overcome in future. Only brief comments on this problem will be given in Section 8.9.8. The problem of the quantity of information will be discussed first in the following subsection.

8.9.3 BANDWIDTH REQUIREMENTS IN VARIOUS THREE-DIMENSIONAL TELEVISION SCHEMES

The sampling point numbers of various three-dimensional images were computed in Chapter 7. In Section 7.6, for a better understanding of the results obtained, the bandwidth required to transmit various three-dimensional images are shown for assumed values of parameters (see Table 7.3). In this subsection, the computations performed to derive Table 7.3 will be described.

We assume lateral resolution and frame frequency both comparable to those of today's commercial television: N (line number) = 500 and f_F (frame frequency) = 30 pictures/s. However, in some cases the resolution cannot be reduced below the resolution of the human eye; for these cases bandwidth required for transmitting such "unreduced" images will be shown.

We start from Table 7.2. In light of the sampling theorem, when the bandwidth (maximum frequency) is B, we may transmit the brightness information of two sampling points in a time interval 1/B. Hence, if f_F = 30 $[s^{-1}]$, then $B = 1/2 \times 30 \times P\ [s^{-1}] = 15P\ [s^{-1}]$. We further assume in Table 7.2 that

$$N = N_V = N_H = 500, \quad \lambda = 500\text{ nm}, \quad \Omega = \Omega_H = 0.2,$$

$$a = 2\text{ m}, \quad b = 20\text{ cm}, \quad \text{and} \quad w = 40\text{ cm}.$$

(1) *Binocular hologram images.* Suppose we transmit two images in holographic form. Doubling the formula for the off-axis reference beam, we obtain $P = 8 \times 10^6$. Therefore, B = 120 MHz.

(2) *Wide viewing-zone holography without information reduction.* This is one of those imaging techniques having essentially the resolution of the human eye. From the assumed parameters, $P = 10^{11}$, hence $B \doteqdot$ 1,500 GHz.

(3) *Eye-piece-type binocular holography.* This is another example of an imaging technique having the resolution of the human eye. From the formula $P = 5.5(A/\lambda)^2$, the sampling point number $P \fallingdotseq 2 \times 10^8$, which leads to $B \fallingdotseq$ 6.0 GHz for the two images.

(4) *Reduced-information holography*. The above-mentioned parameter values lead to B = 600 GHz. [In this case, the assumption $N^2 \ll N'^2$ for Eq. (7.38) does not strictly hold. However, all the minor errors are neglected here.]

(5) *Integral photography*. $P = 2.8 \times 10^9$ leads to B = 42 GHz.

(6) *Multiple photography without vertical parallax*. $P = 5 \times 10^7$ leads to B = 750 MHz.

(7) *Unidirectional holography without vertical parallax*. $P = 2 \times 10^8$ leads to B = 3 GHz.

The above results are tabulated in Table 7.3 in Section 7.6.

Table 7.3 shows that the bandwidth required to transmit autostereoscopic television information is tremendous. However, if we transmit binocular image information plus some additional information concerning the hidden part, the receiver should be able to reconstruct, in principle, a complete autostereoscopic image. This assumption leads to, for example, B = 50 MHz for N = 500. Therefore, the redundancy ratios of IP and holography are 800-fold and 12,000-fold, respectively.

Obviously, the biggest reason for such large redundancy is that the images observed from every direction are transmitted equally so that the reconstructed image may be observed from any direction at the receiving end. If a person-to-person three-dimensional television-telephone communication system is established in future, the bandwidth requirement would be reduced greatly by using "mobile" video cameras which move in accordance with the receiver's eyes.

8.9.4 POSSIBILITY OF THREE-DIMENSIONAL TELEVISION SCHEME IN WHICH HOLOGRAMS ARE TRANSMITTED

For the above reasons, we may conclude that even in the distant future a three-dimensional television system in which holograms are transmitted without information reduction will never be realized. If any holographic image transmission is realized, it will probably be some mixed technique of holography and multiple photography.

When a holographic approach is employed in the display of three-dimensional television, another problem remains unsolved: lack of an animated-hologram display device. In other words, at present we do not have a satisfactory display device which is provided both with a resolution high enough for hologram display and with a response quick enough for television information. Such a display device is a prerequisite for "holographic" television. Presently we have three candidates:

(1) a display device using photochromic glass,
(2) a display device using a very thin oil film (similar to Eidophor),
(3) a display device using amorphous semiconductors (semiconductor glass).

Photochromic glass was described in detail in Section 6.15. It is provided with a satisfactorily high resolution, but the temporal response is rather poor. In the oil-film device, a very thin oil film is scanned by an electron beam to generate deformation by electrostatic fields (26),(28). It features a quick temporal response, but the resolution is poor (less than 100 lines/mm). Amorphous semiconductors have long been investigated with respect to their electrical switching characteristics. Only in the past few years, however, has their property as a reversible optical recording material earnestly been investigated. Each of the above three devices will require long-term efforts before becoming practical for use in three-dimensional television schemes.

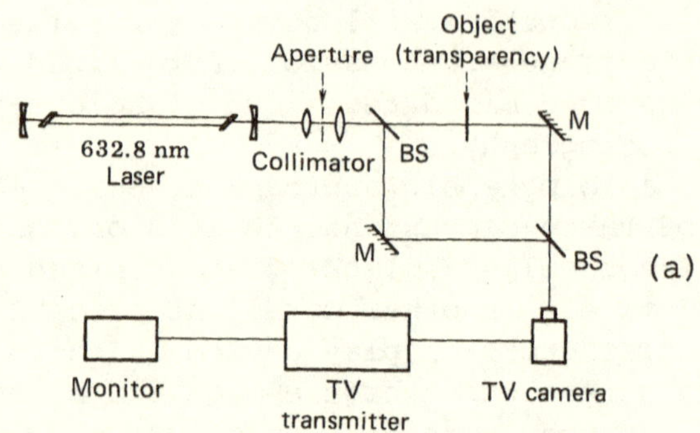

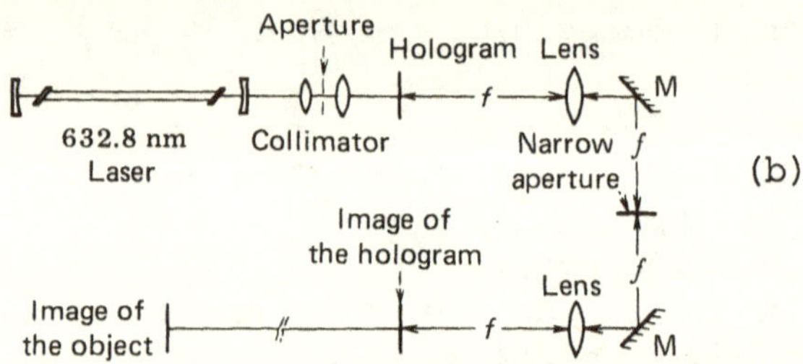

Fig.8.16 Experiment of television transmission of a hologram [after Enloe et al. (29)]. (a) Transmitter. (b) Optical system used in the receiver.

8.9.5 TRANSMISSION EXPERIMENTS OF HOLOGRAMS VIA TELEVISION CHANNELS

Several reports from Bell Laboratories describing transmission of holograms via television channels are found in the literature. Though these works are not directly related to three-dimensional imaging, they will briefly be introduced as a basis for future developments toward three-dimensional television techniques.

A. First Experiment by Enloe et al.

Probably the first experiment of television transmission of holograms was performed by Enloe *et al.*

(29), who transmitted a hologram of a still, planar, black-and-white transparent pattern consisting of four letters, BELL.

The experimental system is shown in Fig.8.16. In the transmitter, the hologram of the object is made upon the screen of a vidicon. Since the vidicon does not have a very high resolution, the angle made by the object wave and the reference wave is chosen as small as 1°, so that the spacing of the fringes is not excessively small. In the receiver, the hologram pattern reproduced upon a cathode-ray tube is once recorded as a photograph, which, after development, is illuminated by ordinary laser light to reconstruct the image. The narrow aperture seen at the right in Fig.8.16b rejects the direct-light and the conjugate-image component which are both directed very close to the wanted component. The other parts are probably self-explanatory. In Ref.(29), a photograph of the reconstructed image (the four letters, BELL) is shown.

B. *Hologram Heterodyne Scanner*

The greatest difficulty in Enloe's experiment was insufficient resolution of the vidicon. A scheme to overcome this difficulty was also proposed by Enloe *et al.*, and was named the *hologram heterodyne scanner* (30).

As described in Section 7.3.1, in ordinary off-axis reference-beam holography, if the reference-beam angle is chosen wide enough so that the spectrum to be recorded (the rectangular spectrum seen in Fig.7.6a) does not interfere with the self-modulation component (the triangular spectrum), a spatial frequency bandwidth of 4W is required where W is the real frequency range to be recorded. This fact greatly increases experimental difficulty because at present even the best commercially available camera tube has a resolution of about 1000 lines. To reduce the required bandwidth, the only method is to reduce the reference-beam angle; the interference will be increased however.

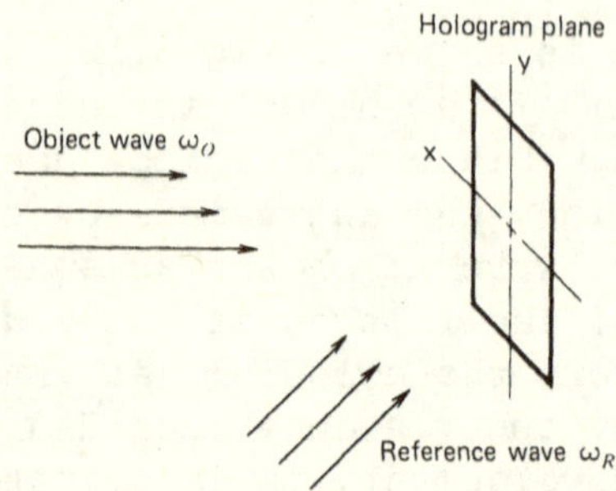

Fig.8.17 Hologram heterodyne scanner.

In the hologram-heterodyne-scanner scheme, the reference beam is reduced to zero (*i.e.*, an on-axis reference beam is used), still avoiding the above interference. Thus the bandwidth can be reduced to W.

(1) *Optical arrangement to be considered.* We assume that on the hologram plane (or on the screen of the camera tube in the present case), the object and reference waves, having different wavelengths

$$O(x,y) = O_0(x,y)\exp[j\omega_O t+j\phi_O(x,y)] \tag{8.3}$$

$$R(x,y) = R_0 \exp[j\omega_R t-jax-j\phi_R], \tag{8.4}$$

are incident as shown in Fig.8.17. The object and reference waves are assumed to be incident normally and obliquely, respectively. The added amplitude upon the screen plane

$$I = O_0^2+R_0^2+2R_0O_0(x,y)\cos[(\omega_O-\omega_R)t+\phi_O(x,y)+ax+\phi_R]. \tag{8.5}$$

(2) *Scanning of the screen by a thin reference beam.* We first consider a method that is not practical but good for understanding the following, more practical schemes.

If we assume $\omega_O = \omega_R$ in Eq.(8.5) as in ordinary holography,

$$I = O_0^2+R_0^2+2R_0O_0(x,y)\cos[\phi_O(x,y)+ax+\phi_R]. \tag{8.6}$$

The first, second, and third terms represent the triangular spectrum, the spikelike spectrum at $f = 0$, and the rectangular spectrum around $f = \theta_R/\lambda$ as shown in Fig.7.6, respectively.

One possible method for reducing the reference-beam angle, still avoiding interference, will be as follows. Instead of an ordinary camera tube, we use a phototube having uniform sensitivity over the entire screen area. A movable pinhole mechanism is placed just in front of the screen; the pinhole mechanically scans the screen surface in the x direction. Then a signal will be generated from the term $\cos[\phi_O(x,y)+ax]$ in Eq.(8.6), the first and second terms being eliminated. Therefore, the reference-beam angle may be reduced to λW, resulting in a bandwidth of 2W as shown in Fig.7.6b.

In practice, however, the necessary scanning speed can never be obtained with a mechanical scanning scheme. Enloe *et al.* state that a more practical method might be to use a pencillike reference beam and scan the screen surface optically.

(3) *Heterodyne method.* Pencil-beam scanning is an interesting idea but not quite practical because the beam must be focused to a very small spot, comparable to the fringe spacing. Therefore, it seems that the best way will be to go back to Eq.(8.5) and again consider the case in which $\omega_O \neq \omega_R$. Two beams of such different frequencies may be obtained as two sidebands of a modulated laser beam, or from two different oscillation modes of a mode-locked laser, or from two lasers mutually mode-locked. Anyway, if $\omega_O \neq \omega_R$, we may take out only the third term of Eq.(8.5) by filtering out the frequency components around $\omega = |\omega_O-\omega_R|$, even when $a = 0$, that is, $\theta_R = 0$. This is the basic idea of the hologram-heterodyne-scanner scheme.

C. *On-Axis Reference-Beam Method Using Multiple Exposures*

Almost concurrently with the above proposal of Enloe *et al.*, Burckhardt proposed two on-axis reference-

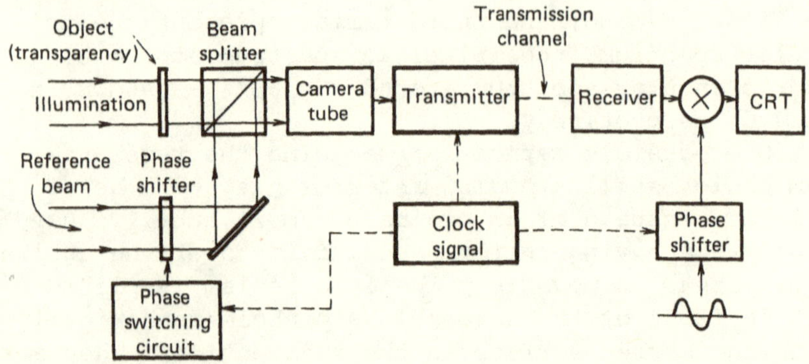

Fig.8.18 On-axis reference-beam method using multiple exposures [after Burckhardt (32)].

beam methods using multiple exposures to reduce the resolution requirement of the camera tube (31)(32). In the first proposal (31), four exposures for a picture are required; whereas in the second (32) the required number is reduced to three, and it is proved that three is the minimum required number. In this item the second proposal will be described.

The setup of the proposed system is shown in Fig. 8.18. Both the object and reference beams are normally incident upon the screen of the camera tube. First, a hologram of the object is made upon the screen, scanned, converted into an electrical signal, and transmitted. At the receiver, this signal is multiplied by a sinusoidal waveform, and displayed on a cathode-ray tube. Next, the transmitter sends out a new hologram signal with a reference beam whose phase is advanced by 120°. The receiver again displays the hologram after multiplying the signal by a sinusoidal wave, whose phase is also advanced by 120°. The same process is repeated for further 120° advances of both the reference-beam phase in the transmitter and the sinusoidal-wave phase in the receiver. Thus, the image displayed finally on the cathode-ray tube can be made equivalent to one obtained with an off-axis reference scheme, as shown in the following.

We assume that the initial phase of the reference beam is defined so that R_0 of the first hologram is pure real in Eq.(8.6). Then the optical intensity of the first hologram may be expressed as

$$I = (O_0+R_0)(O_0^*+R_0)$$

$$= O_0O_0^*+R_0^2+O_0R_0+O_0^*R_0. \tag{8.7}$$

If the transmission channel is linear, the received signal

$$U_1 = K_1 I \quad (K_1 \text{ const}). \tag{8.8}$$

After multiplication by the sinusoidal waveform, the signal will be expressed as

$$U_1' = K_1 I \cos \omega t. \tag{8.9}$$

If the display on the cathode-ray tube is also linear and the hologram is scanned in the x direction, the displayed light intensity of the first hologram

$$I_{c1} = K_0+K_2U_1' \quad (K_0 \text{ and } K_2 \text{ const})$$

$$= K_0+K_1K_2 \cos \omega_s x(O_0O_0^*+R_0^2+O_0R_0+O^*R_0) \tag{8.10}$$

where ω_s is the spatial frequency in the x direction corresponding to the temporal modulating frequency ω.

The second and third hologram patterns displayed upon the cathode-ray tube are computed:

I_{c2}: by replacing R_0 by $R_0 \exp(j2\pi/3)$

and $\cos \omega_s x$ by $\cos(\omega_s x+2\pi/3)$,

I_{c3}: by replacing R_0 by $R_0 \exp(j4\pi/3)$

and $\cos \omega_s x$ by $\cos(\omega_s x+4\pi/3)$.

Adding all three terms, we obtain

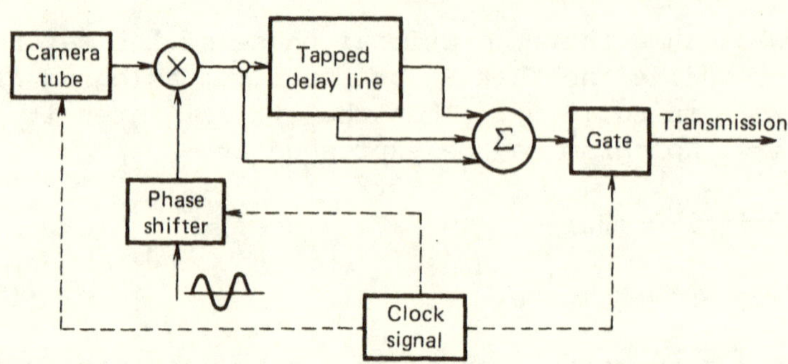

Fig.8.19 Reduction of transmission bandwidth in the on-axis reference-beam method [after Burckhardt (32)].

$$
\begin{aligned}
I_{total} &= I_{c1}+I_{c2}+I_{c3} \\
&= 3K_0+(3/2)K_1K_2O_0R_0 \exp(j\omega_s x) \\
&\quad +(3/2)K_1K_2O_0^*R_0 \exp(-j\omega_s x),
\end{aligned} \tag{8.11}
$$

which has a form identical to one obtained in an off-axis reference-beam case. Thus it is shown that even when a low-resolution camera tube is used, information equivalent to an off-axis reference-beam hologram can be transmitted if we increase the number of exposures.

In the above scheme, however, the required bandwidth of the transmitting channel is not reduced effectively. The spatial frequency bandwidth to be transmitted in an on-axis reference beam scheme is W, whereas in the above scheme a transmission bandwidth corresponding to 3W is required because the frame frequency must be tripled. On the other hand, since an off-axis reference-beam hologram is finally obtained in the above scheme, the intrinsic lower limit of the transmission bandwidth should be a value corresponding to 2W, as seen in Fig.7.6b. Theoretically, we should be able to reduce 3W to 2W.

A setup for such a reduction of the transmission bandwidth is shown in Fig.8.19 (32). In this scheme, the synthesis process which was originally performed in the receiver is completed in the transmitter, and a synthesized signal requiring the minimum bandwidth is transmitted.

D. *Experiment by Berrang*

The experiment of the system shown in Fig.8.18 was performed by Berrang. In Ref.(33), a picture of the image reconstructed at the receiver end is shown.

8.9.6 POSSIBILITY OF THREE-DIMENSIONAL TELEVISION SCHEME IN WHICH INTEGRAL PHOTOGRAPHS ARE TRANSMITTED

The biggest difficulty in the direct transmission of an integral-photograph signal is its tremendous quantity of information. No one has ever tried an experiment of this kind.

When we reduce the information of IP, we finally reach multiple photography such as LS pictures and multiple projection-type displays. A typical reduction ratio may be found by comparing two bandwidths shown in Table 7.3: 42,000 and 750 MHz.

8.9.7 TELEVISION TRANSMISSION OF VARIFOCAL MIRROR IMAGES

There is no essential difficulty at present in transmitting animated three-dimensional images recorded by the varifocal mirror technique. The first experiment was performed by King and Berry of Bell Laboratories (13).

Figure 8.20 shows the setup of their experiment. When varifocal mirror VM1 vibrates, the position of the virtual image of the object moves upward and downward in the figure. Since the lens L of the camera tube has a large diameter, only a portion of the virtual image at a constant distance can be focused sharply upon the

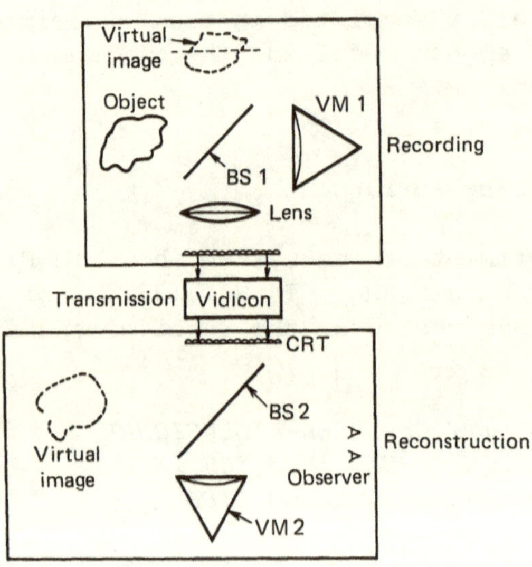

Fig.8.20 Television transmission of varifocal mirror images [after King and Berry (13)].

screen of the camera tube, while the other portions are blurred due to defocus. In the receiver, a similar optical setup is used to observe the cross sections of the object through VM2, which vibrates synchronously with VM1.

In the present state of the art, the varifocal mirror technique is the only practical method of realizing three-dimensional television. However, it has the following difficulties which may not easily be overcome.

In the first place, if the vibrating frequency of the varifocal mirror and the frame frequency of the television system are denoted by f_M and f_T, respectively, the maximum depth resolution

$$N_D = f_T/f_M. \quad (8.12)$$

The vibrating frequency f_M can never be reduced below 18 Hz which is determined by the after-image duration of human eyes. On the other hand, f_T cannot be raised above 180 Hz due to technical limitations; in most cases 60 Hz is the practical limit. Therefore, depth resolution is usually very poor.

Second, a more essential difficulty is the presence of the "phantom-imaging effect," that is, an image part that should not be visible because it is hidden by another part of the object remains visible.

The third difficulty is blurring of the entire image due to the noisy, defocused light component. Since this noise component has a low frequency, it might be removed electronically.

8.9.8 FUTURE PROSPECTS FOR THREE-DIMENSIONAL TELEVISION

A. Classification of Various Three-Dimensional Imaging Techniques in Terms of the Depth Cues Used

Important bases for discussion of future prospects for three-dimensional television techniques are the psychological and physiological backgrounds. As described in Chapter 3, it is accepted that ten depth cues should be considered; among these, the most important are the three physiological cues: accommodation, convergence, and binocular parallax. In addition to the ten cues described in standard textbooks, many specialists accept that a psychological effect which has been named tentatively the *wide-screen effect* in Section 2.8 is also important. The success of Cinerama should be attributed principally to this effect.

The various three-dimensional imaging techniques described in this book may be classified according to the depth cues used, as shown in Table 8.1.

TABLE 8.1 Depth Cues Utilized in Various Three-Dimensional Imaging Techniques

	Accommodation	Convergence	Binocular parallax	Wide-screen effect
Wide-viewing-zone holography	o	o	o	x
Eye-piece holography	o	o	o	o
Projection-type holography	x	o	o	x
Holographic stereogram	Δ	o	o	x
Integral photography	Δ	o	o	x
Lenticular-sheet three-dimensional pictures	x	o	o	x
Projection-type three-dimensional pictures	x	o	o	x
Cinerama	x	x	x	o

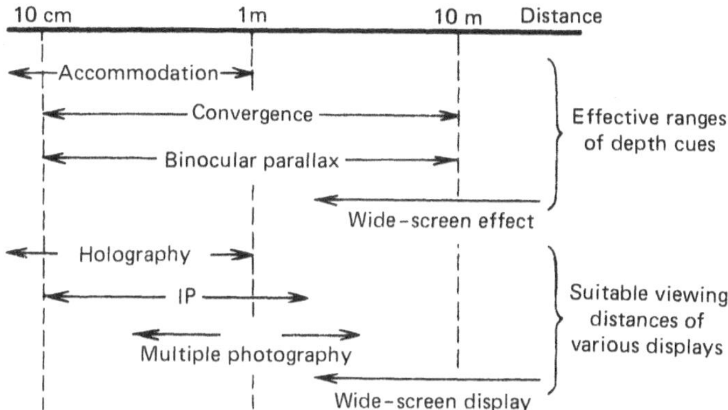

Fig.8.21 The effective ranges of four major depth cues and suitable viewing distances of various three-dimensional displays.

B. Future Prospects of Three-Dimensional Television Broadcasting

We first consider the future of three-dimensional television broadcasts. The effective ranges of the four major cues: accommodation, convergence, binocular parallax, and the wide-screen effect are shown in the upper half of Fig.8.21. We may combine the above effective ranges with Table 8.1 to obtain the lower half of Fig.8.21, which shows the suitable viewing distances of various images.

We may predict something about the future of three-dimensional television broadcasting on the basis of Fig.8.21. It was stated in Section 2.8 that developments of most parallax-type three-dimensional movies were interrupted as a direct consequence of the appearance of Cinerama. It is quite possible that television broadcasting will follow a similar path in the near future.

It is natural to predict that rising standards of living will bring forth enlarged living rooms, in which people will watch television from more distant spots, typically 5 m. This distance is much shorter than the average viewing distance in movie theaters, but is in the range where the wide-screen effect becomes the most effective cue (see Fig.8.21) and the wide-screen display is suitable.

Therefore, the author predicts that the next generation of television broadcasting will feature high-resolution (2000 lines, for example), wide-screen (1-2 m) display giving only illusory depth sensation but much easier to realize than the parallax-oriented ones. Truly three-dimensional television of very high quality will then follow the wide-screen system, but in the distant future.

C. Future Prospects of Other Applications

Therefore, three-dimensional television techniques will first be applied to educational, medical, and information-display purposes. In these fields, binocular stereoscopic systems will be the forerunners. It is extremely difficult to predict the next stage; it seems that the projection-type holography scheme is a promising candidate in various applications.

8.10 OTHER THREE-DIMENSIONAL IMAGING TECHNIQUES

Finally, some proposals and experiments that have appeared quite recently will briefly be described.

8.10.1 USE OF LUMINESCENCE IN CRYSTALS

Lewis proposed using a luminescence phenomenon in a transparent, cubic crystal (34). The basic arrangement is illustrated in Fig.8.22. The crystal is CaF_2 doped with 0.1% Er^{3+}; this material emits light when pumped by two near-infrared wavelengths: 830 and 1540 nm. Therefore, by focusing those lights toward a single spot inside the cube as shown in Fig.8.22, a spatial display may be obtained.

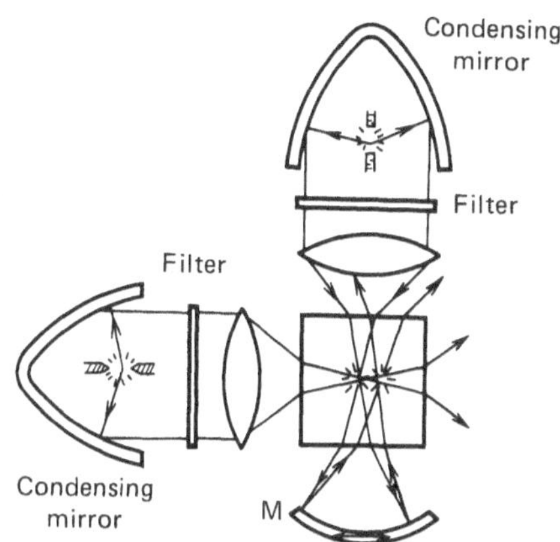

Fig.8.22 A three-dimensional display using the luminescence in a crystal.

In the experiment reported, the display device uses a crystal of $2 \times 2 \times 2$ cm^3. Lewis states that bigger displays may be constructed.

8.10.2 OPTICAL SECTIONING USING A SPATIAL FILTER

Gerritsen *et al.* reported that by using interferometer techniques and a spatial filter, specific cross sections of an object may be recorded optically, and they suggested that this method may be applied to transmission of three-dimensional images (35).

In their scheme, an interferometer generates a light beam that focuses at a specific depth resulting in a striped pattern. The object is illuminated by this beam, and a photograph is recorded. This photograph is later analyzed using a special filter to extract pictures at various depths.

8.10.3 ACOUSTICAL HOLOGRAPHY

The acoustical holography technique now constitutes a substantial, independent field of engineering. In this book, descriptions of this field is entirely omitted because of length restrictions.

8.10.4 USE OF VISUAL ILLUSIONS

The best-known example is Cinerama (see Section 2.8). There are many other proposals. An example is that of Weiner who showed that when a plane picture is projected upon a screen tilted backward and observed through a frame, depth sensation is obtained (36). This illusion stems from the fact that most scenery consists of a closer lower part (the ground or floor, for example) and a more distant upper part.

References

(1) Waters, J. P., Holographic Image Synthesis Utilizing Theoretical Methods, *Appl. Phys. Lett. 9,* No. 11, 405-407 (Dec. 1, 1966).

(2) Lohmann, A. W., and Paris, D. P., Binary Fraunhofer Hologram Generated by Computer, *Appl. Opt. 6,* No. 10, 1739-1748 (Oct. 1967).

(3) Lesem, L. B., Hirsch, P., and Jordan, J. P., Jr., The Kinoform: A New Wavefront Reconstruction Device, *IBM Res. Develop. 13,* No. 3, 150-155 (March 1969).

(4) King, M. C., Noll, A. M., and Berr, D. H., A New Approach to Computer-Generated Holography, *Appl. Opt. 9,* No. 2, 471-475 (Feb. 1970).

(5) Chutjian, A., and Collier, R. J., Recording and Reconstructing Three-Dimensional Images of Computer-Generated Subjects by Lippmann Integral Photography, *Appl. Opt.* 7, No. 1, 99-101 (Jan. 1968).

(6) Saunders, B. G., Stereoscopic Drawing by Computer - Is It Orthoscopic? *Appl. Opt.* 7, No. 8, 1499-1504 (Aug. 1968).

(7) Ortony, A., A System for Stereo Viewing, *The Computer J. 14,* No. 2, 140-144 (May 1971).

(8) Kageura, K., Binocular Stereoscopic Display of Computer Outputs, Graduation Thesis, June 1971, Department of Electronic Engineering, University of Tokyo.

(9) Traub, A. C., Three-Dimensional Display, U.S. Patent 3,493,290, Feb. 3, 1970.

(10) Traub, A. C., A New 3-Dimensional Display Technique, *The Mitre Documents*, No. M-68-4, May 1968.
(11) Traub, A. C., Stereoscopic Display Using Varifocal Mirror Oscillations, *Appl. Opt. 6,* No. 6, 1085-1087 (June 1967).
(12) Rawson, E. G., Three-Dimensional Computer-Generated Movies Using a Varifocal Mirror, *Appl. Opt. 7,* No. 8, 1505-1511 (Aug. 1968).
(13) King, M. C., and Berry, D. H., Varifocal Mirror Technique for Video Transmission of Three-Dimensional Images, *Appl. Opt. 9,* No. 9, 2035-2039 (Sept. 1970).
(14) Kasahara, T. *et al.,* Stereo-Radiography Using Holographic Techniques, *J.J.A.P. 8,* No. 1, 124-125 (Jan. 1969).
(15) Groh, G., and Kock, M., Three-Dimensional Display of X-ray Images by Means of Holography, *Appl. Opt. 9,* No. 3, 775-777 (March 1970).
(16) Yamada, S., and Kaku, E., private communication.
(17) Gobar, D., Three-Dimensional Picture Projection, U.S. Patent 3,479,111 (filed Aug. 24, 1967; patented Nov. 18, 1969).
(18) Nomura, H., Three-Dimensional Image Recording Using Holography, M.S. Dissertation, Feb. 1972, Department of Electronic Engineering, University of Tokyo.
(19) Mengel, L. I., 3-Dimensional TV System, *Electronics,* 45, 128 (Oct. 1958).
(20) Johnston, H. R., Hermanson, C. A., and Hull, H. L., Stereo Television for Remote Control, *Electronics,* 120-122 (Jan. 1951).
(21) Three-Dimensional TV for Industry, Television Newsletter, May 1955.
(22) Low-Cost Three-Dimensional TV, *Electronics,* 196-198 (July 1953).
(23) Yamaguchi, Y., and Fukushima, K., Stereoscopic Television Without Glasses, *NHK Monthly Technical Report 4,* 132-135 (May 1961).
(24) Okoshi, T., Three-Dimensional Imaging and Three-Dimensional Television, *J. Inst. Electron. Commun. Eng. Jap. 51,* No. 10, 1249-1257 (Oct. 1968).
(25) Noda, M., *et al.,* Stereoscopic X-ray Television

for Medical Use, *J. Inst. Telev. Eng. Jap. 25,* No. 5, 388-398 (May 1971).

(26) Baumann, E., The Fischer Large-Screen Projection System (Eidophor), *J. SMPTE 60,* 344-356 (April 1953).

(27) Herman, S., Principles of Binocular Three-Dimensional Display with Applications to Television, *J. SMPTE 80,* No. 7, 539-544 (July 1971).

(28) Leith, E. N., Upatnieks, J., Hildebrand, B. P., and Haines, K., Requirements for a Wavefront-Reconstruction Television Facsimile System, *J. SMPTE 74,* 893-896 (Oct. 1965).

(29) Enloe, L. H., Murphy, J. A., and Rubinstein, C. B., Hologram Transmission Via Television, *Bell Syst. Tech. J. 45,* No. 2, 335-339 (Feb. 1966).

(30) Enloe, L. H., Jakes, W. C., Jr., and Rubinstein, C. B., Hologram Heterodyne Scanners, *Bell Syst. Tech. J. 47,* No. 9, 1875-1882 (Nov. 1968).

(31) Burckhardt, C. B., and Doherty, E. T., Formation of Carrier Frequency Holograms With an On-Axis Reference Beam, *Appl. Opt. 7,* No. 6, 1191-1192 (June 1968).

(32) Burckhardt, C. B., and Enloe, L. H., Television Transmission of Holograms With Reduced Resolution Requirements on the Camera Tube, *Bell Syst. Tech. J. 48,* No. 5, 1529-1535 (May-June 1969).

(33) Berrang, J. E., Television Transmission of Holograms Using a Narrowband Video Signal, *Bell Syst. Tech. J. 49,* No. 5, 879-887 (May-June 1970).

(34) Lewis, J. D., *et al.*, A True Three-Dimensional Display, *IEEE Trans. Electron Devices ED-18,* No. 9, 724-732 (Sept. 1971).

(35) Gerritsen, H. J., and Horwitz, B., Experimental Demonstration of Optical Sectioning by Spatial Filtering and Its Possible Use in Transmission of Three-Dimensional Pictures, *Appl. Opt. 10,* No. 4, 862-867 (April 1971).

(36) Weiner, M. L., The Development of a New Three-Dimensional Photographic Process Through Utilization of a Horizontal-Depth Projection Screen Method Without the Use of Stereoscopic Vision, *Appl. Opt. 8,* No. 6, 1225-1228 (June 1969).

9 Conclusion

Reviewing the entire volume of this book, we can improve Table 1.1 classifying various three-dimensional imaging techniques to a more complete form. The improved, more detailed table is shown as Table 9.1.

The classification employed in this table is based upon the methods used in three-dimensional imaging. Another possible classification might be one based upon applications; for example, the still-picture display, movies, television, microscopy, x-ray imaging, acoustical imaging, and so forth. If we take the technical classification as the abscissa and the applicational classification as the ordinate, we may obtain a matrix-like table. Such work, however, is left for the reader's final exercise. Readers will find, if they try this exercise, at least two or three seeds of new ideas or new equipment at some of the cross points in the matrix.

TABLE 9.1 Classification of Various Three-Dimensional Imaging Techniques Described in This Book (Note that the classification in the left columns is more or less arbitrary)

Classification			Light Used	Systems
(in the wide sense)	Binocular imaging		Incoherent	Binocular viewer Parallax stereogram Lenticular-sheet binocular picture Polarization method[a] Two-color method[a]
			Coherent	Eye-piece holography
	dimensional imaging	Multiple-picture recording	Incoherent	Lenticular-sheet three-dimensional picture Projection-type three-dimensional displays, using Regular direction-selective (d-s) screen Unidirectionally d-s screen Radially d-s screen Holographic d-s screen
			Coherent	Projection-type holography Sampling-type holography
			Mixed	Holographic stereogram

Three-dimensional imaging	Autostereoscopic three-	Continuous parallax	Incoherent	Integral photography Use of luminescence in crystals Projection onto smoke screen[a]
			Coherent	Holography: Computer-generated hologram Image hologram Unidirectional holography Color holography
			Mixed	Holocoder-hologram method White-light-reconstruction holography Lippmann-Bragg holography
	Sectioning		Incoherent	Varifocal mirror technique Computer-generated integral photography Moiré topography[a]
			Coherent	Use of spatial filter
	Illusions		Incoherent	Use of wide screen (Cinerama) Use of visual illusions

[a]Not described in this book.

Index

U

V

W

X

Y

A 6
B 7
C 8
D 9
E 0
F 1
G 2
H 3
I 4
J 5

A NOTE ON THE TYPE

The main text is in Courier, a monospace type originally created by Howard Kettler for IBM in 1955. The type was later redrawn by Adrian Frutiger in 1961 for IBM's "Selectric" series typewriters, becoming ubiquitous in the pre-digital era. The front matter and colophon are in Times Ten, a Linotype redesign of Times Roman, and the diagram notations are in Frutiger's Univers.

www.ingramcontent.com/pod-product-compliance
Lightning Source LLC
LaVergne TN
LVHW050917080826
845145LV00001B/113

* 9 7 8 0 9 8 2 2 2 5 1 4 1 *